Damaging Effects of Weapons and Ammunition

Damaging Effects of Weapons and Ammunition

Igor A. Balagansky
Novosibirsk State Technical University
Russia

WILEY

This edition first published 2022

Registered Office
John Wiley & Sons, Inc., 111 River Street, Hoboken, NJ 07030, USA

Editorial Office
111 River Street, Hoboken, NJ 07030, USA

For details of our global editorial offices, customer services, and more information about Wiley products visit us at www.wiley.com.

Wiley also publishes its books in a variety of electronic formats and by print-on-demand. Some content that appears in standard print versions of this book may not be available in other formats.

Library of Congress Cataloging-in-Publication Data

Names: Balagansky, I. A. (Igor Andreevich), 1952– author.
Title: Damaging effects of weapons and ammunition / Igor A. Balagansky.
Description: Hoboken, NJ : Wiley, [2022] | Includes bibliographical references and index.
Identifiers: LCCN 2022001002 (print) | LCCN 2022001003 (ebook) | ISBN 9781119779537 (cloth) | ISBN 9781119779544 (adobe pdf) | ISBN 9781119779551 (epub)
Subjects: LCSH: Ammunition–Evaluation. | Explosives, Military–Evaluation. | Ballistics.
Classification: LCC UF700 .B35 2022 (print) | LCC UF700 (ebook) | DDC 623.4–dc23/eng/20220214
LC record available at https://lccn.loc.gov/2022001002
LC ebook record available at https://lccn.loc.gov/2022001003

Cover Design: Wiley
Cover Image: © Taro Hama @ e-kamakura/Getty Images

Set in 9.5/12.5pt STIXTwoText by Straive, Pondicherry, India

10 9 8 7 6 5 4 3 2 1

Dedicated

to

My Teachers

Contents

Preface

There are currently only a few publications available that provide information on the damaging effects of weapons and ammunition in a systematic and up-to-date manner. The author has attempted to fill this void.

This book contains a description of the physical fundamentals of the effects of weapons and ammunition. The introduction provides the basic concepts of the theory of combat effectiveness of weapons. The first chapter deals with the designs and effects of fragmentation munitions; the second chapter talks about munitions with shaped charges. The third chapter describes high-explosive weapons and ammunition; the fourth chapter is devoted to penetration weapons and ammunition. Since the study of the explosion and impact processes nowadays is impossible without extensive application of numerical simulation, the author found it necessary to devote a separate chapter to these issues (Chapter 5). It does not aim to teach how to create software systems for numerical modeling – a large number of those exist already – rather, it is important to learn how to use existing systems competently.

Calculation formulas for estimating parameters of damage fields and their interaction with different objects are provided. Damage criteria are provided, too; vulnerability characteristics of objects in relation to different damaging factors are given, as well as formulas for evaluation of safe distances and resistance of different objects and structures to explosion and impact action.

When writing this book, the author has used the experience of teaching the relevant disciplines at the Novosibirsk State Technical University, Bauman Moscow State Technical University, Ustinov Baltic State Technical University, and Zhukovsky Military Aviation Technical University in Russia. It also reflects the results of the author's own research.

The book does not cover designs and operation of weapon control and guidance systems, as they are separate and extensive subject for study. The dimensions used are not always given in the international unit system SI, since specific artillery units are generally used in cited literature and classical publications on the effects of weapons and ammunition.

The author hopes that readers will understand the units used, which will help them to get acquainted with those publications in the future. To facilitate the unit translations, tables with ratios and necessary explanations are provided in the appendix. It is assumed that readers are familiar with the basics of probability theory, explosion physics, and gas dynamics.

The textbook is aimed at students, postgraduate students, and professionals, which specialize in the study of explosive phenomena and explosion safety. In addition, it will be useful for other professionals in the development of protective measures against terrorism and explosion-related technological accidents.

I gratefully acknowledge the support of my son, Alexei, in preparing this book, especially for the help in improving the style of my English.

I

Introduction

I.1 Ammunition Types and Characteristics of Their Damaging Effect

I.1.1 Basic Concepts and Definitions

The term "weapons and ammunition" includes a wide range of different elements: from small arms ammunition and hand grenades to artillery and mortar projectiles, aircraft bombs and depth charges, missile and torpedo warheads, as well as engineering and sea mines designed to destroy enemy personnel and equipment, destroy structures and fortifications, and perform special tasks.

Damaging effect is defined as the ability of ammunition to damage a target provided that it is already close to it and all its elements have operated without failure. The effectiveness of the damaging effect of ammunition on the target and its combat effectiveness should not, therefore, be confused. Obviously, the combat effectiveness depends not only on the effectiveness of the ammunition near a target but also on the accuracy of ammunition delivery, the reliability of all its elements (the fuse in particular), the ability to withstand the enemy's defensive actions, and many other factors.

As an example, you might recall the history of thermonuclear weapons. The first thermonuclear device was exploded in the United States, yet the first thermonuclear bomb was created in the USSR. The device created in the United States had such a weight (62 tons) that it was meaningless to think about delivering it to a real target. It had a damaging effect, but its combat effectiveness was completely absent.

Damaging Effects of Weapons and Ammunition, First Edition. Igor A. Balagansky.

I.1.2 Types of Ammunition and Their Damaging Effects

In order to ensure a damaging effect on targets, it is necessary to exert any action on them. It is customary to highlight:

- action of fragment flow;
- penetrating action;
- shaped charge action;
- action of a shock wave and explosion products;
- incendiary action;
- action of penetrating and electromagnetic radiation of a nuclear explosion.

All ammunition is divided into the following types, depending on its main effects:

- fragmentation (with natural, controlled fragmentation, and preformed fragments; rod warheads; gunpowder and explosive shrapnel; case-shot);
- penetrating (small arms ammunition, armor-piercing, and concrete-piercing);
- shaped charges (with single or tandem shaped charges; explosion-formed projectiles);
- high explosive (blast) ammunition, including volumetric explosion ammunition;
- incendiary;
- nuclear (based on fission or fusion);
- unified by type of action.

Ammunition that does not damage targets but serves as a countermeasure to protect or reduce the damage from the enemy's ammunition can be identified as a separate group. For the same purpose, active protection of armored vehicles is developed to destroy enemy ammunitions as it approaches the target.

There are also many auxiliary and special-purpose munitions: lighting, smoke, and others that will not be considered here.

As we can see, there are just a few ways for ammunition to act on targets, despite the huge variety of weapon and ammunition designs. For the most part, all of them are determined either by the kinetic energy of the munition itself or by the chemical energy of the explosive that the munition is equipped with. In both cases, the equations that describe the behavior of materials during impulse loading determine the parameters of the field of damage and the processes of damage itself.

In this textbook, we will examine in some detail the effects of fragmentation, high explosive, penetrating ammunition, and ammunition with shaped charges.

I.2 Generalized Characteristics of the Damaging Effect

The damaging effects of weapons and ammunition can be described by different parameters. For example, for fragmentation ammunition, such characteristics are sometimes used as the number of effective fragments, the distance at which a fragment is able to pierce a barrier of a given thickness. The effect of ammunition with shaped charges is assessed by the thickness of the pierced armor. To assess the effect of high-explosive ammunition, the value of overpressure at the front of the shock wave is used, and so on. These vastly different parameters do not allow for sufficient characterization of the damaging effect and, thus, to compare various ammunition types with each other. This makes the task of selecting and justifying the design parameters of the newly developed munitions even more challenging.

To solve these problems, the so-called generalized characteristics of the damaging effect are used, which contain information about the design parameters of the ammunition, the particular indicators of their damaging effect, the vulnerability of the target, the conditions for the use of ammunition, and the degree of damage to the target.

I.2.1 Degrees of Damage

In the case of land-based targets, the minimum time during which an affected object cannot function as a combat unit is used as a numerical value of the damage degree. Three main degrees of damage are used, depending on the time the object remains nonfunctioning:

- *A* – time sufficient to solve the operation's objectives;
- *B* – time sufficient to address the daily objectives;
- *C* – when the object is suppressed for the duration of a battle.

Note that for different targets and different combat missions, the times corresponding to different degrees of damage may vary significantly and, in some cases, additional degrees of damage are introduced (e.g. for armored vehicles, an additional *D* degree may be used). Sometimes the *A* degree is associated with the total destruction of the target, *B* with its partial failure, and *C* with temporary suppression.

The degree of damage is established in the statement of design and is determined for military equipment by the duration of disruption of the combat function and by the minimum required hospitalization and treatment time for the personnel (Table I.1).

Table I.1 Degrees of damage of targets.

Target	**Military equipment**				**Personnel**		
Degree of damage	*A*	*B*	*C*	*D*	*A*	*B*	*C*
Duration of disruption	7 days	12–24 hours	3 hours	30 minutes	2–3 months	1 month	7 days

Source: From Balagansky et al. [1].

I.2.2 Contact and Remote Ammunition

All ammunition can be divided into ammunition with an impact or contact effect that can only damage a target after a direct hit (shaped charges, armor-piercing, concrete-piercing) and those with a remote effect that can damage a target when exploded at some distance from it (fragmentation, high-explosive ammunition).

I.2.3 Generalized Characteristics of Contact Ammunition

For contact ammunition, the exhaustive characteristic that defines their destructive effect on single small-size targets is the conditional damage law $G(m)$, which is understood to be the probability of damaging a target that is hit by m projectiles [2]. The conditional damage law simultaneously determines both the destructive power and the degree of vulnerability of the target to the projectile attack. The $G(m)$ function only makes sense for integer m, but for clarity, all points in the smooth curve are usually connected, as shown in Figure I.1.

A convenient numerical characteristic of the conditional damage law is the average number of hits required for destroying a target:

$$\omega = \sum_{m=0}^{\infty} [1 - G(m)]. \tag{I.1}$$

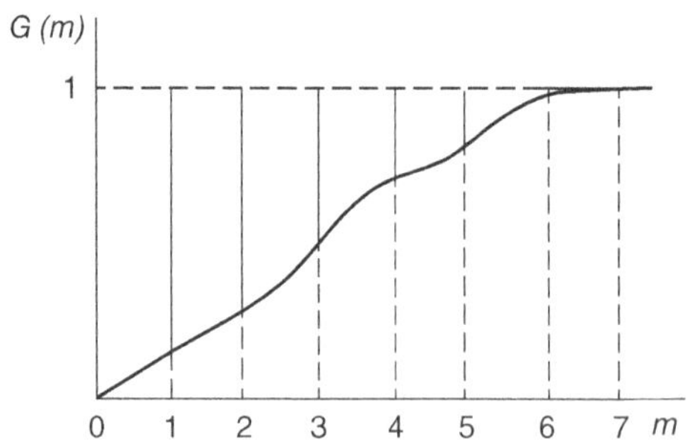

Figure I.1 Type of conditional damage low. *Source:* From Wentzel [2].

The higher the ordinates of the damage law values are, the lower the value of ω. The value of ω, being an integral characteristic of the conditional damage law, fully determines the destructive effect of the projectile and the vulnerability of the target. The more powerful the projectile and the more vulnerable the target is, the lower the average number of hits is required.

I.2.4 The Accumulation of Damage

The conditional damage law is particularly simple if there is no accumulation of damage. Accumulation of damage is the phenomenon where the projectiles "help each other" to damage the target, i.e. when the target can be damaged by the combined action of two or more projectiles (or other hitting elements), neither of which, used separately, would damage the target. For example, if an airplane target has fuel tanks with an inert filler, it often requires at least two projectiles to destroy such tanks, the first of which penetrates the tank and the second ignites the fuel spill. Strictly speaking, the accumulation of damage always takes place. However, in most cases, it is negligible and can be ignored.

Suppose there is no accumulation of damage and the projectiles damage the target independently of each other. Let's denote r the probability of damaging the target if one projectile hits it. Then the law of damage $G(m)$ – the probability of damaging the target when m projectiles hit it – is equal to the probability that at least one of the projectiles will damage the target:

$$G(m) = 1 - (1 - r)^m. \tag{I.2}$$

Such a law of damage is called a degree law. So, if there is no accumulation of damage for some target, its law of damage is a degree law, where r is the probability of damaging the target in one hit.

The value of r can be interpreted as the relative area of vulnerable target components. Indeed, let the target consist of only two types of components: certainly vulnerable and absolutely non-vulnerable. Then the probability of damage is equal to the probability that a projectile hitting the target will hit its vulnerable components. If we consider the distribution of projectiles hitting the target to be approximately uniform, which is true for small targets, the r-value will be equal to the relative area of vulnerable components.

It can be demonstrated that for a target that has no accumulation of damage, the average number of hits required is the value inverse of the relative area of its vulnerable components:

$$\omega = \frac{1}{r}. \tag{I.3}$$

The value of the full target area (S_t) divided by the average number of hits required (ω) is called the vulnerable area of the target S_v. Typically, during calculations, we consider the projection of the full or vulnerable target area to the direction of the projectile's approach. Then, depending on the prevailing direction of target firing, the full or vulnerable area of the target is referred to as the average full or vulnerable area of the target. The probability of damaging the target, in this case, can be calculated as the probability of hitting the vulnerable target area S_v.

I.2.5 Generalized Characteristics of the Damaging Effect of Remote Ammunition

Remote ammunition not only affects targets with a direct hit but also when it explodes at some distance from the target. The target is damaged either by the products of the explosion and shock wave (high-explosive ammunition) or by high-velocity fragments (fragmentation ammunition).

The main characteristic that determines the effectiveness of such ammunition is the coordinate law of damage $G(x, y, z)$. The coordinate law of damage is a functional relationship between the probability of the target damage and coordinates of the explosion point of the ammunition relative to the target [2].

The simplest type of coordinate law is for high-explosive ammunition, the effect of which is estimated by the target destruction radius R_d. Hence, for high-explosive ammunition, the coordinate law is expressed as a simple stepwise dependence on the distance to the target R. If the projectile exploded at a distance $R \leq R_d$ from the target, then $G(R) = 1$, and at $R > R_d$, $G(R) = 0$. The area of space around the target, inside which $G(R) = 1$, is called the area of unconditional target damage. The boundaries of this area are lines or surfaces equidistant to the target contours (Figure I.2).

It is more difficult to calculate the coordinate law of damage $G(x, y, z)$ for fragmentation ammunition since the fact of damaging the target is accidental. After all, the number of fragments hitting the target is random at the given breakpoint position, and there is a certain probability that none of the fragments will hit the target or, if hit, will be unable to damage its vital components.

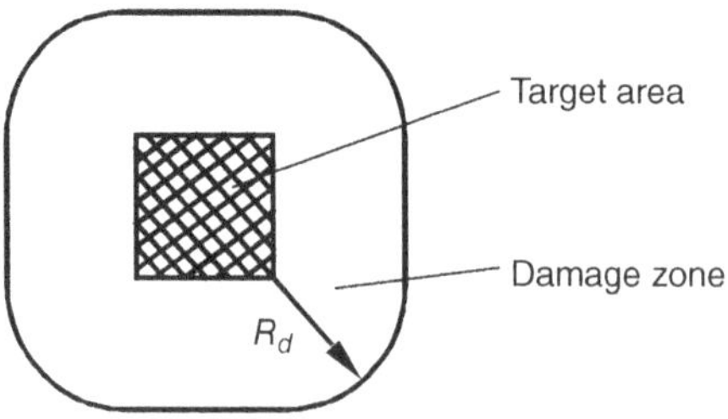

Figure I.2 Zone of unconditional target damage by the blast action. *Source:* From Wentzel [2].

Let's consider a component, the failure of which leads to the destruction of the whole target. Obviously, at some given mutual position of the target and the projectile to calculate the probability of damaging the component g, you need to know the number of fragments n that hit it and the probability of its destruction when hitting one fragment p_1. If the number of hitting fragments were known precisely, the probability of damaging the component g could be determined by the formula

$$g(n) = 1 - (1 - p_1)^n. \tag{I.4}$$

In reality, a different number of fragments may hit a component of the target. Therefore, to calculate the probability of damage of a given component, you need to know the law of distribution of hits, i.e. the probability that a certain number of fragments will hit a given area.

Experimental data in full compliance with probability theory suggest that the law of distribution of the number of fragments hitting the components, whose angular sizes are small compared with the width of the sector of the fragment field, is close to the Poisson's law [2]. In this case, the Poisson's law formula is as follows

$$p_n = \frac{n^{-n}}{n!} e^{-<n>}, \tag{I.5}$$

where p_n is the probability that exactly n fragments will hit the component; n is a random number of hitting fragments; $<n>$ is an expected value of the number of fragments that fit into the component area.

Having carried out the corresponding transformations and typed the designation $<m> = <n> p_1$, we will get the expression for the coordinate law of the damage of the component area

$$g(x, y, z) = 1 - e^{-<m>(x,y,z)}. \tag{I.6}$$

The coordinate law of damage the complete target will be written in the same way:

$$G(x, y, z) = 1 - e^{-<m>(x,y,z)}, \tag{I.7}$$

where $<m>$ is the expected value of the number of fragments damaging the target.

In case of a flat scattering of burst points (firing at surface targets), the coordinate law will be determined by two coordinates of the burst point on the plane:

$$G(x, y) = 1 - e^{-<m>(x,y)}. \tag{I.8}$$

Studies have shown that coordinate laws of damage vary depending on the nature of the target, ammunition capacity, and other conditions. When considering these laws, three different areas can be identified around the center of the target (Figure I.3). One of these areas (γ_d) is characterized by the fact that a round of ammunition bursting at any point within it always leads to the destruction of the

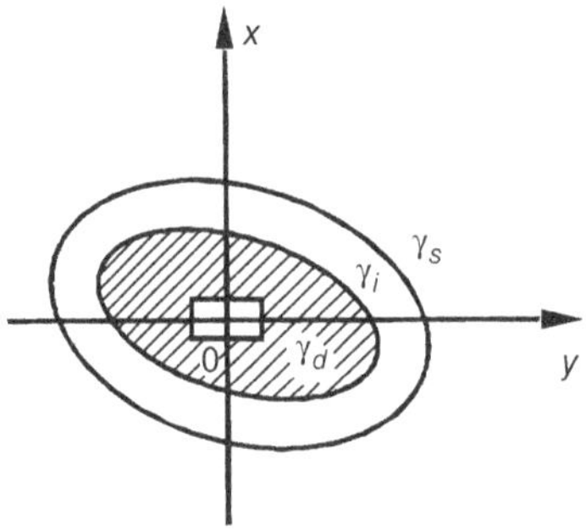

Figure I.3 Areas surrounding the center of the target. *Source:* From Fendrikov and Yakovlev [3].

target. This area is called a reliable damage area. This is followed by an area (γ_i) where it does not necessarily result in the target being damaged. This is referred to as the area of unreliable damage. Then there is an area (γ_s) where ammunition does not damage the target at all; this area is called the safe area.

I.2.6 Specified Zone of Target Damage

The form of the coordinate law can be simplified by artificially expanding the area of reliable damage at the expense of the area of unreliable damage and completely eliminating the area of unreliable damage from consideration. The obtained extended reliable damage area is referred to as the specified damage zone of the target, which is characterized by the area S_{sp}, and its sizes – by the specified target sizes [3]. For all points of the specified zone according to its definition $G(x, y) = 1$, and outside this zone $G(x, y) = 0$, in other words, in this case, the coordinate law has a stepwise graph.

By definition, the area of a specified damage zone can be determined as follows:

$$S_{sp} = \iint_S G(x,y)dS. \tag{I.9}$$

The integration is performed over the whole area, within which the coordinate law does not turn zero.

For nuclear warheads, the specified damage zone is a circle with an area $S_{sp} = \pi R^2$, where R is the radius of the specified damage zone.

For conventional ammunition, the specified damage zone is often represented as a rectangle, and therefore, its area is $S_{sp} = 4l_x l_y$, where the $2l_x$ and $2l_y$ are the depth and front of the specified target sizes.

The probability of damaging an elementary target with several shots when using the concept of a specified damage zone can be calculated as the probability of at least one ammunition hitting this zone. In this case, the probability is calculated using the same formulas as when assessing the effectiveness of contact ammunition.

1.3 Evaluation of the Effectiveness of Shooting

1.3.1 The Concepts of Combat Effectiveness of Weapons

A detailed study of the theory and methods of assessing the combat effectiveness of various types of weapons and ammunition is not included in this textbook. However, consideration of the basic concepts will certainly be useful for a better understanding of what follows. Using simple models of combat operations, we will illustrate how generalized characteristics of the damaging effect are used in assessing the approximate combat effectiveness.

As a rule, the combat effectiveness of a weapon is assessed by the amount of damage inflicted on enemy objects (targets). The effectiveness of firing depends on a large number of different factors: type, size, mobility, and vulnerability of the target; characteristics of the dispersion and destruction effect of ammunition; the amount of ammunition (the number of guns, missiles, batteries, divisions, aircrafts, etc.) involved in the destruction of targets; methods of firing; reliability of weapons; time of staying in position; counteraction of the enemy, etc. In particular, according to the army data, 6 seconds after firing begins, the first soldier already reaches a shelter, and 11 seconds afterward, the last soldier leaves the exposed area.

1.3.2 Classification of Targets, Typical Efficiency Indicators

To assess the effectiveness of the shooting, the whole variety of targets can be represented by three typical kinds [2].

A single target is an elementary, usually small target that provides certain functions (aircraft, ship, tank). The task of firing is its damage. As an indicator of efficiency, the probability of damage $W = P(A)$ is often used, where A – the damage of the target.

A group target is a target consisting of several single targets, such as the position of an anti-aircraft missile system, a tank convoy, or a group of aircraft or ships united by a common combat mission. When firing, it is necessary to damage the group as a whole and thus prevent it from fulfilling its combat mission. According to the experience of World War II, the enemy's refusal to continue the attack is directly related to the number of losses. Loss of 50–60% tanks means refusing to continue the attack with the probability of 0.90–0.95. The average number of damaged targets in the group $M_d = M[K_d]$, where the random value of K_d is the number of damaged targets, is usually taken as an indicator of the shooting efficiency. In some cases, a more specific task is set before a group target is fired, for example, to disable the enemy's antiaircraft missile system. Damaging the position of the antiaircraft missile system may mean either damaging all missile launchers or damaging the target illumination radar. Then the indicator of the effectiveness of firing on a group target will be the probability of completing the task A: $W = P(A)$.

An area target is a target consisting of a set of objects distributed within a certain area in an obscure way, for example an accumulation of personnel and battle equipment. Typical for an area target is that it is not the individual objects that are targeted, but the entire area as a whole. As an efficiency indicator, the expected value of the portion of the damaging area $M = M[U]$ is used, where $U = S_d/S_t$ is the ratio of the damaging area to the full target area.

I.3.3 Dispersion During Shooting

There is always dispersion in any type of firing, both with unguided and guided missiles. The main reasons for the dispersion of projectiles during firing are an inaccurate determination of target coordinates; aiming error; the influence of meteorological factors (wind, change of atmospheric pressure, humidity); fluctuations of the launcher; production tolerances during ammunition manufacturing.

All errors affecting the deviation of the projectile from the target can be divided into systematic and random errors. Systematic errors from shot to shot do not change, they can be measured and taken into account later (e.g. the deviation of the aim point from the center of the projectile dispersion) and the effect of low atmospheric pressure when firing at altitudes other than sea level. Random errors cannot be measured, as they vary from shot to shot. As a result of the combined effect of all firing errors, the actual trajectory of the projectile never coincides with the calculated trajectory, and the point of hit (or blast) of the projectile inevitably deviates from the calculated point to which the projectile was directed. This phenomenon is called "dispersion."

The law of distribution of random values characterizing the point of hit (or blast) of ammunition is called the law of dispersion. For contact ammunition or remote one with flat dispersion, this law is presented as the law of distribution of the two coordinates (x, y) of points of hitting. Usually, the distribution is given as the value of the probability density $\varphi(x, y)$. The value $\varphi(x, y)\cdot dx\cdot dy$ is the probability of hitting an area with dimensions $dx\cdot dy$ adjacent to the point with coordinates (x, y) (Figure I.4).

Similarly, for projectiles with volumetric dispersion, the dispersion law is the distribution of the three coordinates of the blast point (x, y, z) and is characterized by the probability density $\varphi(x, y, z)$, with the value of $\varphi(x, y, z)\cdot dx\cdot dy\cdot dz$ being the probability of the projectile blast in the elementary volume $dx\cdot dy\cdot dz$ adjacent to the point (x, y, z).

Let's consider the case of flat dispersion as a simpler one. Imagine shooting a contact projectile or a remote one with flat dispersion. First of all, we must choose a flat surface on which we will study the dispersion of hit points. This surface is

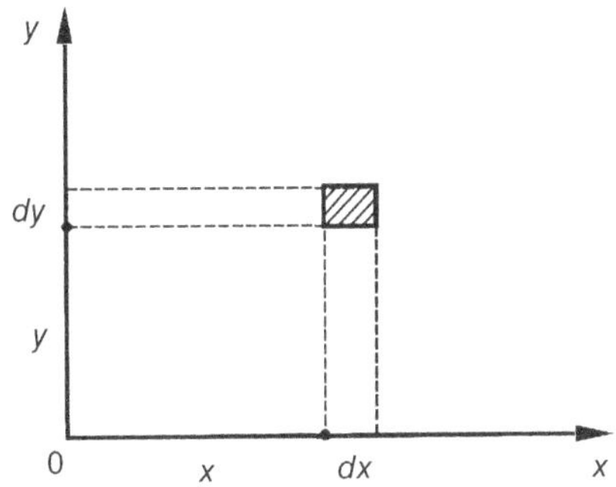

Figure I.4 Area with dimensions of $dx{\cdot}dy$, adjacent to a point at the coordinates (x, y). *Source:* From Wentzel [2].

commonly referred to as a picture plane or a dispersion plane. When shooting at land or sea targets with remote ammunition, this is usually the surface of the ground or sea. When a land or sea target is shot contact projectiles, it is usually considered to be on a vertical dispersion plane. In the case of air targets, the picture plane is most often drawn through the point of hit perpendicular to the vector of the relative velocity at which the projectile meets the target.

When the picture plane Q is fixed, the rectangular $X0Y$ coordinate system is selected. Figure I.5 shows the picture plane and the coordinate system for the case of shooting an air target.

Usually, when a remote projectile is fired at sea or ground targets, the X-axis is at least close to the direction of firing and the dispersion along the X-axis characterizes the deviation along the firing line, the Y-axis thus characterizes the deviation across the firing line.

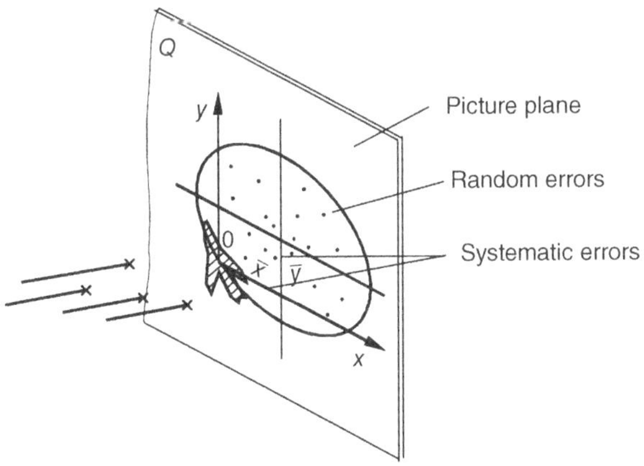

Figure I.5 Setting up the coordinate system on the picture plane. *Source:* From Wentzel [2].

As a law of dispersion, all types of firing and bombing are generally subject to normal law. This is due to the fact that the firing error on each of the axes can be represented as the sum of a large number of elementary errors resulting from various factors. If one of the coordinate axes (usually $0X$) is at least close to the direction of firing, such axes are called the principal axes. In this case, the dispersion law will take the simplest form:

$$\varphi(x,y) = \frac{1}{2\pi\sigma_x\sigma_y} e^{-\frac{1}{2}\left[(x-\bar{x})^2/\sigma_x^2 + (y-\bar{y})^2/\sigma_y^2\right]}, \qquad (I.10)$$

where $\bar{x}, \bar{y}$ are the coordinates of the dispersion center. They characterize a systematic firing errors. If there is no systematic error, these values are equal to zero; σ_x, σ_y – standard deviations in the $0X$, $0Y$ axes, respectively.

In the practice, it is not the values of σ_x, σ_y that are usually used, but the so-called probable (median) deviations of $0X$, $0Y$ axes, which are denoted by E_x and E_y, respectively:

$$E_x = \rho\sqrt{2}\cdot\sigma_x, \quad E_y = \rho\sqrt{2}\cdot\sigma_y \quad \rho \approx 0.477. \qquad (I.11)$$

Median deviations are convenient because they correspond to the principal half-axis of the dispersion ellipse, within which exactly half of all hits lie. The law of dispersion, in this case, takes the following form:

$$\varphi(x,y) = \frac{\rho^2}{\pi E_x E_y} e^{-\rho^2\left[(x-\bar{x})^2/E_x^2 + (y-\bar{y})^2/E_y^2\right]}.$$

It is usually assumed that errors along the firing line (E_x) do not depend on errors across the firing line (E_y). Therefore, these values can be considered independently of each other.

I.3.4 Scheme of Two Groups of Errors

Random shot error consists of several components of random errors: target coordinate error, correction for meteorological and ballistic factors, technical dispersion caused by differences in weight and shape of projectiles, etc. In a single shot, each component of the total shot error is repeated only once, in which case it is said that there is one group of errors. When several shots are fired at the same target, some components of the total shot error caused by common sources are repeated, while other components caused by different sources are not repeated. For example, when shooting at an unobserved target from the same gun with the same scope setting, all the shots will repeat the target coordinate error and will not repeat the error caused by variations in the shape and weight of the projectiles. Please note that in this case, the target coordinate error, although repeated from shot to shot, is an accidental rather than systematic error.

Shot errors that have at least one common source and contain at least one common component are called dependent. Shots fired under these conditions are referred to as dependent shots. In real-world conditions, shots are always dependent, so we have to consider a complex error system consisting of at least two error groups.

If we return to the case of firing from one gun at an unobserved target (uncontrolled firing), we can consider two groups of random errors: data preparation errors (E_{x0}, E_{y0}) – group error, and technical dispersion (B_d, B_s) – individual error. Then the total median error of the shot along the firing line

$$E_x = \sqrt{E_{x0}^2 + B_d^2}, \tag{I.12}$$

across the firing line

$$E_y = \sqrt{E_{y0}^2 + B_s^2}. \tag{I.13}$$

The group error makes the salvo as a whole deviate from the target (Figure I.6). Individual errors create dispersion inside the salvo.

An estimate of the values of aggregated median errors when an artillery division is firing after complete preparation of the initial data can be obtained from Table A.1 in the Appendix.

The degree of dependence between the shots can be characterized by the correlation ratio of the hit points coordinates. In the scheme of two error groups, the correlation ratio for each pair of shots is equal to the ratio of the square of the group error probable deviation to the square of the total error probable deviation:

$$\mu^{(x)} = \frac{E_{x0}^2}{E_{x0}^2 + B_d^2}, \tag{I.14}$$

$$\mu^{(y)} = \frac{E_{y0}^2}{E_{y0}^2 + B_s^2}. \tag{I.15}$$

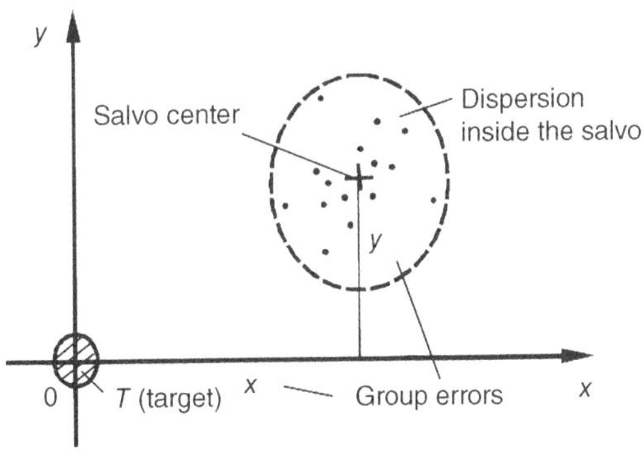

Figure I.6 Effect of the group and individual shooting errors. *Source:* From Wentzel [2].

In practice, most often these two ratios differ slightly and are averaged by the formula

$$\mu = \sqrt{\mu^{(x)}\mu^{(y)}}. \tag{I.16}$$

Theoretically, there can be two extreme cases of dependence between shot errors: the first case is when there is no common error source, i.e. no repeating component ($\mu = 0$); the second case is when all error sources are common, i.e. all components are repeated as many times as the shot is fired ($\mu = 1$). In the first case, the shots will be independent, and in the second case, the shots will be functionally dependent. With concentrated firing and independent shots, the ammunition hits into different points randomly spread around the center of aim, and with functionally dependent shots, all ammunition hits into the same point at a random distance from the center of aim.

I.3.5 Probability of Damaging a Single Target

The probability of damaging a single target W is one of the key indicators of combat effectiveness in the theory of firing, which underlies the calculation of many other indicators. A single target is damaged by one or more shots. Most methods of estimating the effectiveness of multiple shots are based on a preliminary determination of the probability of a single shot damaging the target, so the methods of calculating this probability greatly contribute to the simplicity and accuracy of estimating the effectiveness of the shooting in more complex cases.

I.3.5.1 Damaging the Target with a Single Shot

Let us first consider damaging the target with a single shot. The single-shot probability of damaging the target W_1 is defined as the probability of a complex event, which consists of hitting the target and damaging it if the target is hit:

$$W_1 = p_1 G(1), \tag{I.17}$$

where p_1 is the probability of hitting the target with one shot; $G(1)$ is the probability of damaging the target with one shot (the value of the conditional damage law $G(m)$ at $m = 1$).

With a known average required number of hits ω for a target that has no damage accumulation,

$$W_1 = \frac{p_1}{\omega}. \tag{I.18}$$

The probability of hitting the target with one shot p_1 can be calculated as follows. If the target is projected on the dispersion plane by lines parallel to the direction of firing, then p_1 is the probability that the point of a hit on the picture plane will be

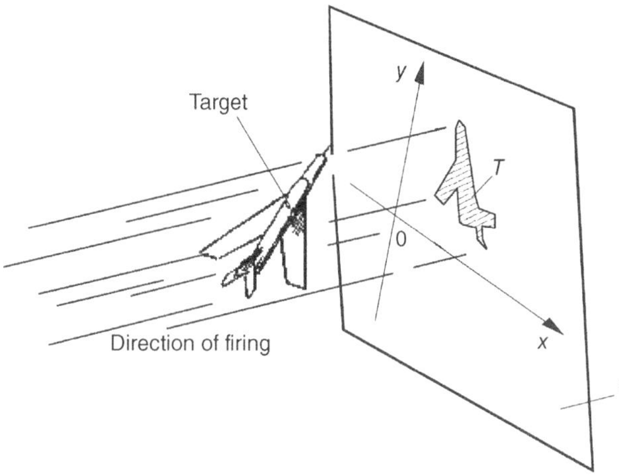

Figure I.7 Projecting of a target on a picture plane. *Source:* From Wentzel [2].

within some region of the target's projection (Figure I.7). This probability is expressed with the formula

$$p_1 = \iint_{T} \varphi(x,y)dxdy, \tag{I.19}$$

where $\varphi(x, y)$ is the law of dispersion (I.11).

In some cases, the integral (I.19) is expressed through known functions. Let us give the corresponding calculation formulas.

1) The target is a rectangle with sides parallel to the principle dispersion axes (Figure I.8).

$$p_1 = \frac{1}{4}\left[\hat{\Phi}\left(\frac{\beta - \bar{x}}{E_x}\right) - \hat{\Phi}\left(\frac{\alpha - \bar{x}}{E_x}\right)\right] \times \left[\hat{\Phi}\left(\frac{\delta - \bar{y}}{E_y}\right) - \hat{\Phi}\left(\frac{\gamma - \bar{y}}{E_y}\right)\right], \tag{I.20}$$

where α, β, γ, δ – coordinates of target's boundaries on $0x$ and $0y$ axes; $\hat{\Phi}$ is the normalized function of Laplace:

$$\hat{\Phi}(x) = \frac{2\rho}{\sqrt{\pi}} \int_0^x e^{-\rho^2 t^2} dt \tag{I.21}$$

(the values of this function for different values of the argument are given in the Appendix, Table A.2).

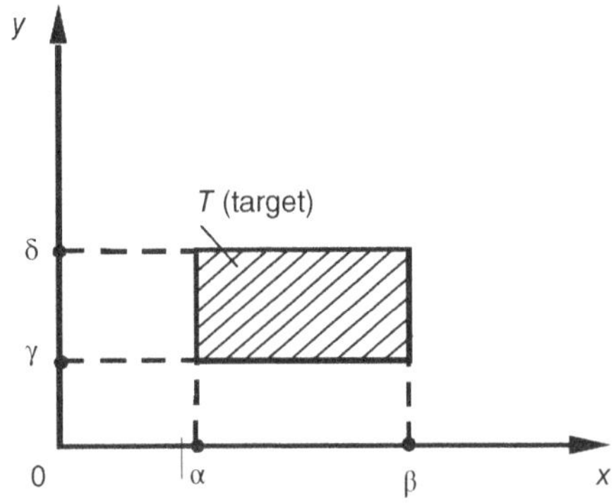

Figure I.8 Coordinates of target boundaries. *Source:* From Wentzel [2].

2) Target – circle, dispersion – circular ($E_x = E_y = E$). If the dispersion center coincides with the target center (there is no systematic error), the probability of hitting within the circle of radius r_t is expressed by the formula

$$p_1 = 1 - e^{-\rho^2 (r_t/E)^2}. \tag{I.22}$$

If the value and configuration of the vulnerable target area S_v are known, the probability of damaging the target can be calculated simply as the probability of hitting the vulnerable target area.

Let us now consider what will change if the target is damaged by remote warheads. The probability of damaging the target, in this case, will be determined by the following formula:

$$W_1 = \iint_S \varphi(x,y) G(x,y) dxdy, \tag{I.23}$$

where $G(x, y)$ – coordinate law of damage.

If the area of the specified damage zone S_{sp} and its specified sizes $2l_x$ and $2l_y$ are known, the probability of damaging can be defined simply as a probability of hitting the specified target area S_{sp}, for example, by the formula (I.20).

Thus, whether by contact or remote ammunition, the probability of damaging a single target can always be calculated as the probability of hitting either a vulnerable area of the target S_v or the specified damage zone of the target S_{sp}.

I.3.5.2 Damaging the Target with Multiple Shots

When shooting with multiple shots, the probability of damaging the target is affected by the dependence of the shots due to repeated errors.

In the extreme case of functionally dependent shots ($\mu = 1$) with concentrated firing, all ammunition hits the same point, and therefore, the probability of damaging the target with several shots W_n is equal to the probability of damaging with one shot W_1:

$$W_n = W_1.$$

The other extreme case of shooting is independent shots ($\mu = 0$). In this case, the probability of damaging the target with n shots will be

$$W_n = 1 - \prod_{i=1}^{n}(1 - W_i), \tag{I.24}$$

where W_i is the probability of hitting the target with an i-shot.

With the equal probability of damaging the target with each of the n independent shots,

$$W_n = 1 - (1 - W_1)^n. \tag{I.25}$$

Example

Five hunters shoot one duck at a time and independently of each other. Each hunter would kill a duck with a probability $W_1 = 0.2$. How likely would all five hunters be to kill a duck?

Solution

Since the shots are independent, the probability can be calculated by the formula (I.25):

$$W_5 = 1 - (1 - 0.2)^5 = 0.672.$$

To calculate the probability of the target damage with dependent shots, the approximate methods of calculation are usually used, in which the dependence of shots is either ignored or taken into account approximately. One of the simplest methods of taking into account the dependence of the shots is to calculate the probability of damaging the target using the approximate formula

$$W_n = W_1 + \left(W_n^{(n)} - W_1\right)\sqrt{1 - \mu^2}, \tag{I.26}$$

where W_1, $W_n^{(n)}$ – the probability of target damage at one and n independent shots, respectively; μ – correlation ratio of shots (see I.16).

The relative error of determining the probability of damaging the target with this method at $nW_1 \approx 0.5$–5 does not exceed several percents.

Example

One of the hunters from the previous example takes five shots at a duck in a row. The correlation ratio between the shots is $\mu = 0.5$. What's the equal probability of damaging the duck?

Solution

Shots are dependent. The value of $nW_1 = 5 \cdot 0.2 = 1$. So, we can use the formula (I.26). Let's put the calculated values from the previous example into the formula

$$W_5 = 0.2 + (0.672 - 0.2)\sqrt{1 - 0.5^2} = 0.608.$$

The dependence of the shots always leads to a reduced probability of damaging the target compared to the case of independent shots. This effect is more pronounced with more shots fired and a bigger correlation ratio μ. With a small number of shots ($n = 2$–4) and relatively small correlation ratio values ($\mu < 0.5$), the correction for the dependence of shots is small and the probability of damaging the target can be calculated as in the case of independent shots. If there are a large number of shots and their correlation is significant, correction for the dependence of shots should be taken into account.

An extremely important issue in efficiency theory is to establish a rational relationship between the group error of firing (E_{x0}, E_{y0}) and individual or technical dispersion (B_d, B_s). The optimal characteristics of the individual dispersion must be commensurate with the characteristics of the aiming error ($E_{x0} \cong B_d$); ($E_{y0} \cong B_s$). Unfortunately, this is not always the case in existing weapon systems; in particular, for artillery weapon systems and small-caliber unguided aircraft missiles, the group error of firing often far exceeds the individual dispersion.

The optimal ratio between individual and group firing errors can be achieved either by increasing the aiming accuracy or by increasing the individual dispersion of the ammunition. In the practical case of a firing situation, the effectiveness of the firing was increased by up to five times when the technical dispersion was increased to its optimum value without increasing the aiming accuracy.

Another way to increase the effectiveness of the dependent shots, which is widely used for ground artillery, is to shoot with an artificial dispersion of the shots, which is achieved by shooting one target at several aiming positions.

I.3.6 Evaluation of the Effectiveness of Firing on a Group Target

Most often the task of shooting at a group target is to damage the largest possible number of units in the group. As an indicator of the effectiveness of firing in this case, the average number of damaged units from the group is used.

$$M_d = M[X_d], \tag{I.27}$$

where the random value of X_d is the number of units damaged.

We will deduce the most general formula for M_d, suitable for all cases without exception when shooting at a group target. For this purpose, let us consider a group

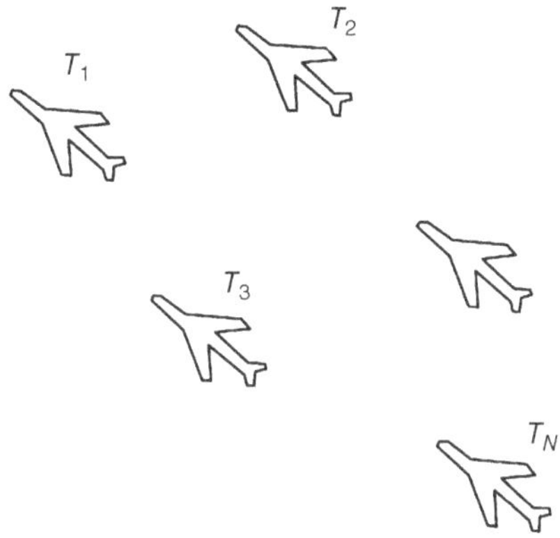

Figure I.9 Group target. *Source:* From Wentzel [2].

target consisting of N units of T_1, T_2, ..., T_N (Figure I.9). This target is to be fired with any ammunition (contact or remote), consisting of any number of shots – dependent or independent, under constant or variable conditions. The aim may be to target individual units or a group as a whole.

Let's deduce the formula for the expected value of the number of damaged units. For this purpose, let us present the total number of damaged units X_d as a sum of N random values:

$$X_d = X_1 + X_2 + \cdots + X_N = \sum_{i=1}^{N} X_i. \tag{I.28}$$

Each i-th unit of T_i has its own random value of X_i, which we will define as follows:

- if the unit of T_i is damaged, $X_i = 1$;
- if unit T_i is not damaged, $X_i = 0$.

It is not difficult to make sure that the total number of damaged targets X_i simply equals the sum of all X_i values. According to the theorem on the summation of expected values:

$$M[X_d] = M[X_1] + M[X_2] + \cdots + M[X_N] = \sum_{i=1}^{N} M[X_i]. \tag{I.29}$$

Let's denote the probability of damaging the i-th unit in the whole shooting as W_i. Then by defining the expected value

$$M[X_i] = W_i \cdot 1 + (1 - W_i)\, 0 = W_i.$$

By substituting (I.29), we get

$$M[X_d] = W_1 + W_2 + \cdots + W_N = \sum_{i=1}^{N} W_i,$$

or finally

$$M_d = \sum_{i=1}^{N} W_i, \tag{I.30}$$

i.e. the average number of damaged units within a group target is equal to the sum of probabilities of damaging all individual units.

To use this formula, you must first calculate the probability of each target being damaged from the group by all the shots fired, and then sum up these probabilities.

Example

Five independent shots are fired at a group of five ducks. The whole group is fired as a single unit without observing the results or transfer of fire. Each shot aimed at a group can kill no more than one duck. The chances of killing the first, second, third, fourth, and fifth ducks in one shot are, respectively,

$$p_1 = 0.1; \quad p_2 = 0.2; \quad p_3 = 0.2; \quad p_4 = 0.1; \quad p_5 = 0.1.$$

It is necessary to determine the average number of killed ducks as a result of the whole shooting.

Solution

Find the probability of killing individual ducks for the whole shooting:

$$W_1 = W_4 = W_5 = 1 - (1 - 0.1)^5 = 0.410,$$

$$W_2 = W_3 = 1 - (1 - 0.2)^5 = 0.672.$$

From here

$$M_d = 3 \cdot 0.410 + 2 \cdot 0.672 = 2.574.$$

Formula (I.30) is perfectly universal and valid for any kind of group target and any way of organizing the shooting. However, the probability W_i is simply calculated only if the firing mode of each individual unit is independent of whether other units are damaged or not. This condition is broken, for example when it is possible to transfer fire from one damaged unit to another undamaged unit.

I.3.7 Evaluation of the Effectiveness of Firing at Area Target

As an indicator of the effectiveness of firing at an area target, an expected value of the fraction of the damaged area is often used.

$$M = M[U], \tag{I.31}$$

where $U = S_d/S_t$ is the ratio of the damaged area to the full target area.

In general, if no assumptions are made about the target, damage area, and firing conditions, the task of evaluating the effectiveness of firing at an area target becomes quite difficult. However, taking into account the overall low accuracy of the information we have about the area target, we can make a number of assumptions and significantly simplify the problem. In particular, it makes no sense to enter into the calculation of the exact configuration of the target and the damage area, and it is possible to replace both areas with rectangles.

Let's accept the following assumptions:

- the target is a rectangle with dimensions T_x, T_y, with sides parallel to the principal axes of dispersion;
- the damage zone is also a rectangle with the dimensions of L_x, L_y, with the sides parallel to the principal dispersion axes.

Let's imagine the process of firing at an area target as if each time when the target T is shot, the damage area of L is reset onto the target. The position of zone L with respect to the target T is characterized by one random point O_1, the epicenter of the explosion, which may take one or another position with respect to the origin of coordinates – the aiming point O. If the origin of coordinates O coincides with the center of the target, it is said about aiming at the center of the target; if it does not coincide, it is said about a takeaway aiming point from the center of the target.

Depending on which position the O_1 point will take as a result of the shooting, the damage zone L will cover one or another part of the target area (Figure I.10). For several shots, the damaged target area depends on the concrete location of the centers of all damage zones. The total damaged area S_d does not equal the sum of the areas damaged by the individual shots, as the damage zones may overlap. Where there are overlaps, the increase in the destructive effect is usually neglected in target zones that are covered twice, three times, or more, and the damaged area S_d is taken to be the area covered by at least one damage zone. By dividing the damaged area of S_d by the target area of S_t, we obtain a portion of the damaged area.

$$U = \frac{S_d}{S_t}.$$

This random value characterizes the success of firing at an area target.

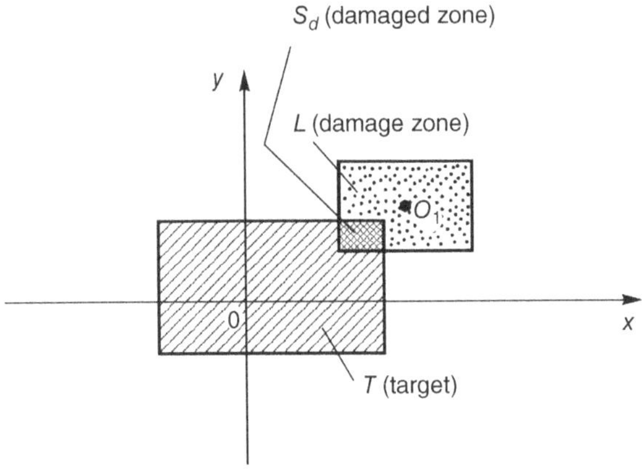

Figure I.10 The mutual position of the target area and its damaged zone. *Source:* From Wentzel [2].

Thus, we are interested in the average fraction (expected value) of the damaged area $M = M[U]$. The methods of calculating of this value are different depending on whether one or more shots are fired at the target.

I.3.7.1 Fraction of Damage U with One Shot

Consider the area target T, on which a single shot is fired, with damage area L (Figure I.11). Let us condition all linear dimensions to express not in meters, but in probable deviations on corresponding axes. When it is necessary to express them in meters, we will mark it with a special symbol (m):

$$T_x = \frac{T_x^{(m)}}{E_x};\quad T_y = \frac{T_y^{(m)}}{E_y};\quad L_x = \frac{L_x^{(m)}}{E_x};\quad L_y = \frac{L_y^{(m)}}{E_y};$$
$$a_x = \frac{a_x^{(m)}}{E_x};\quad a_y = \frac{a_y^{(m)}}{E_y}. \tag{I.32}$$

So, one shot is considered, in which the damage zone L is reset on target T so that the random point O_1 (the epicenter of the explosion) is dispersed around the origin of coordinates O according to normal law. The probable deviations, chosen as units, are equal to one. It is necessary to find the average damaged fraction U.

Consider a random overlapping area S_d of the damage zone L with the target T (see the shaded area in Figure I.12) and the damaged fraction U:

$$U = \frac{S_d}{S_t}. \tag{I.33}$$

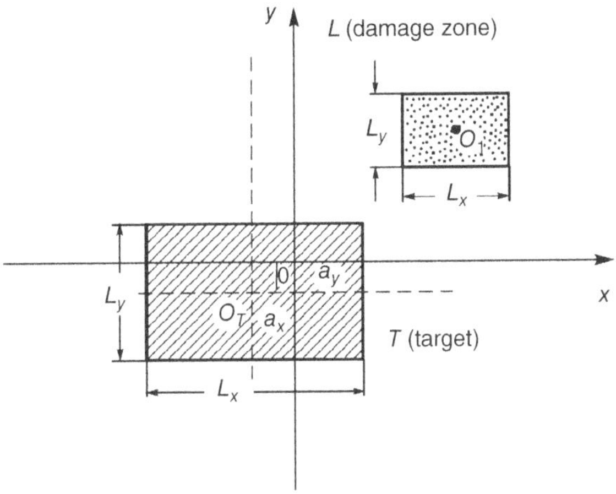

Figure I.11 Mutual position and sizes of the target and the damage zone. *Source:* From Wentzel [2].

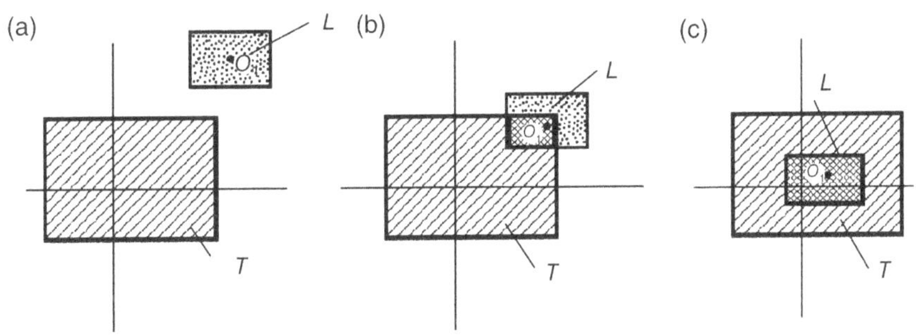

Figure I.12 Random area overlapping the damage zone with the target. *Source:* From Wentzel [2].

The value of U depends on the random position of the epicenter O_1. If point O_1 lies far away from the target (Figure I.12a), the damaged fraction will be equal to zero. If point O_1 lies as shown in Figure I.12b, then the damage zone and the target will overlap partially. Finally, if the damage zone lies as shown in Figure I.12c, the maximum possible area will be covered. In this case (when the damage zone is smaller than the target), this maximum fraction is equal to the ratio of the damage zone area to the target area:

$$u_{\max} = \frac{S_L}{S_t} = \frac{L_x L_y}{T_x T_y}.$$

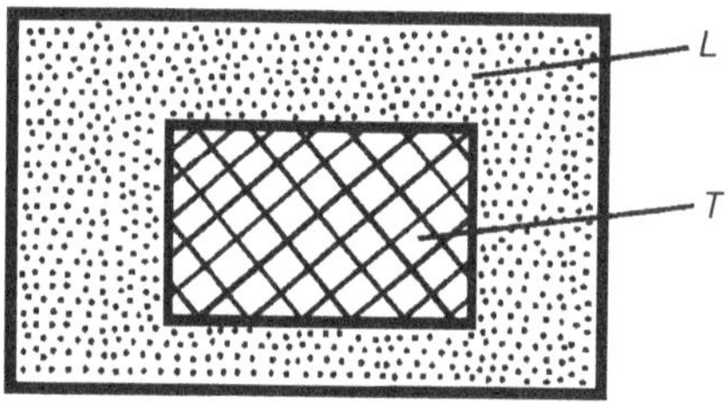

Figure I.13 Complete coverage of the target area by a damage zone. *Source:* From Wentzel [2].

There is another case when the damage zone is large and the target is small and all can be covered by the damage zone (Figure I.13). In this case,

$$u_{max} = 1.$$

One or another, at any ratio of damage zone and target size, there is some maximal value of u_{max} of damaged fraction U.

A random value U is a so-called mixed type value that has separate values with finite probabilities other than zero and intervals where the distribution function is continuous and only a certain probability density corresponds to each individual value. The distribution function (integral distribution law) for such random variables has breaks (jumps) in several points, and in the intervals between them grow continuously [4].

Remember that the integral law of distribution of the portion of the damaged area at one shot determines the probability of the occurrence that the portion of the damaged area U will be less than that specified by argument u.

$$F(u) = P(U < u)$$

Figure I.14 shows the form of the value distribution function U – a fraction of the damaged area at one shot. The extremes 0 and u_{max} values of U have probabilities other than zero. At these points, the distribution function has breaks (jumps). The jumps are equal to the probabilities of these values:

$$p_0 = P(U = 0), \quad p_m = P(U = u_{max}).$$

For intermediate values $0 < u < u_{max}$, the distribution function is continuous and the probability of each individual value is zero.

At a single shot, it is not difficult to build an exact law of U value distribution.

For example, let's look at the size ratio of the target T and the damage zone L (dashed rectangle), as shown in Figure I.15. It is clear that the probability of p_0 is nothing more than the probability of the epicenter O_1 hitting the zone shaded by Figure I.15 – the outer part of rectangle $A'B'C'D'$.

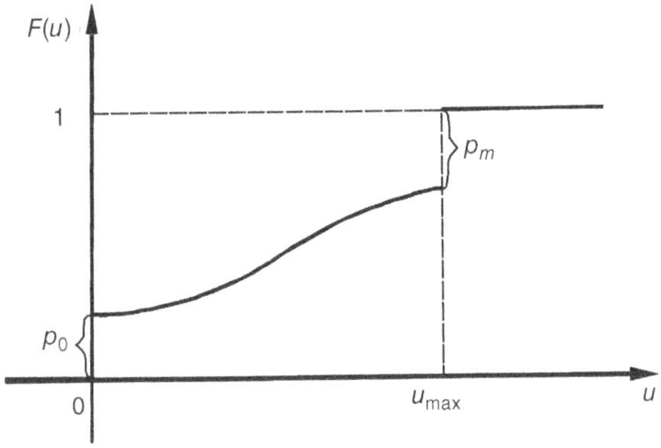

Figure I.14 The function of the damaged fraction distribution with one shot. *Source:* From Wentzel [2].

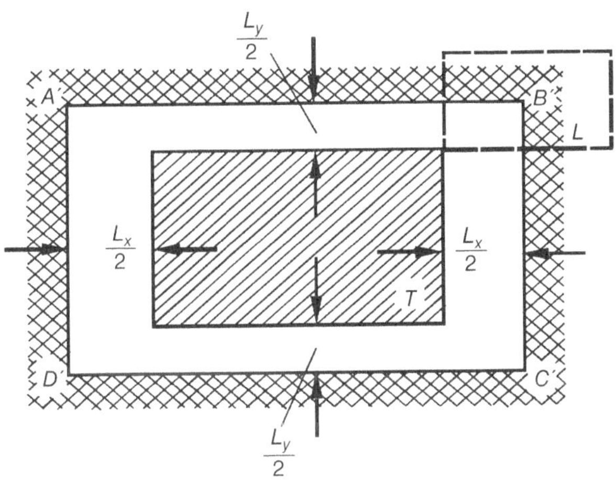

Figure I.15 Illustration for creating a distribution law of a portion of the damaged area. *Source:* From Wentzel [2].

Similarly, the probability p_m (Figure I.16) can be found as the probability of O_1 hitting the shaded area of *ABCD* corresponding to the maximum overlapping area (in this case $L_x \times L_y$).

A similar geometrical formation can be used to find the distribution function $F(u)$ for any intermediate value $0 < u < u_{max}$. To do this, you need to draw a series of curves (geometric places of the center of the damage zone) corresponding to the same damage fraction u. The probability of obtaining the damaged fraction less

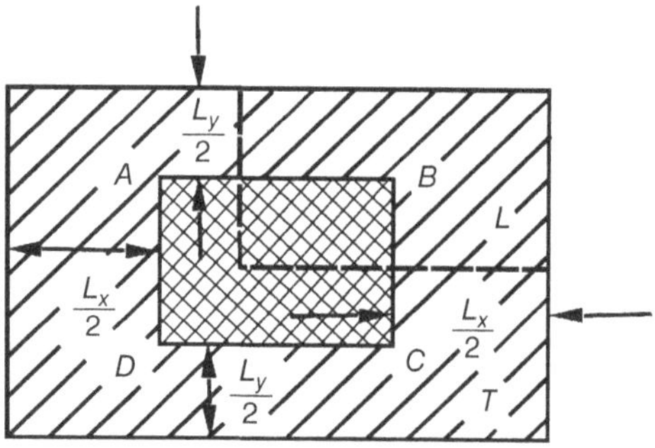

Figure I.16 Illustration for creating a distribution law of portion of the damaged area. *Source:* From Wentzel [2].

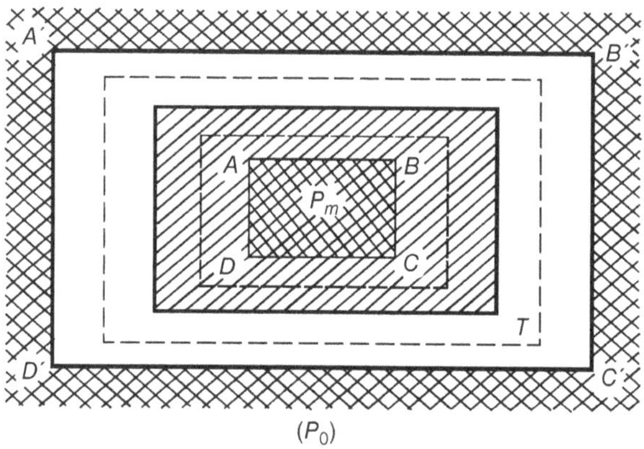

Figure I.17 Creation of the law of distribution of the fraction of damaged area for the intermediate values $0 < u < u_{max}$. *Source:* From Wentzel [2].

than u is the probability of the point hitting O_1 to the outer part of the corresponding curve.

However, establishing the exact shapes of these curves is a rather difficult task, and, taking into account that the rectangular configuration of the target and the damage zone is chosen enough freely, it is possible to replace all such curves with rectangles (see the dotted line in Figure I.17). The probability of the point O_1 hitting the outer part of each of such rectangles corresponding to the given value of the damaged fraction u is calculated by the formula

$$F(u) = 1 - \frac{1}{4}\left[\hat{\Phi}(\beta_u) - \hat{\Phi}(\alpha_u)\right]\left[\hat{\Phi}(\delta_u) - \hat{\Phi}(\gamma_u)\right], \tag{I.34}$$

where α_u, β_u, γ_u, δ_u – are the coordinates limiting the rectangle corresponding to this value of u in the 0x and 0y axes (Figure I.18)

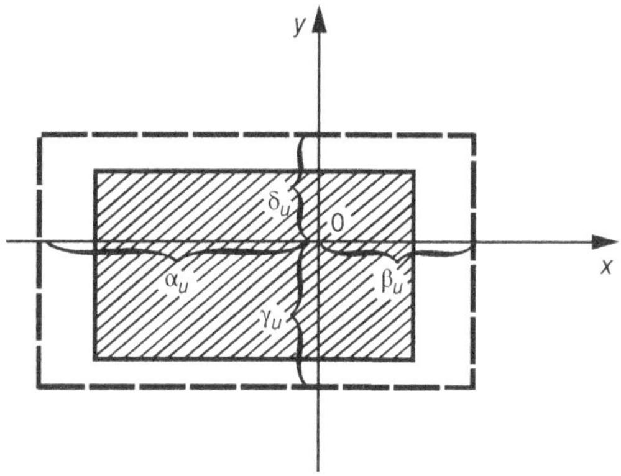

Figure I.18 Determination of probability obtaining a given fraction of damaged $F(u)$ using the normalized Laplace function. *Source:* From Wentzel [2].

For a rough approximation of the function $F(u)$, four points are almost enough: two boundary and two middle points, for example for $u = u_{\max}/3$ and $u = 2u_{\max}/3$.

Example

One shot is fired at a target with dimensions $T_x = 2$; $T_y = 1$. Damage zone has dimensions $L_x = L_y = 3$. There is no offset of the aim point. We need to draw on four points the function $F(u)$ of distribution of damage fraction U.

Solution

The creation of zones is shown in Figure I.19. The epicenter hitting in the zone (p_m) – the inner part of $ABCD$ rectangle – corresponds to covering the whole target $(u = 1)$. Hitting in the zone (p_0) (outer part of $ABCD$ rectangle) means no target coverage $(u = 0)$. We find the probability of hitting in the zones (p_0) and (p_m), as well as the outer parts of the dotted rectangles corresponding to $u = u_{\max}/3$ and $u = 2u_{\max}/3$

$$p_0 = 1 - \hat{\Phi}(3.5) \cdot \hat{\Phi}(4) = 0.025,$$
$$p_m = \hat{\Phi}(2) \cdot \hat{\Phi}(2.5) = 0.746,$$
$$F(1/3u_{\max}) = 1 - \hat{\Phi}(3.33) \cdot \hat{\Phi}(3.17) = 0.056,$$
$$F(2/3u_{\max}) = 1 - \hat{\Phi}(2.67) \cdot \hat{\Phi}(2.83) = 0.125.$$

The graph of the function $F(u)$, drawn on four points, is shown in Figure I.20.

Knowing the distribution function $F(u)$ of value U for one shot, it is easy to find its average value (expected value) $M = M[U]$. For random variables of mixed type, the expected value consists of two parts: sum and integral. The sum is applied to those values that have nonzero probabilities (i.e. where the distribution function makes a jump) and the integral to the area where the distribution function is continuous.

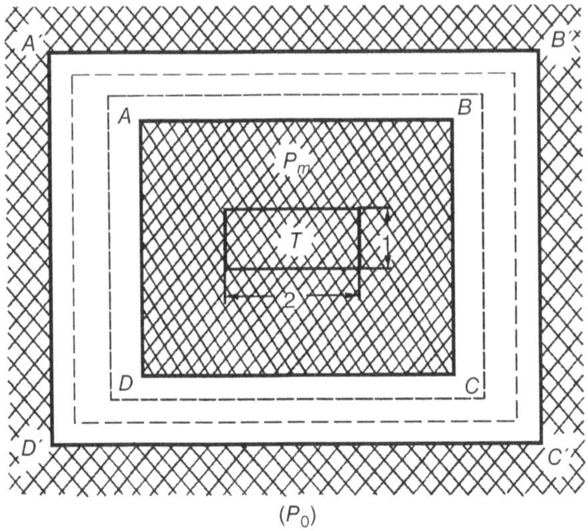

Figure I.19 Illustration to the example: the creation of the law of distribution of the fraction of damaged area. *Source:* From Wentzel [2].

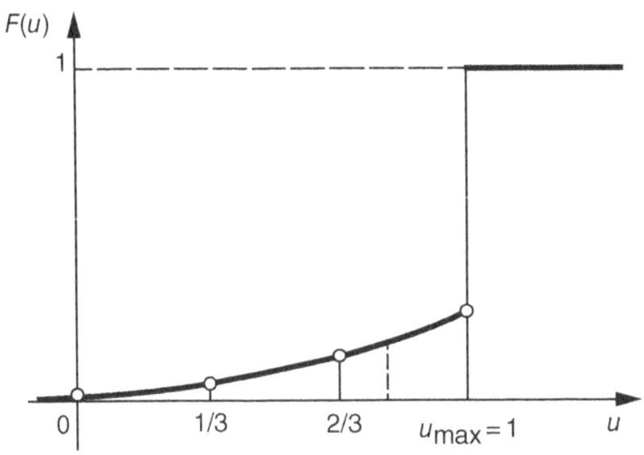

Figure I.20 Illustration to the example: function of distribution of the damaged area. *Source:* From Wentzel [2].

In our case, there are two jumps in the $F(u)$ distribution function: at 0 and u_{max}. Hence,

$$M[U] = u_{max}p_m + \int_0^{u_{max}} u(F(u))'_u du. \tag{I.35}$$

Knowing the function of $F(u)$ distribution of U value for one shot, it is easy to find the probability R_u that for one shot, the damaged fraction U will be not less than the set value u:

$$R_u = 1 - F(u).$$

Example

For the previous example, we need to find the probability that at least 80% of the target area will be damaged.

Solution

According to the graph in Figure I.20 at $u = 0.8$, we have

$$F(0.8) \approx 0.16, \quad R_{0.8} = 1 - 0.16 = 0.84.$$

This approximate method of finding the law of distribution of the damage fraction is suitable only for the case of one shot and cannot be extended to the case of several shots at the same area target.

I.3.7.2 Determining the Average Damage Fraction at Multiple Shots

There are methods to estimate the effectiveness of damaging area targets, suitable for both single and multiple shots. We won't be considering them here.

Let us consider here only an approximate method for determining the average fraction of the area damaged by several independent shots if the damage fraction by one shot is known. With an increasing number of shots n, the average fraction of the damaged area M_d grows not in proportion to the number of shots n, but more slowly. There are two reasons for this:

- with several shots fired, individual projectile areas may overlap;
- the average M_n fraction of damaging cannot exceed one at any number of shots and at unlimited increasing n can only approach to one asymptotically.

The exact relationship between M_n and the number of shots n is quite complex. However, numerous calculations have shown that for practical needs, it can in most cases be replaced by function like

$$M_n = 1 - (1 - M)^n, \tag{I.36}$$

where M is the average damaged fraction of a single shot.

Formula (I.36) is not accurate enough if the target size is large compared to the dispersion area of the projectiles.

I.4 Self-assessment Questions

1) What is the difference between the terms "effectiveness of the damaging effect" and "battle effectiveness"?
2) List the main types of ammunition and their damaging effects.

3) Describe the main types of degrees of damage.
4) Name and describe the generalized characteristics of damaging effects for contact ammunition.
5) What is the accumulation of damage?
6) Name and describe the generalized characteristics of the damaging effects of remote munitions.
7) How do the specified target sizes differ from their actual sizes for contact and remote ammunition?
8) List the kinds of targets and typical efficiency indicators of their damage.
9) List the main reasons for the dispersion of the ammunition when firing.
10) What is a picture plane and how is it built?
11) What's the dispersion law of ammunition when firing?
12) What are the similarities and the differences between systematic and group shooting errors?
13) What kind of shots are defined as dependent and why?
14) Why are probable deviations rather than standard deviations used in efficiency calculations?
15) How are generalized damage characteristics used in calculations of battle effectiveness?

References

1 Balagansky, I.A. and Merzhievsky, L.A. (2004). *Effects of Means of Destruction and Ammunition*. Novosibirsk: Novosibirsk State Technical University [in Russian].

2 Wentzel, E.S. (1964). *Introduction to Operations Research*. Moscow: Soviet Radio [in Russian].

3 Fendrikov, N.M. and Yakovlev, V.I. (1971). *Methods of Calculation of Combat Efficiency of Armament*. Moscow: Voyenizdat [in Russian].

4 Wentzel, E.C. (1969). *Probability Theory*. Moscow: Nauka [in Russian].

1

Fragmentation Ammunitions

1.1 Basic Concepts and Definitions. General Information

1.1.1 Classification of Fragmentation Ammunition

Fragmentation and high-explosive fragmentation ammunition represent one of the broadest and most developed classes of ammunition designed to damage practically all types of targets except concrete, underwater, and heavily armored ones. In a modern interpretation, fragmentation weapons are defined as weapons that attack a target with a high-speed flow of a large number of inert elements of the same type. These can be fragments of natural fragmentation, controlled fragmentation, and preformed elements, which are thrown, as a rule, by the explosion of a high-explosive charge.

Fragmentation ammunition is a remotely acting ammunition that can damage targets with a fragment flow at some distance from the explosion location, which is called a kill interval. They are characterized by a large variety of designs due to the multitude of combat missions and a variety of means of delivery to the target.

Since there are many types and forms of fragmentation weapons, and they are continuously being developed, it is not possible to provide a complete overall classification. Some of the partial classifications that can be identified include:

- by type of target;
- by type of weapon;
- by fragment field configuration;
- by the presence of fragment field control;
- by the way damage elements are formed;
- by the location of the explosion (ground, air);
- by individual design features.

According to the type of target, multipurpose (universal) fragmentation munitions (e.g. blast fragmentation artillery projectiles with natural fragmentation of

Damaging Effects of Weapons and Ammunition, First Edition. Igor A. Balagansky.

medium calibers) and specialized projectiles designed to attack targets of a certain class are distinguished. The latter primarily include antipersonnel ammunition designed to destroy non-protected and weakly protected personnel.

In the classification by type of weapon, the main determining factor is the maximum overload during the action. For fragmentation munitions (antipersonnel fragmentation grenade launchers, small caliber fragmentation bombs, cluster munitions, fragmentation warheads, etc.), the overload when fired is a determining factor, while for high-explosive fragmentation projectiles, the overload when interacting with the ground is also an important indicator. The value of the maximum overload to a significant degree determines the structural appearance of a weapon, including the wall thickness of the body, and, consequently, the filling of the explosive, the possibility of using a controlled fragmentation, or preformed elements, etc.

1.1.2 High-explosive Fragmentation Projectiles of Field Artillery

Field artillery projectiles include conventional high-explosive fragmentation projectiles or fragmentation-cumulative-type cluster munitions, and high-precision high-explosive fragmentation projectiles. The main Russian field artillery systems and their characteristics are presented in Table 1.1 (towed ordnance) and in Table 1.2 (mobile ordnance), where D_{max} – a maximum distance of fire; S – a rate

Table 1.1 Towed guns and mortars.

Ordnance	D_{max}, km	S, shots/min	B_d/D_{max}	Q, kg	M_{or}, t	n
82-mm mortar 2B14-1 "Podnos"	4.27	24	1/95	3.1	0.039	120
82-mm automatic mortar 2B9M "Vasilec"	4.27	120	1/95	3.1	0.62	300
120 mm transportable mortar 2C12 "Sani"	7.1	12	1/250	16.0	0.21	48
120-mm gun 2B16 "Nona-K"	8.8 (FBP) 7.0 (FBM) 12.8 (ARFBM)	8	1/342	17.3 (FBP) 16.0 (FBM)	1.2	80
122-mm howitzer D-30A	15.3	6–7	1/546	21.8	3.2	80
152-mm gun-howitzer D-20	17.4	5–6	1/290	43.56	5.6	60

Table 1.1 (Continued)

Ordnance	D_{max}, km	S, shots/min	B_d/D_{max}	Q, kg	M_{or}, t	n
152-mm howitzer 2A65 “Msta-B”	24.7 (FBP) 28.5 (FBP/GG)	8	1/363	43.56	7.0	60
152-mm gun 2A36 “Hyacinth-B”	28.5 (FBP) 33.0 (ARP)	6	1/267	46.0	9.8	60

Source: From Odintsov [1].

Table 1.2 Mobile guns and mortars.

Ordnance	D_{max}, km	S, shots/min	B_d/D_{max}	Q, kg	M_{or}, t	n
120-mm Mobile mortar 2C9 “Nona-S”	7.1 (FBM) 8.8 (FBP) 12.8 (ARFBM)	10	1/342	16.0 (FBM) 17.3 (FBP)	8.5	25
120-mm Mobile mortar 2C23 “Nona-SVC”	7.1 (FBM) 8.8 (FBP) 12.8 (ARFBM)	10	1/342	16.0 (FBM) 17.3 (FBP)	14.5	30
122 mm Mobile howitzer 2C1 “Gvozdica”	15.2	5	1/546	21.8	15.7	40
152-mm Mobile howitzer 2C3 “Akacia”	17.3	3	1/290	43.56	27.5	40
152-mm Mobile gun 2C5 “Hyacinth-C”	28.4 (FBP) 33.0 (ARP)	6	1/267	46.0	28.2	30
152-mm Mobile howitzer 2C19 “Msta-C”	24.7 (FBP) 28.5 (FBP/GG)	8	1/374	43.56	42.5	50
203-mm Mobile gun 2C7 (2C7M) “Pion” (“Malka”)	37.5 (FBP) 47.5 (ARP)	15 (2.5)	1/222	110.0	46.0	4 (8)
240-mm Mobile mortar 2C4 “Tulpan”	9.65 (FBM) 19.0 (ARFBM)	0.9	1/420	130.7	27.5	20

Source: From Odintsov [1].

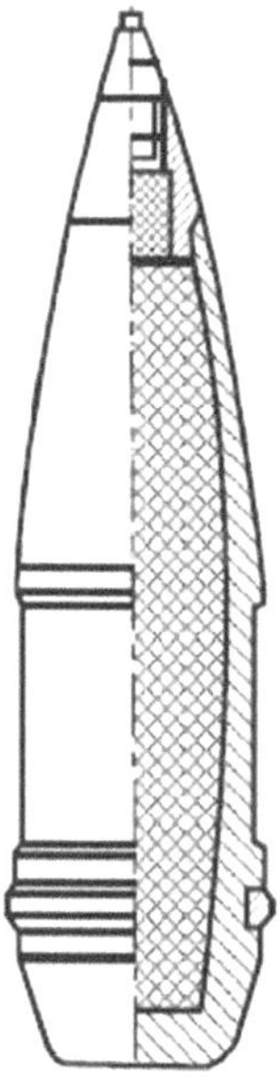

Figure 1.1 152-mm fragmentation blast projectile "Grif." *Source:* From Balagansky et al. [2].

of fire; B_d/D_{max} – individual error per maximum distance; Q – projectile mass; M_{or}, – ordnance mass; n – allowance of ammunition; FBP – fragmentation blast projectile; FBM – fragmentation blast mine; ARP – active–reactive projectile; ARFBM – active–reactive fragmentation blast mine; FBP/GG – fragmentation blast projectile with a gas generator.

The trend is clearly seen in the formation of two classes of artillery weapons – powerful long-range systems of 152 and 203 mm calibers intended for the armament of the main forces and 120 mm caliber systems belonging to the class of assault guns and intended for the armament of mobile forces.

Natural fragmentation projectiles are the most common and universal type of artillery ammunition. Their main advantages are simplicity of design and manufacture, relatively low cost, high power at firing, and penetration into a barrier. The main disadvantage is their insufficient damage effect due to the unsatisfactory fragmentation of their shells.

A characteristic representative of the FBP is a Russian 152-mm fragmentation blast projectile "Grif" (Figure 1.1). The projectile is designed for D-20, ML-20, 2C3 field guns, and can also be used for firing from 2A65 and 2C19 systems. The maximum distance of fire for this projectile is 17.4 km. The mass of the projectile is 43.56 kg, length is 709 mm (4.66 calibers), the maximum initial velocity is 655 m/s.

The dominant trend in the development of field artillery systems of the main force and its ammunition is the increasing distance of fire. For the new generation of 155-mm NATO guns with 52 caliber barrel length, the required distance is 40–50 km. Methods for increasing the distance directly related to the projectile include the use of:

- active–reactive projectiles;
- projectiles with reduced aerodynamic resistance;
- sub-caliber projectiles with separable sabots;
- direct-flow jet propulsion projectiles.

Figure 1.2 shows the 152 mm active–reactive FBP 3OF22. The aerodynamic resistance of the ERFB/BB projectile (Figure 1.3) is reduced by giving it a full-ogive shape and by reducing the bottom resistance with the use of a gas generator. The 6.13 caliber length projectile has no cylindrical surfaces at all, and its centering in the barrel is maintained by the protrusions on the body.

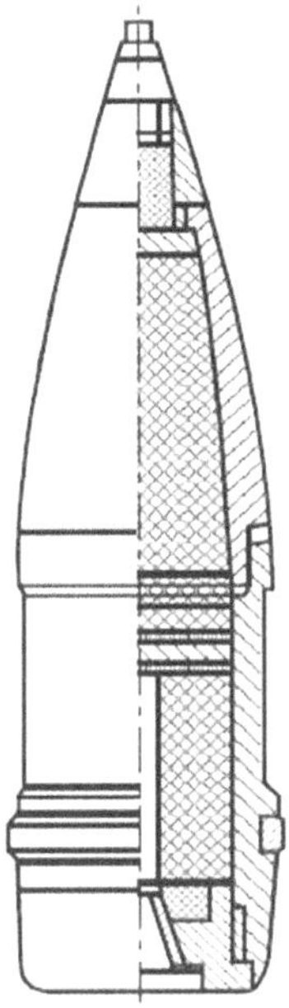

Figure 1.2 152 mm active reactive FBP 3OF22. *Source:* From Balagansky et al. [2].

This projectile has a maximum range of 38–40 km. At present, the ERFB/BB full-ogive-shaped projectiles are standard in all 155-mm ordnances of NATO and several other countries.

1.1.3 Brief Description of Other Classes of Fragmentation Ammunition

Along with FBP, there are many other types of fragmentation ammunition, their appearance and characteristics determined by their combat missions. We will not even try to describe all these types here.

The British anti-vehicle mine HB-876 (Figure 1.4) is an example of a design that turns into a stationary vertical position after falling to the ground. After firing from the cassette, the mine is lowered using a parachute, which is detached after landing. The spring feet on the body of the mine bend to the sides to ensure the mine's required vertical orientation. In addition to the circular field of damage of the meniscus elements, the mine's damage effect is also provided by a hemispherical shell located on the upper end of the high-explosive charge.

Figure 1.3 155-mm full-ogive shape projectile ERFB: 1 – fuze; 2 – body; 3 – protrusions; 4 – gas generator. *Source:* From Balagansky et al. [2].

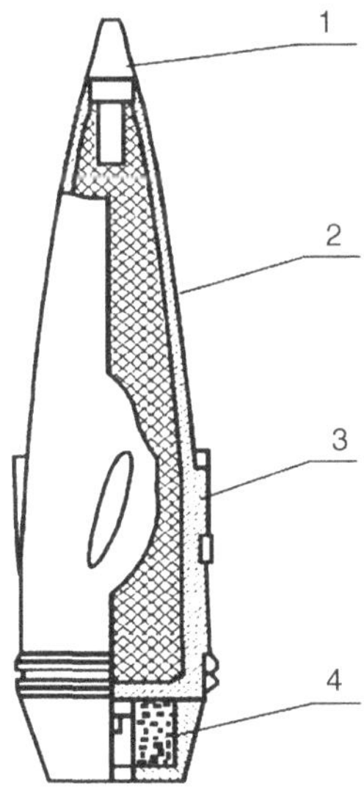

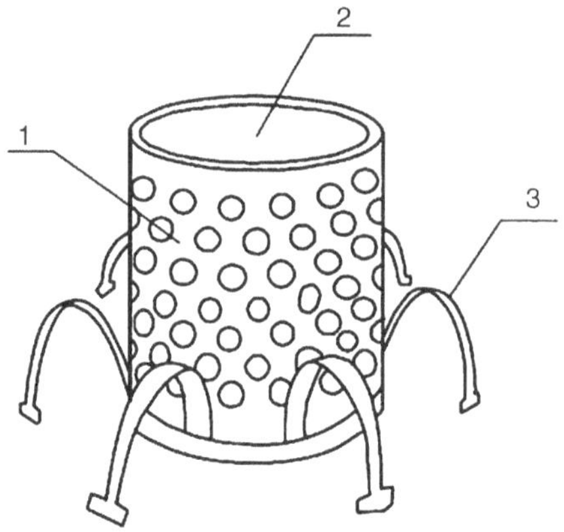

Figure 1.4 British anti-vehicle mine HB-876: 1 – fragmentation shell with meniscus damaging elements; 2 – hemispherical shell, which creates explosively formed penetrator; 3 – spring feet of the vertical orientation mechanism. *Source:* From Balagansky et al. [2].

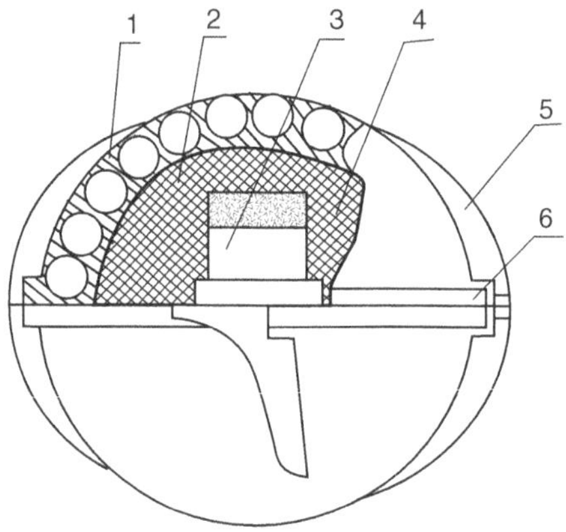

Figure 1.5 Russian antipersonnel submunition SHOAB-0.5: 1 – shell with preformed fragments; 2 – explosive charge; 3 – inertial fuse; 4 – detonator; 5 – aerodynamic tides; 6 – connecting belt. *Source:* From Balagansky et al. [2].

The advantage of ammunition with preformed fragments is its high efficiency provided by the stable characteristics of the fragmentation field. In modern designs, preformed fragments are typically compact bodies (sphere, cube, short cylinder) made of steel or tungsten-based heavy alloy.

An example of a munition with steel spheres in an aluminum alloy shell is a SHOAB-0.5 spherical submunition (Figure 1.5). Typical examples of ammunition with preformed fragments can be hand grenades (Figure 1.6), engineering mines with circular and directional fields, warheads of antiaircraft-guided missiles, etc.

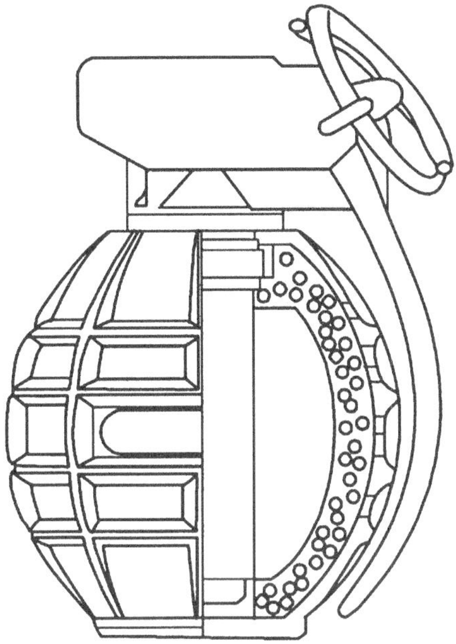

Figure 1.6 German hand grenade M-DN21 (weight 225 g, fragment mass 0.45 g). *Source:* From Balagansky et al. [2].

It should be noted that the use of a light binder for preformed fragments (Figure 1.5) and the absence of the supporting shell leads to a rapid "blowing out" of the binder material by the products of detonation, the stream of the products of detonation into the resulting gaps, and, as a consequence, a strong decreasing of fragments velocity. In such cases, binders with a density comparable to the density of the fragment material should be used. At the same time, despite a decrease in the explosive filling rate, a significant increase in the fragment velocity is observed.

Another type of fragmentation warhead is the continuous rod warhead. Warheads of this type use a long rod "folded" (connected end-to-end) in a bundle surrounding the high-explosive filler. When the warhead functions, these rods expand to form a complete "hoop" centered at the munition's axis. This circle expands as a continuous rod until its circumference is approximately equal to the length of the unfolded rod, after which it will fragment. Figure 1.7 shows the design and function of a continuous rod warhead. The ring of fragmentation is highly effective against aerial targets, and as a result, continuous rod warheads are often found in antiaircraft munitions.

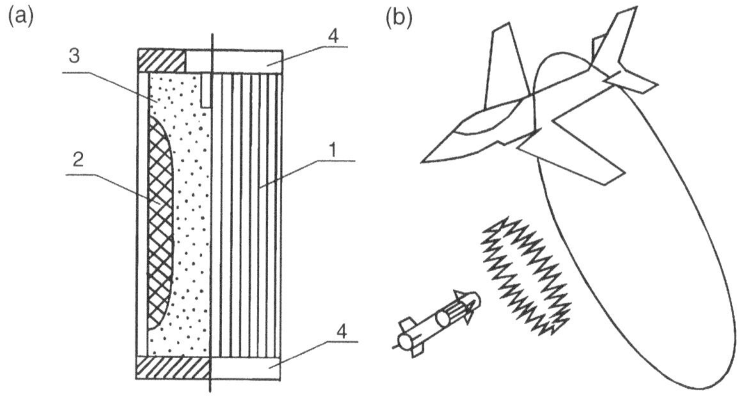

Figure 1.7 Continuous rod warhead of an antiaircraft-guided missile and its effect on a target (a, b): 1 – set of rods; 2 – damper; 3 – HE charge; 4 – covers. *Source:* From Odintsov [3].

1.2 The Mechanics of High-speed Deformation and Destruction of Shells Under the Action of an Explosion

Crushing of a projectile or warhead shell into fragments is a complex, rapid process of transforming the energy of the explosion products into the work of deformation, shell destruction, and throwing of fragments. This process can be roughly divided into the following stages: initiation of the explosive charge; detonation and formation of the explosion products; loading and throwing of the shell by the explosion products with the simultaneous onset of plastic deformation of the metal; the movement of the expanding shell with simultaneous initiation and growth of microcracks and micro-damage; formation of main macrocracks with their subsequent exit to the outer surface of the shell; full growth of cracks to the entire shell thickness with the formation of fragments and the expiration of the explosion products in through cracks; the movement of fragments in the air; their interaction with the target.

To date, the destruction of cylindrical shells loaded with an explosion has been thoroughly studied. A significant contribution to the study of the behavior of the explosive destruction of the shells to fragments was made by J. Rinehart, J. Pearson, N. Mott, S. Hoggat, R. Recht, J. Taylor, D. Grady, M. Held, V. Odintsov, E. Gryaznov, A. Gerasimov, et al.

The process of explosive expansion of a cylindrical shell can be seen in Figure 1.8. For the description of the process, we will introduce the following designations: b_0 – initial external radius of the shell; b_c – radius of crack formation (occurrence of cracks); b_f – radius of destruction (occurrence of through cracks

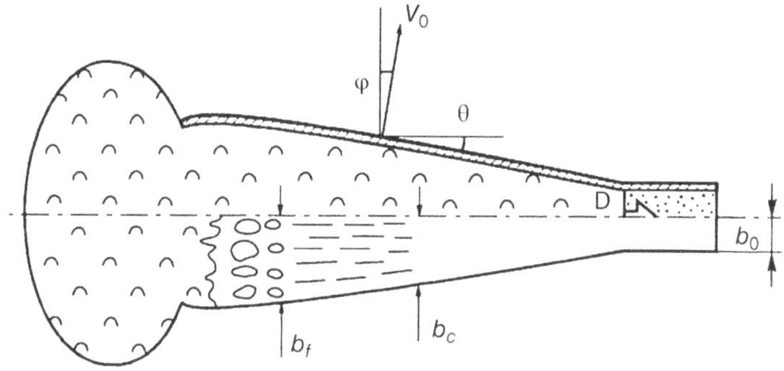

Figure 1.8 Scheme of fragmentation shell throwing with detonation products. *Source:* From Odintsov [4].

fixed on the break of products of detonation on the external surface); D – detonation velocity of HE charge; V_0 – velocity of the shell expansion; φ – the angle between the direction of vector V_0 and the plane perpendicular to the axis of HE charge; θ – the angle of rotation of the shell $\theta = 2\varphi$.

The dynamic deformation of the shells prior to failure exceeds the limit deformations obtained as a result of static tests. This phenomenon – the dynamic peak of plasticity – is explained by shell deformation conditions under the influence of contact explosive loading and material properties.

The most homogeneous loading regime can be realized by the central initiation of spherical shells.

The maximum velocity V_0 obtained by fragments for cases of axial and spherical symmetry can be calculated using the Pokrovskiy–Gurney formulas:

- for axial symmetry

$$V_0 = 0.5\varphi_0 D\sqrt{\frac{\alpha}{2-\alpha}}, \tag{1.1}$$

- for spherical symmetry

$$V_0 = 0.5\varphi_0 D\sqrt{\frac{5\alpha}{10-4\alpha}}, \tag{1.2}$$

where α – fill factor

$$\alpha = \frac{m_{HE}}{(M + m_{HE})};$$

m_{HE} – mass of HE charge; M – mass of shell; D – velocity of detonation.

For a shell of natural fragmentation $\varphi_0 \cong 0.98$, for a shell of controlled fragmentation $\varphi_0 \cong 0.90$–0.95, for a shell with preformed elements $\varphi_0 \cong 0.80$–0.85.

The approximate value of the angle of declination of a fragment field is assumed to be equal to

$$\varphi = \arcsin\left(\frac{V_0}{2D}\right). \tag{1.3}$$

When thick-walled cylindrical shells are blown up, two independent systems of cracks are formed in their walls: inside the wall and on the inner surface. The first system is a system of tear-off cracks oriented mainly in the radial direction, i.e. orthogonally maximum tensile stress. Deviations of cracks from this direction can be caused by local non-homogeneity of the stress field caused by both non-homogeneous material structure and the mutual influence of cracks. The second system of cracks comprises shear cracks oriented mainly along the main shear direction, i.e. at an angle of 45° to the radial direction. This fracture mechanism is associated with the fact that on the inner surface of the shell for a significant time interval, the state of unequal compression is realized, which prevents the formation of a tear-off fracture.

The expansion of the shell during an explosion leads to the formation and branching of cracks on the outer surface of the shell and the plastic flow of the material in the area adjacent to the inner surface of the shell. At high deformations and their high velocities, the plastic flow process loses homogeneity and the deformation localization zones (localization strips) with a thickness of 1–20 μm occur, located mainly along the planes of maximum tangential stresses. The localization strips are the regions of the easiest crack propagation.

Figure 1.9 shows the configurations of fragments and fracture surfaces formed by the fracture of shells. Areas of brittle radial fracture, internal microcracks, and

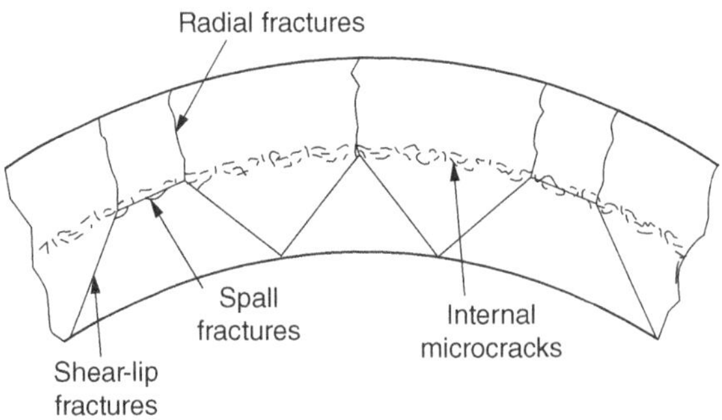

Figure 1.9 Configurations of fragments and fracture surfaces resulting from the shell fracturing process. *Source:* From Balagansky et al. [5]. Reprinted with permission of Elsevier.

shear-lip fracture can be identified in the cross-section of the fragments. Spall fracture surfaces leading to shell thickness division do not normally form in cylindrical shells.

Separation of the shell along its thickness caused by spall fracture is realized in rarefaction waves under certain conditions. The possibility of the occurrence of spall directly depends on the value of the pressure gradient behind the shock wave front and can be implemented by means of a special loading scheme. Spall fracture is typical for spherical shells at central initiation, which is associated with the high intensity of the rarefaction waves inside of such shells at blast loading.

The important role of the method of explosive loading on the behavior of the destruction of the shells to fragments and the formation of fragment fields should be noted.

To illustrate the above, we will present the results of numerical simulation of the process of explosive loading of two variants of the spherical shell with the central (a) and peripheral (b) initiation of detonation in the explosive charge [5]. For the spherical shell, the influence of the method of loading is shown evidently.

Modeling was performed using the ODVAX software package. The results are presented for comparison in pairs for both variants at appropriate moments in time. The figures show the configurations at the initial stage and at such moments in time when at least one variant in the shell is implemented tensile stress. The presence of tensile stresses was controlled in the calculation process by the value of the average stress (first invariant of the stress tensor). The possibility of the appearance of a spall was estimated with the use of the simplest criterion of destruction – maximal tensile stress:

$$\sigma = \frac{1}{3}(\sigma_{11} + \sigma_{22} + \sigma_{33}).$$

Tensile stresses exceeding the spall strength of the shell material were output in the form of isolines. Particles with corresponding parameters were drawn as isolines. Spall strength value for medium carbon steel:

$$\sigma^* \cong 2\,\text{GPa}.$$

Figures 1.10–1.20 shows configurations of flow fields and isolines of only tensile stresses in different moments of time after initiation of HE charges (TG 36/64) in a spherical shell of 85 mm diameter with central (a) and peripheral (b) initiation. In Figures 1.10–1.15, one can observe the evolution of the detonation front in HE. During the passage of the first wave pulse in the shell at the central initiation (a), the state of tree-axial non-uniform tension is realized with the maximum value of the tensile stress oriented perpendicular to the shell surface, reaching 6.5 GPa, which is much higher than the spall strength of carbon steel. The area of tensile stress at the time $t = 7.5$ μs covers more than half of the middle surface of the shell.

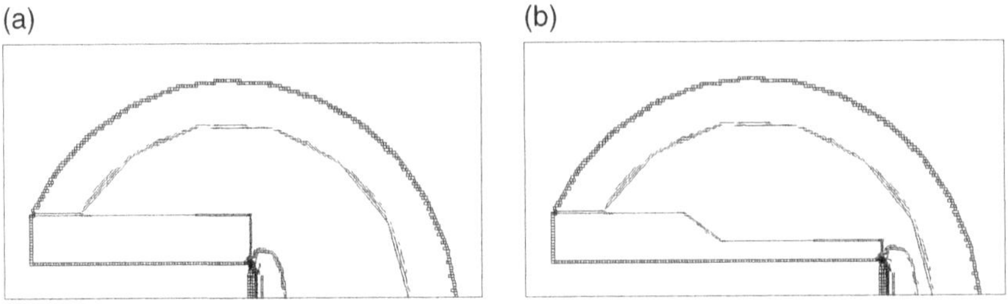

Figure 1.10 Configurations of flow fields at time $t = 1\ \mu s$ since the initiation of the HE charges in shells with a central (a) and peripheral (b) initiation. *Source:* From Balagansky et al. [5]. Reprinted with permission of Elsevier.

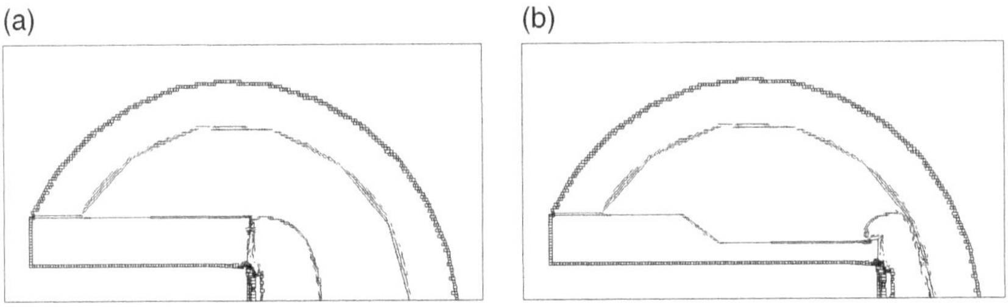

Figure 1.11 Configurations of flow fields at time $t = 2\ \mu s$ since the initiation of the HE charges in shells with a central (a) and peripheral (b) initiation. *Source:* From Balagansky et al. [5]. Reprinted with permission of Elsevier.

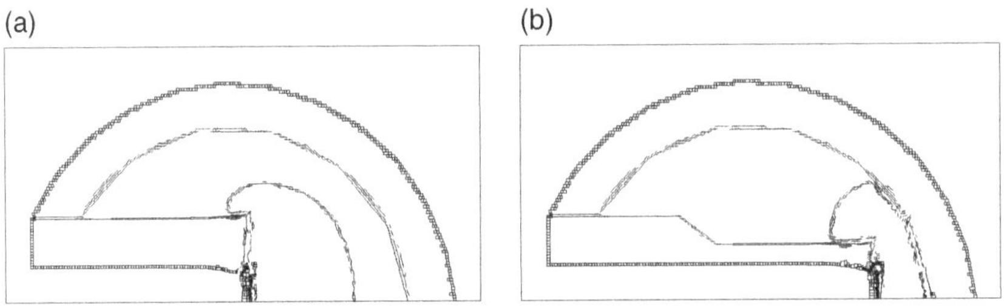

Figure 1.12 Configurations of flow fields at time $t = 3\ \mu s$ since the initiation of the HE charges in shells with a central (a) and peripheral (b) initiation. *Source:* From Balagansky et al. [5]. Reprinted with permission of Elsevier.

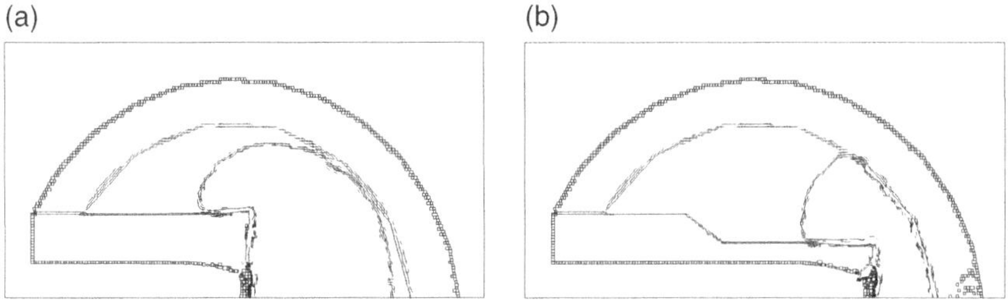

Figure 1.13 Configurations of flow fields at time t = 4 μs since the initiation of the HE charges in shells with a central (a) and peripheral (b) initiation. *Source:* From Balagansky et al. [5]. Reprinted with permission of Elsevier.

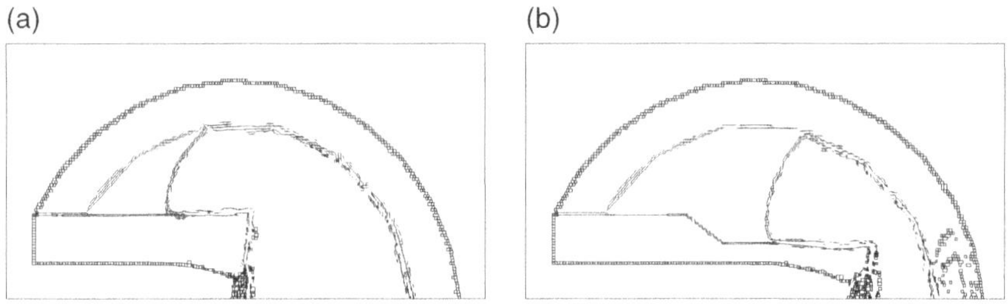

Figure 1.14 Configurations of flow fields at time t = 5 μs since the initiation of the HE charges in shells with a central (a) and peripheral (b) initiation. *Source:* From Balagansky et al. [5]. Reprinted with permission of Elsevier.

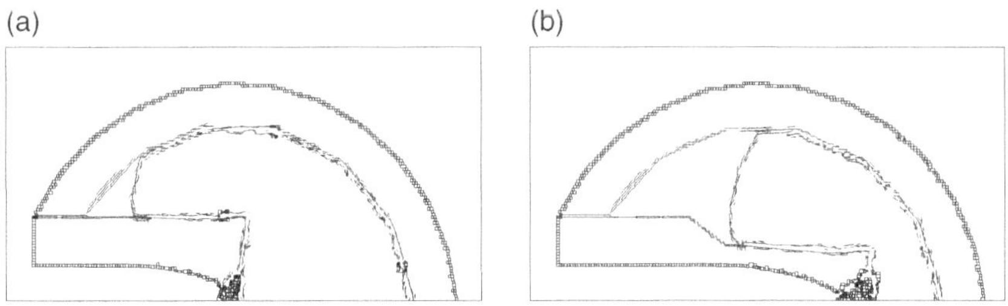

Figure 1.15 Configurations of flow fields at time t = 6 μs since the initiation of the HE charges in shells with a central (a) and peripheral (b) initiation. *Source:* From Balagansky et al. [5]. Reprinted with permission of Elsevier.

(a) (b)

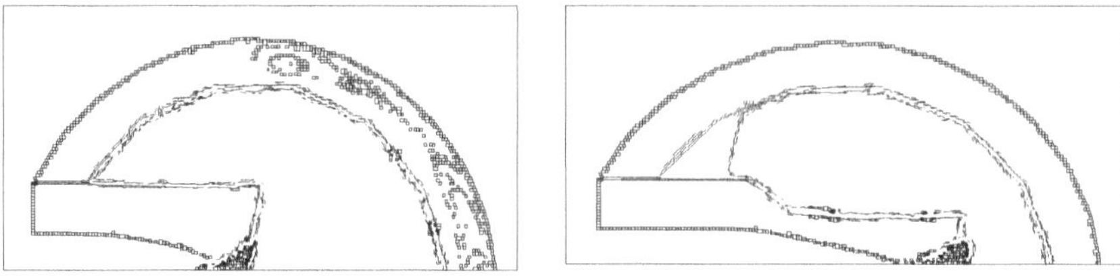

Figure 1.16 Configurations of flow fields at time t = 7.5 μs since the initiation of the HE charges in shells with a central (a) and peripheral (b) initiation. *Source:* From Balagansky et al. [5]. Reprinted with permission of Elsevier.

(a) (b)

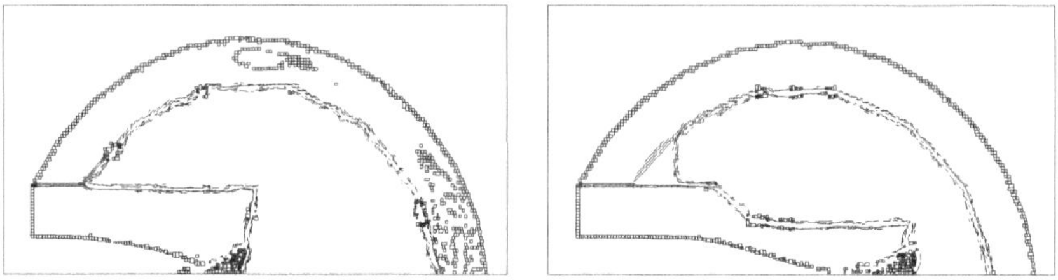

Figure 1.17 Configurations of flow fields at time t = 8 μs since the initiation of the HE charges in shells with a central (a) and peripheral (b) initiation. *Source:* From Balagansky et al. [5]. Reprinted with permission of Elsevier.

(a) (b)

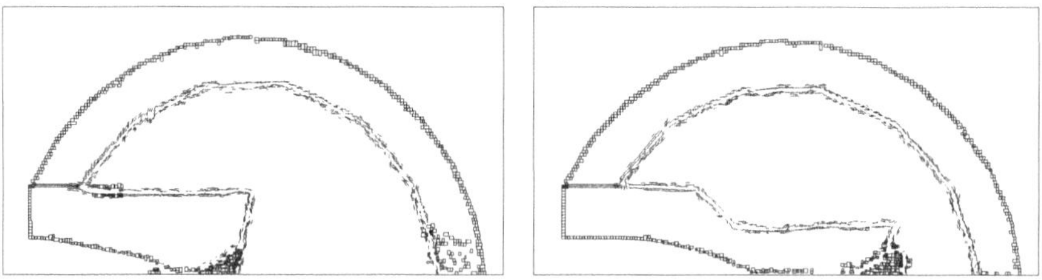

Figure 1.18 Configurations of flow fields at time t = 9 μs since the initiation of the HE charges in shells with a central (a) and peripheral (b) initiation. *Source:* From Balagansky et al. [5]. Reprinted with permission of Elsevier.

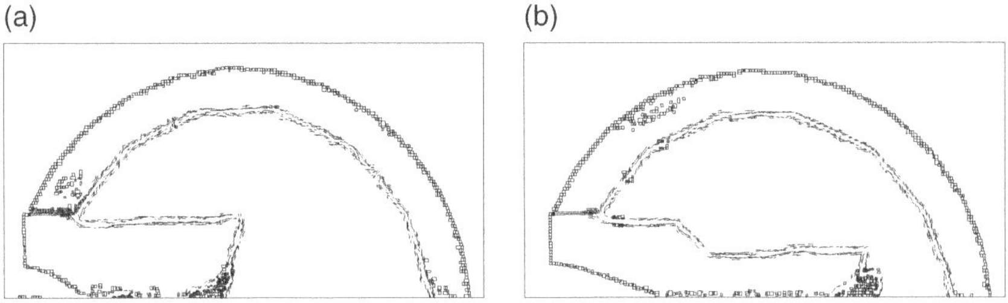

Figure 1.19 Configurations of flow fields at time t = 11 μs since the initiation of the HE charges in shells with a central (a) and peripheral (b) initiation. *Source:* From Balagansky et al. [5]. Reprinted with permission of Elsevier.

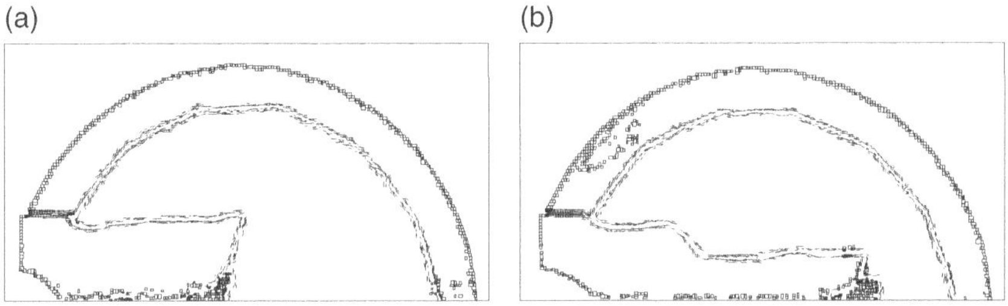

Figure 1.20 Configurations of flow fields at time t = 12 μs since the initiation of the HE charges in shells with a central (a) and peripheral (b) initiation. *Source:* From Balagansky et al. [5]. Reprinted with permission of Elsevier.

It should be noted that there is no way to obtain a state of tree-axial tensile strength in a solid material, except during its explosive loading. It is important that in the entire tensile stress area, the orientation of the maximum stress is perpendicular to the median surface of the shell.

For a shell with peripheral initiation (b), a completely different picture is observed. The areas covered by the tensile stress are significantly smaller in this case. The direction of action of the maximum stress does not coincide with the normal to the middle surface of the shell, which leads to a significant change in the behavior of destruction. Thus, with a change of position of an initiation point from the center to the periphery, it is necessary to expect a decrease in the intensity of spall phenomena. The results of modeling correspond well to the experimental data.

The final type of fragments will be determined by which of the competing destruction mechanisms prevails. Through brittle cracks are typical for brittle material shells (cast iron). For very ductile shells, a shear fracture is more prevalent.

1.3 Modeling the Processes of Explosive Fragmentation of Shells Using Standard Samples

1.3.1 The Basic Theorem of the Dimensional Theory

Similarity and Dimension Theory developed by L.I. Sedov [6] is a powerful tool for the analysis of continuous media mechanics processes. The combination of the conclusions of dimensional and similarity theory with the general analysis of the physical phenomena mechanism can serve as a useful theoretical method of research. Physical laws, established theoretically or directly from experience, are functional dependencies between the values characterizing the phenomenon under study. The numerical values of these dimensional values depend on the choice of the system of units of measurement, which is in no way related to the essence of the phenomenon. Therefore, the functional dependencies expressing physical regularities that do not depend on the system of units of measurement should possess some special structure. Let us have a dimensional value a, which is a function of independent dimensional values a_1, a_2, ..., a_n.

$$a = f(a_1, a_2, ..., a_k, a_{k+1}, ..., a_n); \tag{1.4}$$

Some of these parameters may be variables in the process under consideration, others may be constants.

Let us clarify the structure of the function (1.4) in the assumption that this function represents some physical law independent of the choice of the system of units. Assume that among the dimensional quantities a_1, a_2, ..., a_n, the first k quantities ($k \leq n$) have independent dimensions. Independence of dimensions means that a formula expressing the dimensionality of one of the quantities cannot be represented as a combination in the form of a degree monomial from the formulae of dimensions for other quantities. For example, the dimensions of length L, speed L/T, and energy ML^2/T^2 are independent; the dimensions of length L, speed L/T, and acceleration L/T^2 are dependent. Among mechanical quantities, there are usually no more than three with independent dimensions. According to the π – theorem, using a relative system of units of measurement, the ratio (1.4) can be represented as follows

$$P = f\,(1, 1..., P_1, ..., P_{n-k}). \tag{1.5}$$

Thus, the relation between $n + 1$ dimensional quantities a, a_1, ..., a_n, independent of the choice of the system of units of measurement, takes the form of the relation between $n + 1 - k$ quantities $P, P_1, \ldots, P_{n-k}$, representing dimensionless combinations of $n + 1$ dimensional quantities. Thus, the system of defining parameters should possess the property of completeness.

Any physical relationship between dimensional quantities can be formulated as a relationship between dimensionless quantities. This, in fact, is the source of useful applications of the similarity theory and dimensions method to the investigation of mechanical problems.

1.3.2 Dimensional Analysis for Fragmentation Processes

Let us apply in practice ideas of dimensional theory for the estimation of fragmentation of shells into fragments [4].

The fragmentation mechanisms are determined by grain size, substructures, and inclusions. Therefore, the problem of evaluating the fragmentation of ammunition shells into fragments has two sides: the choice of steel grade and heat treatment required to obtain some predicted quality of fragmentation, and the calculation of the total number of fragments and their distribution in mass groups and angles of propagation.

The main factors influencing fragmentation are the chemical composition of steel, grain size, physical and mechanical properties.

1.3.2.1 Chemical Composition

For steels, carbon content is a major determining factor. The following steel classes are distinguished by this indicator: $C < 0.3\%$ – low-carbon steels; $0.3 < C < 0.7\%$ – medium-carbon steels; $C > 0.7\%$ – high-carbon steels. As the carbon content increases, the fragmentation increases. An especially intensive increase in fragmentation occurs when the carbon content is more than 0.83% when the perlite–ferrite structure of steel turns into perlite–cementite. Inclusions of brittle cementite in steel are hotbeds of destruction and contribute to more intensive fragmentation. Other elements having the same effect as additives are silicon, phosphorus, and sulfur.

1.3.2.2 Grain Size

Grinding ferrite grain leads to higher resistance to brittle fracture, including lower cold breaking threshold and simultaneously to higher yield strength. As the grain size increases, the fragility of steels increases.

The physical and mechanical properties of metal depend on its chemical composition and grain size. Standard static mechanical properties include: σ_s – yield strength; δ – elongation at break; a_H – impact strength.

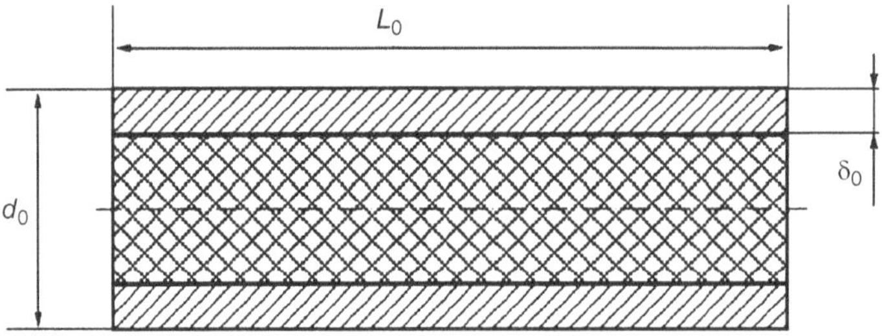

Figure 1.21 Cylindrical shell filled with explosive. *Source:* From Odintsov [4].

The most indicative characteristic is the specific work of fracture E_s (J/m^2). Of the values mentioned above, the impact strength a_H, having the same dimension, is usually used.

Taking yield strength σ_s and specific work of fracture E_s as the main characteristics of a metal, we will perform a dimensional analysis for the fragmentation process. Let us consider the fracture of a cylindrical shell (Figure 1.21) with sizes d_0, δ_0, L_0. In the process of fragmentation, a spectrum of fragments with some characteristic mass m_c is formed. This mass is a function of geometric characteristics d_0, δ_0, L_0, properties of HE ρ_0, D and properties of the metal γ_0, σ_s, E_s, i.e.

$$m_c = f(d_0, \delta_0, L_0, \rho_0, D, \gamma_0, \sigma_s, E_s). \quad (1.6)$$

Here ρ_0 – explosive density; D – detonation velocity; γ_0 – metal density.

This ratio connects $n+1 = 9$ dimensional parameters. According to the π-theorem it can be replaced by a ratio connecting $n+1-k$ of dimensionless parameters (k – number of independent dimensions, in this case, $k = 3$, i.e. $n+1-k = 6$). Five of the six dimensionless variables can be easily found by known analogs of the mechanical problems. This includes the following complexes:

$$\frac{m_c}{\gamma_0 \delta_0^3}; \quad \frac{\delta_0}{d_0}; \quad \frac{L_0}{d_0 - 2\delta_0}; \quad \frac{\rho_0}{\gamma_0}; \quad \frac{\rho_0 D^2}{\sigma_s}. \quad (1.7)$$

The basic dimensionless parameter of fragmentation F, including characteristics of a combination "HE-metal," including specific work of fracture, we will form by rules of the theory of dimensions. In general view

$$F = \rho_0^{\alpha}(d_0 - 2\delta_0)^{\beta} D^{\gamma} \sigma_s^{\delta} E_s^{\varepsilon}, \quad (1.8)$$

Here, dim $\rho_0 = ML^{-3}$; dim $(d_0 - 2\delta_0) = L$; dim $D = LT^{-1}$;
dim $\sigma_s = ML^{-1}T^{-2}$; dim $E_s = MT^{-2}$.
Then

$$\dim F = (ML^{-3})^{\alpha}(L)^{\beta}(LT^{-1})^{\gamma}(ML^{-1}T^{-2})^{\delta}(MT^{-2})^{\varepsilon} = M^{(\alpha+\delta+\varepsilon)} L^{(-3\alpha+\beta+\gamma-\delta)} T^{(-\gamma-2\delta-2\varepsilon)}$$

As for the dimensionless value F dim $F = M^0L^0T^0$, then we get a system of equations connecting the five parameters α, β, γ, δ, ε:

$$\alpha + \delta + \varepsilon = 0; \quad -3\alpha + \beta + \gamma - \delta = 0; \quad -\gamma - 2\delta - 2\varepsilon = 0. \tag{1.9}$$

Let us involve in addition following known physical ratios: $n_\theta \sim (d_0/2 - \delta_0)^{1/4}$ – experimental data; $n_\theta \sim V_0$ – Pokrovsky–Reinhart–Pearson ratio; $V_0 \sim D$ – formula of Pokrovsky–Stanyukovich.

From here we get $n_\theta \sim (d_0/2 - \delta_0)^{1/4}D$ and consequently, comparing it with the original expression for F, we get $\beta = \gamma/4$. For completing the system of equations, we take the index of degree α at density ρ_0 equal to 1. In this case, solving the system, we will find

$$\alpha = 1; \quad \beta = 1/2; \quad \gamma = 2; \quad \delta = -1/2; \quad \varepsilon = -1/2$$

and therefore,

$$F = \rho_0 D^2 \left(\frac{d_0}{2} - \delta_0\right)^{1/2} / \left(E_S^{1/2}\sigma_S^{1/2}\right). \tag{1.10}$$

In the first approximation, we exclude relatively insignificant dimensionless parameters ρ_0/γ_0, $\rho_0 D^2/\sigma_s$, and introducing designations $\delta_0/d_0 = \delta_d$,

$$\frac{L_0}{d_0 - 2\delta_0} = \frac{L_0}{d_a} = \lambda_0, \quad \text{we get} \quad \frac{m_c}{\gamma_0 \delta_0^3} = f(\delta_d, \lambda_0, F). \tag{1.11}$$

During experimental investigations of the fragmentation of cylinders, it is practical to limit the number of defining dimensionless parameters. Usually, the elongation of the cylinder is fixed, i.e. $\lambda_0 =$ const, and λ_0 is determined by certain physical conditions. In this case, the changes of all fragmentation characteristics in the field of arguments δ_d and F are investigated.

Being part of a dimensionless combination of F of linear size d_a (inner diameter of the shell) indicates the presence of a scaling effect at fragmentation, which manifests itself in the intensification of the fragmentation when the diameter (caliber) of the cylinder increases.

Fragmentation spectrums provide important information about the physics of fracture and the behavior of materials, allow for sufficiently substantiated comparison and selection of materials for a particular purpose, and therefore still retain a significant role.

1.3.3 Ratios for the Total Number of Fragments

There are many known formulas that can be used to calculate the total number of fragments produced by a fragmentation projectile explosion. These ratios are mainly empirical and are of little importance in practice because they require the substitution of different empirical coefficients when used. Here, we will give only one formula for calculating the total number of fragments with a mass more than 0.5 g $N_{0.5}$, proposed by V.A. Odintsov.

$$N_{0.5} = K\frac{\alpha}{\psi}d_0^2D^2,$$

where K is the factor depending on the geometry of the shell (for FBP $K = 70$–100); α, ψ – fill factor of HE and narrowing at the break of the shell material (both in fractions or as percentages); d_0 – outer diameter of the shell (caliber) in decimeters; D – detonation velocity of HE, km/s.

The structure of the formula shows how the total number of fragments changes as the design parameters of the ammunition change.

1.3.4 Standard Fragmentation Cylinders

Standard fragmentation cylinders are designed to investigate the fragmentation of ammunition shell material (mainly steel) and to determine the throwing and fracture properties of HE [7]. In Russia, two basic schemes of standard cylinders are used nowadays: closed (a) and open (b), which are shown in Figure 1.22. For each

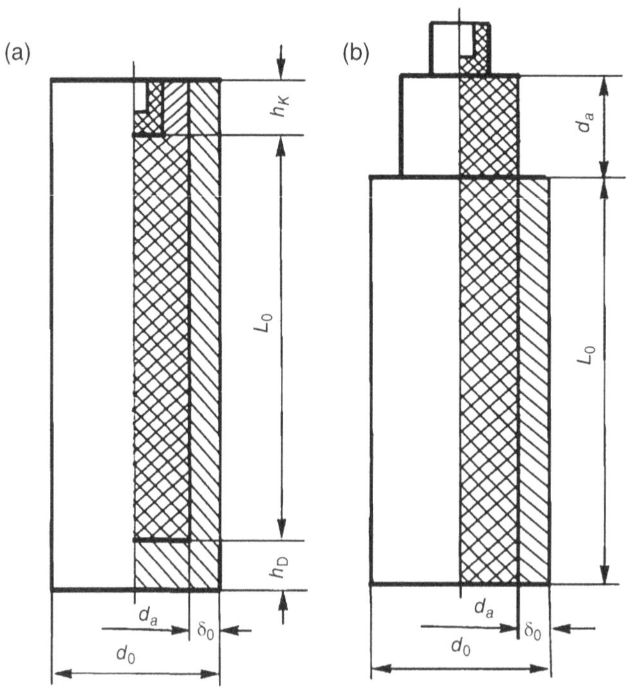

Figure 1.22 Schemes of standard cylinders: (a) closed, (b) open. *Source:* From Odintsov [7].

type of cylinder, depending on the internal diameter, there are three modifications: small, medium, and main. The main one is a cylinder with an internal diameter of $d_a = 40$ mm. This choice is based on the following considerations:

1) according to the results of numerical simulation at $d_a = 40$ mm, $\delta_d = 1/6$ in the average thickness zone of the cylinder wall, the specific tensile impulse $i_c = 5$ GPa μs is realized, which is sufficient to form an internal microcrack zone, which makes it possible to check the action on the cylinder rarefaction wave effects;
2) for a number of materials, it is still possible to determine the characteristics of the fracture mechanics of the cylinder material, in particular, the crack resistance of K_{1C} and impact strength of a_H;
3) the lower limit of $d_a \geq 40$ mm results from the condition of complete detonation of HE charge, especially for mixed HEs with a high content of aluminum powder, potassium perchlorate, etc.;
4) the upper limit of $d_a \leq 40$ mm is determined by the conditions of an explosion of the main cylinder in laboratory vacuum chambers, the maximum mass of explosives charge for which does not exceed 400–500 g.

The length of the chamber of the main cylinder and, respectively, the open cylinder is $L_0 = 160$ mm, the total length of the closed cylinder $L = 200$ mm. External diameter d_0, numberings of the main cylinders, and their masses M depending on the relative thickness of the cylinder wall δ_d are presented in Table 1.3.

Specifications of typical explosives and masses of charges C for cylinder No. 12 are given in Table 1.4.

An important methodological aspect of testing standard cylinders is the selection of the lower limit m_s of the masses of collected fragments. Theoretically, the mass m_s should be determined according to the condition of inclusion in the spectrum of a collection of fragments, giving the smallest fraction. Such a

Table 1.3 Characteristics of standard cylinders.

		Closed cylinder		Open cylinder	
δ_d	d_0	No.	M, g	No.	M, g
1/12	48	9	1065	9-O	695
1/10	50	10	1305	10-O	890
1/8	53.33	11	1730	11-O	1225
1/6	60	12	2660	12-O	1975

Source: From Odintsov [7].

Table 1.4 Specifications of typical explosives and masses of charges.

HE	ρ_0, kg/m^3	D, m/s	$\rho_0 D^2$, GPa	P_{C-J}, GPa	C, g
TNT	1550	6200	59.5	14.9	305
Phlegmatized RDX/Al	1700	7900	106.1	26.5	336
Phlegmatized HMX	1750	8600	129.4	32.4	346

Source: From Odintsov [7].

collection is formed by small shear fragments in the contact zone on the boundary between the shell and the explosion products. Numerous measurements of the width of sliding steps and elongation of contact shear fragments of standard cylinders No. 11, 12 at various combinations of metal-HE and their statistical processing have shown that the value m_s is about 0.25 g. This value is accepted as the mass of fragments m_s for main cylinders ($d_a = 40$ mm). In studies performed in the United States, the m_s value is usually assumed to be 1 grain (0.0648 g).

During experimental processing of spectrums of standard cylinder fragments, it is accepted to classify all fragments into three mass fractions: small fraction $m \in [0, 1\text{ g}]$ with relative content of μ_s, the middle fraction with the mass of fragments $m \in [1, 4\text{ g}]$ with relative content of μ_m, and large fraction with the mass of fragments $m \in [4, \infty\text{ g}]$ with relative content of μ_1 [7].

Evaluation of fragmentation quality can be carried out using the classification diagram constructed for the standard cylinder No. 12 with recalculation of its spectrum for the spectrum of fragmentation blast projectile 100–152 mm caliber with HE filling factor of $\alpha = 0.15$–0.20. The recalculation is made by applying the similarity theory. For each combination of indicators of cylinder $N_{0.25}$, μ_m, with predicted parameters of the fragmentation field of the projectile and, accordingly, the damage D_1 caused by one projectile to the target conglomerate, which includes personnel, non-armored, and lightly armored equipment:

$$D_1 = \sum_{i=1}^{3} P_i S_{sp\,i} C_i,$$

where P_i is the density of i-type targets on the ground, 1/m^2; $S_{sp\,i}$ is the area of a specified damage zone of the target, m^2; C_i is the cost of one i-type target [3].

Constructed on a plane ($N_{0.25} - \mu_m$), the lines of equal damages and the zones between them with sufficient approximation can be approximated by rectangular areas. The classification diagram is presented in Figure 1.23 (class 1 – high-quality fragmentation – corresponds to the condition $N_{0.25} \geq 2000$, $\mu_m \geq 0.45$; class II – qualitative fragmentation – condition $N_{0.25} \geq 1500$, $\mu_m \geq 0.40$; class

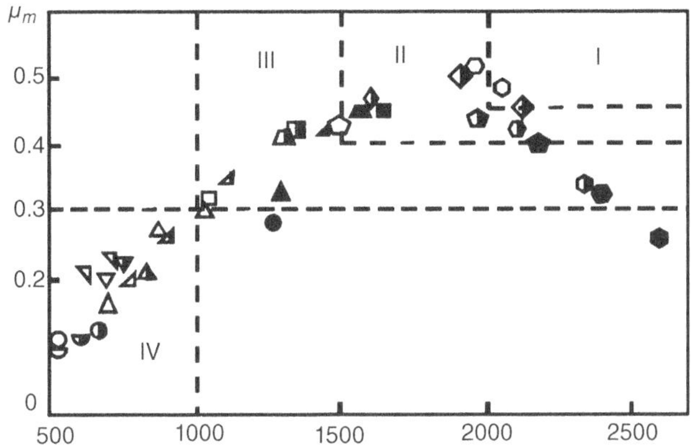

Figure 1.23 Classification diagram of fragmentation quality of standard cylinder No. 12. The number of fragments $N_{0.25}$ has been set along the horizontal axis. *Source:* From Odintsov [3].

III – satisfactory fragmentation – condition $N_{0.25} \geq 1500$, $\mu_m \geq 0.30$; class IV – unsatisfactory fragmentation – $N_{0.25} < 1000$, $\mu_m < 0.30$). The maximum predicted level of number $N_{0.25}$ for standard cylinder No. 12 is 3000 fragments.

1.3.5 The Main Grades of Fragmentation Steels

1.3.5.1 Group of Carbon Steels

Mechanical characteristics of carbon steels are presented in Table. 1.5.

An increase in carbon content in the range of $0.1 < C < 0.6\%$ leads to an improvement in both the numerical and mass characteristics of the spectrum, and the shape of the fragments. The worst fragmentation is done with low-carbon steels

Table 1.5 Mechanical characteristics of carbon steels.

Steel	$\sigma_{0.2}$, MPa	σ_t, MPa	δ, %	ψ, %	a_H, kJ/m^2	HB
10	390	510	33	62	1300	174
20	500	630	19	52	1270	178
35	520	550	4	27	420	254
45	560	680	22	47	670	187
C-60 normalized	670	790	23	39	400	217
C-60 improved	920	1020	18	47	400	285

Source: From Balagansky et al. [2].

10 and 20. Among medium-carbon steels $0.3 < C < 0.7\%$ of the most interested is steel C-60, which is the Russian standard projectile steel. Steel C-60 by its chemical composition corresponds to Steel 60 but has extended limits on the content of phosphorus and sulfur.

1.3.5.2 Siliceous Steels

An example of siliceous steel used for fragmentation ammunition is 60C2 steel. It belongs to the class of spring steels and contains 2% silicon, which increases brittleness. In the United States, AISI-9260 steel of similar composition is used for the production of fragmentation blast projectiles. Siliceous steel with a low alloying element cost provides stable, though not very high advantage over C-60 steel.

1.3.5.3 Chromic Steels

Typical steels in this class are 40X and 45 X 1. The 45 X 1 steel is, like the C-60, standard projectile steel. The presence of chromium in the steel leads to the suppression of radial fracture and the predominant development of shear failure. There is an increased content of large fraction fragments and a significant reduction in the number of small fragments.

1.3.5.4 Silicon–Manganese Steels

This group of steels is currently attracting keen interest. Back in 1970, the high-carbon steel HF-1 with a cement–perlite structure was patented in the United States, its composition approximately corresponds to the Russian grade 110G2C. A significant disadvantage of HF-1 steel is its low ductility, which in some cases does not provide the required strength at firing and impact on the ground. In addition, when equipped with high-blasting explosives, this steel does not provide the necessary number of fragments of medium and large fractions. The most promising material for the manufacture of fragmentation blast projectile bodies is the 80G2C eutectoid steel.

1.3.6 Prospects of Using Manganese Austenitic Steels to Improve Fragmentation Quality

In order to damage easily vulnerable and lightly armored equipment, it is necessary to ensure fragmentation of the shell into large fragments capable of breaking through barriers with steel equivalent of up to 10–20 mm. Analysis of the possibilities of using new steel grades for this purpose has shown that of great interest is the class of austenitic manganese steels with a number of unique properties such as significant hardening under plastic deformation, nonmagnetic, as well as high specific work of fracture. These properties, along with their relatively low cost,

Table 1.6 Chemical composition of steels, %.

Steel grade	Heat treatment	C	Si	Mn	S	P	Cr	Ni	Cu
110G13L	Quenching	1.18	0.8	13.8	0.02	0.02	0.5	0.01	0.01
C-60	Normalization	0.61	0.12	0.75	0.05	0.05	—	0.5	0.3

Source: From Balagansky et al. [2].

make these steels promising to be used as a material for the casings of munitions intended to attack protected targets.

The stable austenitic structure is formed in carbon steels at 0.4–2.0% carbon content and 13–17% manganese content after quenching. The best-known austenitic manganese steel is 110G13L (Hadfield's Manganese Steel). Under normal high-pressure friction conditions, it has an unusually high wear resistance at low hardness. The steel has the typical high toughness and ductility of austenitic steels with sufficiently good strength.

To evaluate the performance of 110G13L steel compared to standard C-60 steel, experiments have been carried out on standard cylinders No. 8 with an outer diameter of 48 mm and a wall thickness of 8 mm made of 110G13L and C-60 steel [2]. The fragmentation of the cylinders and the velocity of the fragments were evaluated. The chemical composition and mechanical properties of steels after heat treatment are given in Tables 1.6 and 1.7.

The goal of the experiments was to investigate the fracture process of shells made of 110G13L and C-60 steels in the same conditions with the following comparison: the distribution of fragments by mass and specific transverse load, the velocity of flight, deformation mechanisms. The experiments were carried out by the explosion of cylinders fitted with different explosives with the subsequent capture of fragments and simultaneous measurement of their velocities using the contact method. The fragments were then extracted from the capture medium, sorted into mass groups, and statistically and geometrically analyzed. The main

Table 1.7 Mechanical properties of steels.

Steel grade	Heat treatment	Regimes of heat treatments	Cooling medium	H_V	σ_b, MPa,	ψ, %
10G13L	Quenching	$T = 1050–1100\,°C$	Water	200	960	25
C-60	Normalization	$T = 800–850\,°C$	Air	196	710	31

Source: From Balagansky et al. [2].

Table 1.8 Main characteristics of the spectra and velocity of propagation of fragments.

Steel grade	HE	$N_{0.25}$, pcs.	$\mu_{<0.25}$, %	$\mu_{0.25-1}$, %	$\mu_{>1}$, %	ξ, $g^{0.66}/cm^2$	<m>, g	V_0, m/s
10G13L	TG 50/50	473	31.24	9.48	59.28	2.22	2.97	960
10G13L	TG 40/60	511	37.16	12.51	50.33	2.27	2.99	1150
C-60	TG 50/50	360	42.17	10.04	47.79	2.01	3.51	870
C-60	TG 40/60	435	45.34	15.37	39.29	2.09	2.33	950

Source: From Balagansky et al. [2].

characteristics of the spectrums and velocity of propagation of fragments for filling with TG 50/50 and TG 40/60 are given in Table 1.8.

In Table 1.8, $\mu_{<0.25}$, $\mu_{0.25-1}$, $\mu_{>1}$ – mass fraction of fragments with masses less than 0.25 g, in the range of 0.25–1 g, and with masses greater than 1 g, respectively; ξ-specific transverse load determined from the ratio $q = \xi \cdot m^{2/3}$, where q – transverse load; m – mass of the fragment.

The analysis of spectrums shows that for shells with austenitic structure, there is an increase in the total number of fragments and redistribution of masses by fractions, indicating an increase in regularity of fragmentation. In addition, there is an increase in the specific transverse load of the fragments and the velocity of their flight. The increase in compactness, the reduction in the percentage of small fragments, and the increase in fragment velocity are due solely to the properties of manganese austenite, such as high ductility and hardenability under plastic deformation.

The special behavior of fracture of the Hadfield steel shell manifests itself in the form of shear fracture. Correspondingly, the increase in fragment velocity is achieved by increasing the shear fracture radius of the shell.

The results provide a better understanding of the fracture of the shells of manganese austenitic steels as a process associated with the structure, affecting the fracture mechanism. Manganese austenitic is more inclined to plastic deformation by the shear mechanism and breaks down into more compact fragments with an increase in the velocity of their flight. The fragments made of Hadfield steel are less destructed in comparison with C-60 steel; the fragment spectrums have a lower percentage of dust and higher regularity of fragmentation.

However, the problem has another aspect. Manganese austenitic steels are difficult to work with due to their extremely high hardenability during plastic deformation. Improved workability can, for example, be achieved by alloying such steels with aluminum.

1.4 Statistical Models of the Fragment Fields and the Fragment Spectra

1.4.1 Fields of Fragment Dispersion, Methods of Controlling the Fields of Dispersion

When a fragmentation projectile explodes, a fragmentation field is formed – a flow of fragments characterized by the direction and speed of movement, as well as density, i.e. the number of fragments per unit of the area they intersect. The density and velocity of the fragment flow are the most important characteristics determining the possibility of hitting the target. Since the target itself can generally be at an entirely arbitrary angle relative to the axis of the ammunition, the task of determining the probability of fragments hitting the target must first know how many fragments fly in this direction. The answer to this question is provided by the so-called law of fragment dispersion. It represents the relationship between the relative number and velocity of fragments flying in each direction relative to the axis of the ammunition. Usually, this direction is given in a spherical coordinate system by two angles: θ, an angle in the equatorial, and φ, an angle in the meridional planes (Figure 1.24). The angle φ is counted from the ammunition axis and can vary from 0 to π. The angle θ varies from 0 to 2π. Thus, the task of finding the law of dispersion is to determine the relative numbers and velocities of fragments flying in the directions determined by the angles φ and θ. This problem can be solved both theoretically and experimentally.

It should be noted that not only the density and velocity of the fragment field but also the behavior of the fragmentation of the ammunition may vary in the angles of

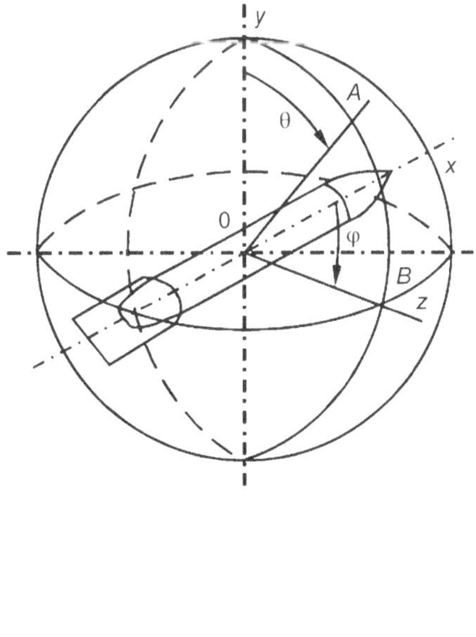

Figure 1.24 The angles θ and φ determining the law of fragment dispersion. *Source:* From Miropolsky et al. [8].

fragments dispersion. We will also consider that the law of fragmentation does not depend on the angles of fragments dispersion.

Two basic classes are distinguished by the shape of a fragment field: axially symmetric fields and non-axially symmetric fields. The first class is the most diverse and includes fields of all barrel ammunition and air bombs, most of the missile warheads, and most of the engineering fragmentation mines. The reference axis coincides with the axis of the warhead, or with the axis of the projectile. The angle is generally counted from the direction of flight of the projectile. A particular case of axially symmetrical ammunition is a warhead with central (spherical) symmetry.

According to the configuration of fragment fields, three main classes can be distinguished (Figure 1.25). Circular and axial fields are axisymmetric and can be described using only one angle φ. A radially directed field is not axially symmetrical and requires a complete description to define it. Fragmentation warheads with circular fields are most common due to their ease of implementation.

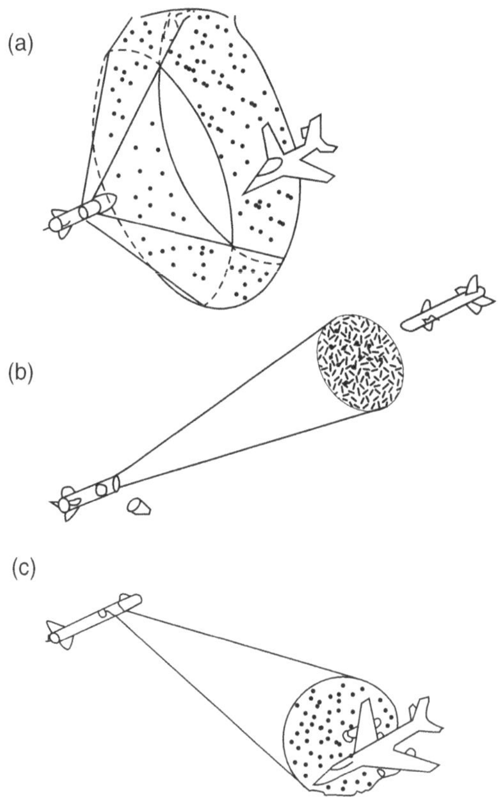

Figure 1.25 Configurations of fragment fields (warheads of guided missiles used as examples): a – circular field; b – axial field; c – radial field axis. *Source:* From Odintsov [3].

Let us introduce a classification of axisymmetric fragment fields depending on the angle of dispersion $\Delta\varphi = \varphi_2 - \varphi_1$, including 80 or 90% of all fragments. The angle of dispersion $\Delta\varphi$ is determined mainly by the shape of the shell and the initiation scheme. For cylindrical shells with elongation $L_0/d_0 = 1.5,...,2.5$ with detonation initiation from the end of charge and from the center of charge, the angle of dispersion $\Delta\varphi$ is about 15° and 25°, respectively. Small angles of dispersion ($\Delta\varphi = 5°, \ldots, 10°$ – fragment fields of "cutting disk" type) are realized by means of shells with a concave shape, two-point initiation on the ends of charge, multipoint synchronous initiation on the axis of charge creating a diverging detonation wave close to the cylindrical one. Large angles of dispersion are achieved by the barrel shape of the warhead or by the coupling of cylindrical and conical volumes.

Experimentally, the angular distributions of the fragments $f(\varphi)$ and the velocity $V_0 = g(\varphi)$ are determined by exploding a warhead in a shield target environment and counting the numbers of holes of the required area in the sectors of duralumin or steel shields corresponding to 5° angular zones (a total of 36 angular zones with numbers $j = 1,2, \ldots, 36$). The test results are recalculated for a whole sphere. The velocity of the fragments in the angular zones is calculated from the distance from the detonation point to the shield measured by the optical high-speed shooting. The schematic of the shield target environment is shown in Figure 1.26.

The law of fragmentation determines the distribution of fragments by the value of their masses. It can be determined experimentally and is statistical in nature. That is explained by the fact that the fragmentation of a shell in each case depends on many different factors, including random ones. Therefore, in order to take into account these random factors, to obtain reliable estimates of parameters,

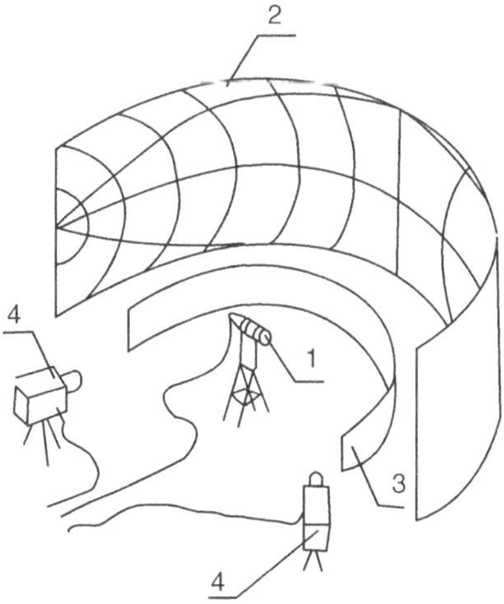

Figure 1.26 *Shield tests*: 1 – projectile; 2 – shield; 3 – ricochet shield; 4 – speed cameras. *Source:* From Andreev et al. [9].

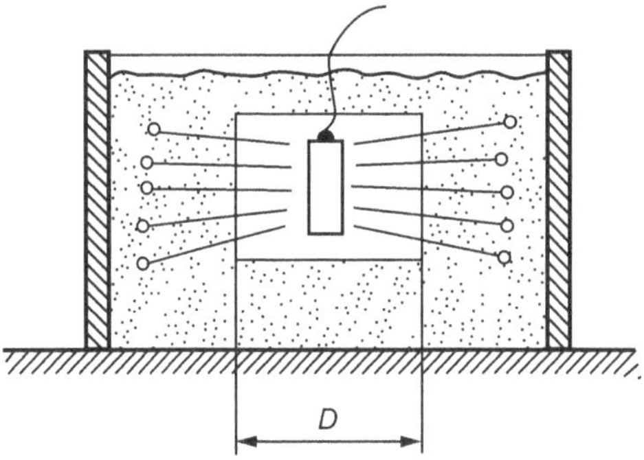

Figure 1.27 Scheme of a blast chamber with a catcher. *Source:* From Odintsov [7].

and to reveal regularities inherent to the ammunition of the type under consideration, it is necessary to make a sufficiently large number of experiments and to process their results using statistical methods.

The characteristics of the fragment spectrum are determined with the help of chamber tests. The essence of experiments is to explode warheads or their models in special pits or blast chambers filled with sand, sawdust, or other fragments catchers (Figure 1.27). The material of the catcher shall not result in the destruction of fragments and thus introduce errors to the experimental results. The fragments collected after the warhead explosion are weighed and sorted by mass intervals on the continuous scale from minimum m_s to maximum m_{max} value.

An example of presentation of the test results is shown in Table 1.9 (open cylinder spectrum $d_a = 25$ mm, $d_0 = 37.5$ mm, $\delta_d = 1/6$, $L_0 = 100$ mm, steel 20/TNT, $M = 484$ g, $N_{0.5} = 189$). The first column ($m \leq m_s$) shows only the total mass of fragments of a given mass group $\hat{M}_1$, which is usually defined as the difference between the initial mass of fragmentation shell M and the total collected mass of fragments of all groups except the first. The difference in mass $M_d = M - \sum \hat{M}_i$ is referred to as shortfall. The relative shortfall in mass $\Delta = M_d/M$ is indicative of the quality of the test and should normally not exceed 0.05.

Table 1.9 Example of presentation of the test results.

m, g	<0.5	0.5–1	1–2	2–3	3–4	4–6	6–8	8–10
$\hat{N}_i$, pcs	—	81	38	17	16	26	8	3
$\hat{M}_i$, g	64.1	56.6	57.6	42.5	55.1	124.7	56.9	26.5

Source: From Odintsov [4].

Based on the results of the analysis, histograms of the distribution “number vs mass” and “mass vs mass” are calculated, as a rule, in the normalized form, when the ordinates are the values of the quantities

$$\hat{f}(m) = \frac{\hat{N}_i}{N_s}\frac{1}{\Delta m_i} \quad \hat{\varphi}(m) = \frac{\hat{M}_i}{M}\frac{1}{\Delta m_i} \tag{1.12}$$

Technical documentation often contains simpler, non-normalized histograms. Their disadvantage is the distortion of the distribution type when moving to groups with large intervals of mass Δm_i. It is obvious that approximation of spectrums by analytical dependencies can be done only with normalized histograms. For the above spectrums, the normalized histograms are shown in Figure 1.28.

The minimum mass of a fragment m_s depends on the mass of a warhead. Usually, m_s is defined as 0.5 g, because fragments of smaller mass may not be effective enough. For small calibers, m_s may be less than 0.5 g. The lower limit is 0.1 g. Such small fragments are very difficult to assemble and separate from the fractions of a destroyed catcher.

In literature, much attention is paid to the evaluation of the data reliability of the chamber tests. It is noted that both the type of catcher medium and the diameter of the air cavity D (see Figure 1.27) in which the explosion is carried out can significantly influence the characteristics of the spectrum. Increasing the diameter of the cavity reduces the average mass of the fragment spectrum. It is recommended to

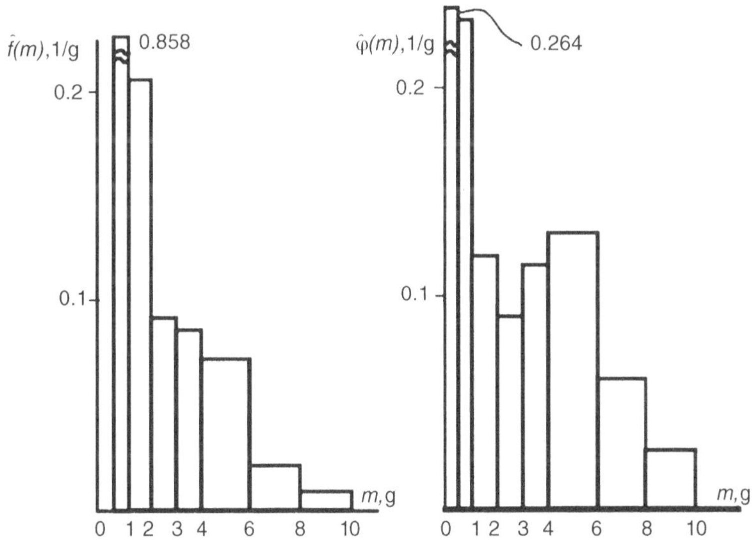

Figure 1.28 Normalized histograms of fragment spectrums. *Source:* From Odintsov [4].

obtain an undistorted distribution of fragments corresponding to the explosion in an unrestricted air environment, choose a ratio of D/d_0 for water catch at least 5, for sandy – at least 10. Here, d_0 is the ammunition caliber.

The modern theory of fragmentation ammunition shows that, all other things being equal, ammunition with shells breaking into fragments of compact form and expected, the optimal mass will be more efficient. Modern technologies for the manufacture of shells and warheads make it relatively easy to solve this problem. One of the early methods to improve the quality of shell fragmentation was a saw-shaped profile applied to the inner surface of the cast bomb shells (Figure 1.29a).

In the process of an HE charge explosion and shell expansion, its destruction starts in places with minimal wall thickness δ_{min}, and as a result, the shell breaks into separate ring-shaped parts. Further fragmentation of the individual rings is random. This method of breaking eliminates the formation of fragments that are too heavy or irregularly shaped.

Similarly, there is the process of forming fragments of shells assembled from separate preformed rings or representing a bar of rectangular cross-section, wound onto a metal body in the form of a spring (Figure 1.29b). Controlled fragmentation of shells can also include the use of plastic "jackets" with shaped notches (Figure 1.29c). Figure 1.29d shows the shell, pretreated by laser, creating narrow zones of microstructure in metal with sharply different strength compared to the base metal.

Regular fracture of shells into fragments of the same mass and shape can be ensured by forming deep cuts on one or both surfaces of the shell walls, which significantly reduces the strength of the walls in the specified sections (Figure 1.30a,c).

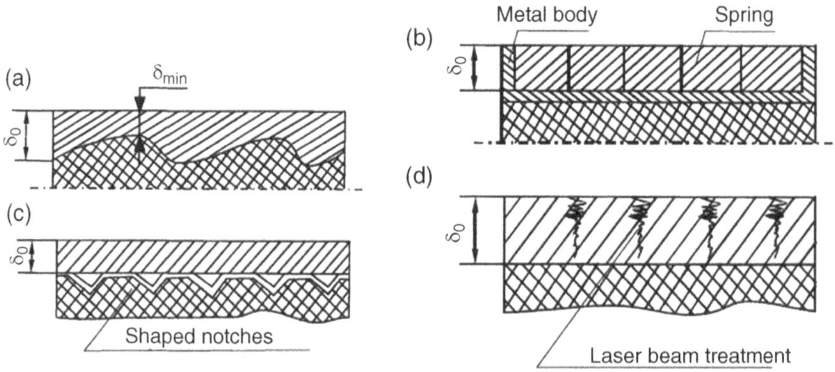

Figure 1.29 Methods of controlled fragmentation of shells into fragments. *Source:* From Miropolsky et al. [8].

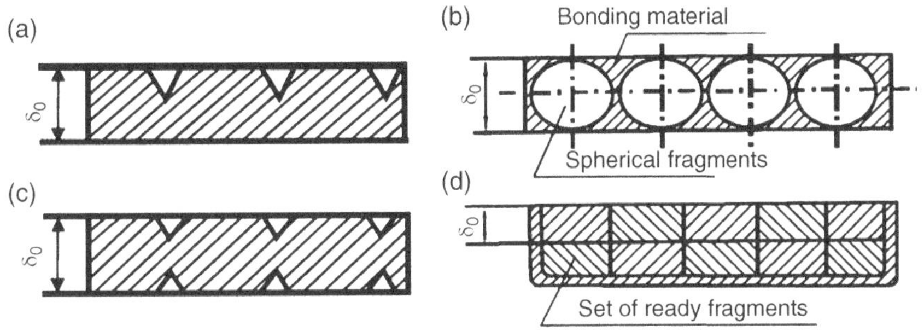

Figure 1.30 Methods of regular fragmentation of shells. *Source:* From Miropolsky et al. [8].

An interesting feature of the external cutting is that the shell often breaks not along the cuts but between them, where rarefaction waves reflected from the surfaces of the cuts meet.

The most perfect way of regular fragmentation of shells is achieved when their walls are sets of preformed fragments of a given mass and shape (Figure 1.30b,d). However, it is necessary to keep in mind that in this case, when a HE charge explodes, the integrity of the shell is broken almost immediately. An intensive flow of explosion products begins in the gaps formed between the fragments so that the fragments themselves can noticeably melt, and their initial speed is significantly reduced.

1.4.2 Laws of Fragment Distribution by Mass

The laws of fragment mass distribution can be applied both in number $F(m)$, $f(m)$, and mass $\Phi(m)$, $\varphi(m)$ forms [4]. The numerical form is intended to calculate the action and the mass form to reveal the resources of metal mass.

1.4.2.1 Numerical Distributions

The function of distributing the number of fragments by mass (integral law of distribution) is introduced as follows:

$$F(m) = \frac{N(< m)}{N_0}, \tag{1.13}$$

where $N(<m)$ is the number of fragments having mass less than m; N_0 is the total number of fragments.

The distribution function has the following general properties:

1) it is a nondecreasing function of its argument, i.e. if $m_2 > m_1$ then $F(m_2) > F(m_1)$.

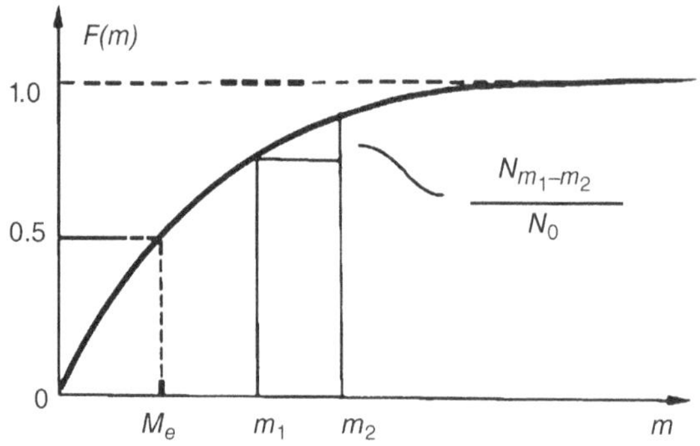

Figure 1.31 Fragments distribution "number by mass" integral law. *Source:* From Balagansky et al. [2].

2) on the left boundary of the definition area, the distribution function is equal to zero; on the right boundary of the definition area (in this case, at $+\infty$), the distribution function is equal to one.

The number of fragments having masses in the interval $m_1 - m_2$ will be determined by the following ratio (Figure 1.31):

$$N_{m_1-m_2} = N_0[F(m_2) - F(m_1)]. \quad (1.14)$$

The median M_e is the value of mass m, for which $F(M_e) = 1/2$.

The density of distribution (differential distribution law)

$$f(m) = \frac{dF(m)}{dm} = \frac{1}{N_0}\frac{dN(<m)}{dm}. \quad (1.15)$$

Unlike the distribution function $F(m)$, which is dimensionless, the distribution density has a dimension of 1/g. Its basic properties are:

1) the distribution density is a nonnegative function, i.e. $f(m) > 0$;
2) the integral from the distribution density within the range from the left to the right boundary of the definition area

$$\int_{m_s}^{m_{\max}} f(m)dm = 1,$$

i.e. the area under the distribution curve is equal to 1.

The number of fragments having masses in the range $m_1 - m_2$ is

$$N_{m_1-m_2} = N_0 \int_{m_1}^{m_2} f(m)dm. \quad (1.16)$$

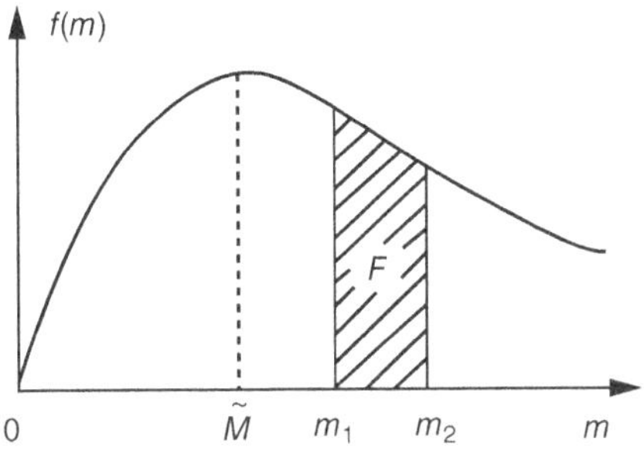

Figure 1.32 Fragment distribution "number by mass" differential law. *Source:* From Balagansky et al. [2].

In other words, the relative number of fragments in each interval is defined as the area F (Figure 1.32).

The mode of the value m is its value $\tilde{M}$, at which the probability density $f(m)$ is maximum. If the distribution curve has one mode, the distribution is called unimodal; if there are several modes, it is polymodal (bimodal, if there are two modes).

The average mass of a fragment

$$\langle m \rangle = \int_0^\infty m \cdot f(m) dm. \tag{1.17}$$

You can show that $<m> = M/N_0$, where M is the mass of the ammunition shell.

1.4.2.2 Mass Distributions

The distribution function $\Phi(m)$ and the distribution density $\varphi(m)$ are introduced by the following relations:

$$\Phi(m) = \frac{M(<m)}{M}; \tag{1.18}$$

$$\varphi(m) = \frac{d\Phi(m)}{dm} = \frac{1}{M} \frac{dM(<m)}{dm}, \tag{1.19}$$

where $M(<m)$ is the total mass of fragments, the mass of each of which is lesser than m.

The relationship between the numerical and mass density of the distribution is determined by the ratio

$$\varphi(m) = \frac{mN_0}{M} f(m). \tag{1.20}$$

The mass of the fragments contained in the interval $m_1 - m_2$ is defined by the formula:

$$M_{m_1-m_2} = M[\Phi(m_2) - \Phi(m_1)] = M\int_{m_1}^{m_2} \varphi(m)dm. \tag{1.21}$$

The mass distribution $\varphi(m)$ is much more informative than the numerical distribution when analyzing the mass balance of fragments. It gives a clear understanding of the size of mass losses for useless fragments. In addition, the mass distribution is much clearer than the numerical distribution, revealing the bimodal behavior of the spectrum.

1.4.3 Analytical Representation of Fragment Distribution Laws

1.4.3.1 Weibull Distribution

The most common way to describe the fracture of materials under explosive and shock loads is to use a Weibull distribution. The numerical form is as follows:

$$F(m) = \frac{N(<m)}{N_0} = 1 - \exp - \left(\frac{m}{m_0}\right)^{\lambda}. \tag{1.22}$$

The distribution is two-parametric. The value m_0 will be further called the characteristic mass of the distribution; the factor λ is a parameter of the fragmentation quality. As the λ increases, fragmentation becomes more homogeneous.

The number of fragments in the interval will be defined as

$$N_{m_1-m_2} = N_0[F(m_2) - F(m_1)] = N_0\left[\exp - \left(\frac{m_2}{m_0}\right)^{\lambda} - \exp - \left(\frac{m_1}{m_0}\right)^{\lambda}\right]. \tag{1.23}$$

Distribution density

$$f(m) = \frac{dF(m)}{dm} = \frac{\lambda}{m_0}\left(\frac{m}{m_0}\right)^{\lambda-1} \exp - \left(\frac{m}{m_0}\right)^{\lambda}. \tag{1.24}$$

The types of distribution density at different values of λ are shown in Figure 1.33a. Figure 1.33b shows the mutual position of characteristic mass m_0, mode $\widetilde{M}$, median M_e, and average mass of a fragment $<m>$.

The obvious advantage of the Weibull law is the possibility of using it to describe fundamentally different types of spectrums.

The average mass of a fragment is calculated using the formula

$$\langle m \rangle = \int_0^{\infty} mf(m)dm = m_0\Gamma\left(1 + \frac{1}{\lambda}\right), \tag{1.25}$$

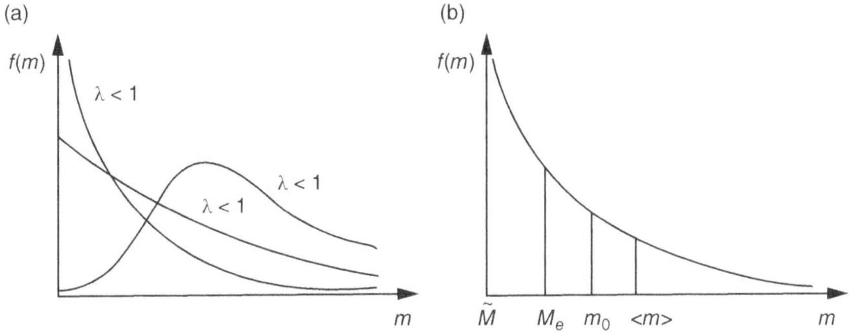

Figure 1.33 Types of distribution density at different λ (a); the relative position of the characteristic mass m_0, mode $\widetilde{M}$, median $M_{e,}$ and average mass of a fragment <m> (b). *Source:* From Balagansky et al. [2].

where $\Gamma(x) = \int_0^{\infty} e^{-t} t^{x-1} dt$ is the gamma function.

The differential law of distribution "mass by mass" looks like

$$\varphi(m) = \frac{\lambda}{m_0 \Gamma(1 + 1/\lambda)} \left(\frac{m}{m_0}\right)^{\lambda} e^{-\left(\frac{m}{m_0}\right)^{\lambda}}. \tag{1.26}$$

1.4.3.2 The Mott Law

In multiple publications, the value $\lambda = 0.5$ is used to describe a wide class of spectrums. The Weibull law at $\lambda = 0.5$ is called the Mott law; thus, the characteristic mass of the distribution is designated by the letter $\mu(m_0 = \mu)$. In this case

$$F(m) = 1 - e^{-(m/\mu)^{1/2}}; \quad f(m) = \frac{1}{2\sqrt{m\mu}} e^{-(m/\mu)^{1/2}};$$

$$\langle m \rangle = \mu\Gamma(3) = 2\mu; \quad N_0 = M/2\mu.$$

The Mott distribution is a one-parameter law.

1.4.4 Distribution of Fragments by Shape

For reliable estimation of the fragment field effect, it is necessary to know the laws of fragment distribution not only by mass but also by shape. The shape characteristic of a fragment is usually taken as the average area of the projection of the fragment (midship section) <S> or other characteristics derived from it:

$$\Phi = \frac{< S >}{Y^{2/3}}, \quad q = \frac{m}{< S >}, \quad \xi_0 = \frac{q}{m^{1/3}} = \frac{m^{2/3}}{< S >},$$

where Φ is a dimensionless fragment shape parameter; q – fragment transverse load; ξ_0 – relative transverse load; $Y = m/\gamma_0$ – fragment volume.

The m and $<S>$ values are correlated. For a fixed mass, the fragments of shear formation have a greater value of the midship section $<S>$ and, consequently, a greater parameter of shape Φ than fragments of brittle fracture (2.0, ..., 2.4 and 1.8, ..., 2.0, respectively). As the mass of a fragment of the given spectrum increases and it moves into the "saber" category, the parameter of the shape increases, i.e. its form becomes worse from the point of view of the damaging effect.

Sometimes changes in the shape parameter for increasing mass are ignored, and Φ is considered constant (ξ_0 = const) for the whole range of masses.

In this case, mass and average midship, and consequently mass and transverse load, turn out to be related not by correlation, but by a functional dependence, for example, in the form of

$$q = \xi_0 m^{1/3}$$

With such an assumption, the value ξ_0 can serve as a characteristic of the compactness of fragments of the whole spectrum of shell fragmentation.

1.5 External Ballistics of Fragments

Let us consider the movement of a fragment in the air under the following assumptions [1]:

- the air density ρ_a along the trajectory is constant;
- the effect of gravity is negligible;
- the midship section area of the fragment in flight is equal to its average value $<S>$;
- the drag coefficient c_x is independent of the fragment velocity.

In reality, the drag coefficient c_x is velocity-dependent and takes its maximum value when the fragment velocity is approximately equal to the speed of sound, so the last assumption is quite a rough approximation.

The equation of fragment motion takes the form

$$m\frac{dV}{dt} = -\frac{\rho_a V^2}{2} < S > c_x,$$

where V is the current fragment velocity.

This equation can be represented as

$$\frac{dV}{dt} = -AV^2, \quad A = \frac{\rho_a < S > c_x}{2m}, \tag{1.27}$$

where A is the ballistic coefficient, 1/m.

The law of change of fragment velocity as a function from the distance x is the most interesting for practical use.

Using the substitution

$$\frac{dV}{dt} = \frac{dV}{dx}\frac{dx}{dt} = V\frac{dV}{dx},$$

it is possible to present the equation of fragment motion in the air in the form of

$$\frac{dV}{dx} = -AV. \tag{1.28}$$

From this equation, it follows, in particular, that the AV value is a loss of velocity of a fragment on the way of 1 m. Integrating this equation at initial condition $V(0) = V_0$, we get

$$\ln\frac{V}{V_0} = -Ax,$$

from here

$$V = V_0 e^{-Ax}. \tag{1.29}$$

This formula is known as the law of exponential attenuation of fragment velocity in flight. From here it is possible to get the expression for the kill interval of the fragment I_k – the distance from the explosion, at which the fragment still retains the velocity needed to damage the target (kill velocity V_k):

$$I_k = \frac{1}{A}\ln\frac{V_0}{V_k}. \tag{1.30}$$

The formula for the ballistic coefficient can be represented as follows:

$$A = \frac{\rho_a c_x \Phi}{2\gamma_0^{2/3} m^{1/3}},$$
$$\Phi = \frac{<S>}{Y^{2/3}} \text{ – fragment shape parameter;} \quad <S> = \frac{S_\Sigma}{4}, \tag{1.31}$$

where S_Σ is the total surface area of the fragment; Y is the volume of the fragment (Table 1.10).

It is more difficult to determine the average midship section area and shape parameters for natural fragmentation fragments. The method of determining the average midship <S>, regulated by a standard, is based on measuring the mass m and two sizes of the largest base of the fragment – length a and width b. According to the E.S. Wentzel model, the fragment is considered to be of a parallelepiped shape with the following sizes $a \times b \times c$ $(a > b > c)$ and therefore thickness

$$c = \frac{Y}{ab} = \frac{m}{\gamma_0 ab},$$

Table 1.10 Values of parameter Φ for bodies of a regular geometrical shape.

Body shape	Φ
Sphere	1.31
Cylinder	
$L/d = 1$	1.38
$L/d = 2$	1.45
Tetrahedron	1.80
Octahedron	1.43
Cube	1.50
Parallelepiped	
1 × 1 × 2	1.58
1 × 1 × 8	2.13
1 × 2 × 2	1.59
1 × 4 × 4	1.89

Source: From Balagansky et al. [2].

from here

$$< S > = \frac{S_\Sigma}{4} = \frac{ab + \frac{m}{\gamma_0}\frac{a+b}{ab}}{2}. \tag{1.32}$$

The real shape of the fragment differs from the parallelepiped the more, the greater the role of shear processes in natural fragmentation.

The shape parameter for such fragments depends on the characteristics of the metal and for fragments of small fractions falls within the range of $\Phi = 1.8$–2.2.

Along with the shape parameter Φ, other metrics are used, also based on the average midship section $<S>$. This includes the transverse load of the fragment

$$q = \frac{m}{< S >} \tag{1.33}$$

and the relative transverse load

$$\xi_0 = \frac{q}{m^{1/3}}. \tag{1.34}$$

Even the developers of fragmentation projectiles do not always fully understand the role of the shape of fragments for their damaging effects. To illustrate the role of the shape of a fragment, consider a simple example.

Example 1.1.
There are two fragments – the first one is a natural fracture fragment and the second one is a spherical fragment. The mass of the fragments in both cases $m = 1$ g; steel equivalent of the target, penetration of which is considered to be damaged, is $h^e_{st} = 3$ mm; the initial velocity of the fragment $V_0 = 1000$ m/s; shape parameter of the fragment: natural fracture $\Phi = 2.8$; sphere $\Phi = 1.31$.

Solution

Let us define kill intervals for each of these fragments. For the steel fragment at ground level, the kill interval can be determined by the experimental formula

$$I_k = 145\frac{m^{1/3}}{c_x\Phi}\lg\frac{V_0}{V_k},$$

where I_k – kill interval, (m); m – mass of the fragment, (g); $c_x = 1.21$ – the drag coefficient of the fragment; Φ – fragment shape parameter; V_k – kill velocity of the fragment, (m/s).

The kill velocity of a fragment under the same assumptions can be calculated by the following formula:

$$V_k = 145\frac{h^e_{st}\Phi}{m^{1/3}},$$

where h^e_{st} the steel equivalent of the target, (mm).

By substituting the initial data in the formulas, we obtain:

- for the spherical fragment $V_k = 570$ m/s; $I_k = 22$ m;
- for the natural fracture fragment $V_k = 1218$ m/s; $I_k = 0$ m, i.e. this fragment does not damage the target at all.

From this simple example, we can see how important the shape of the fragment is. That is why so much attention is paid to the development of fragmentation ammunition with a controlled fracture and preformed fragments.

1.6 Kinds of the Damaging Effect of Fragments

The effect of fragments on the target, depending on its type, can be evaluated according to one of three concepts: the concept of piercing equivalent barriers (thickness-piercing concept); the criterion concept; and the synthetic concept representing the combination of the former ones.

The thickness-piercing concept is based on replacing a target with a steel or duralumin barrier (the equivalent of the target) whose penetration is considered equivalent to target damage. The formula of the Geodesy Research Institute for the pair "steel–steel" is based on the above approach.

$$h = 2.1\, q \lg \left(1 + 6.5V^2\right), \tag{1.35}$$

where h (mm) is the thickness of the pierced barrier, q (g/cm^2) is the transverse load of the fragment, V (km/s) is the velocity of the fragment.

Nonmetallic equivalent barriers are used to evaluate the personnel effect, including 1-in. (25.4 mm) thick, dry pine boards, 180-mm thick NATO standard plasticine targets, etc.

The criterion approach is most appropriate in cases where the damage of the target cannot be considered a simple piercing of the barrier, for example in cases of physically complex processes of penetration with subsequent ignition, initiation, and behind-the-barrier action. An important advantage of the criterion approach is the possibility of building probabilistic models of target damage, set, as a rule, by functions $p = f(K)$.

As the criteria, *full kinetic energy* $W_{\mathrm{cr}} = \dfrac{mV^2}{2}$, *specific energy* $E_{\mathrm{cr}} = \dfrac{mV^2}{2 < S >}$, *specific impulse* $i_{\mathrm{cr}} = \dfrac{mV}{< S >}$, are most often used.

The main disadvantage of the criterion of full kinetic energy is that it does not consider the shape of the fragment.

The specific energy criterion is used to evaluate the damaging effect on all classes of targets, including those that are lightly armored. A typical and not experimentally confirmed feature of the specific energy criterion is the predicted possibility of a decrease in fragment mass due to a minor increase in velocity.

In this sense, the specific impulse criterion is an intermediate position between the criteria of full and specific energy.

1.6.1 Ignition Effect of Fragments

The ignition effect plays an especially important role in the action of fragment flows on aerodynamic targets: the area of fuel tanks on airplanes is 30–60% of the vulnerable target area.

The defining parameter for ignition is the specific impulse of a fragment. At $i \leq 0.16$ kgf/cm^2, the probability of ignition is equal to zero. With the increasing specific impulse, this probability increases, and at $i \geq 2.5$ kgf/cm^2 reaches almost one. With the increase in flight altitude of the aircraft due to the decrease in pressure and temperature of the environment and the associated decrease in the partial pressure of fuel vapor, the concentration of the mixture, oxygen in the mixture and the environment, the probability of fuel ignition decreases and at an altitude of about 16 km becomes practically equal to zero.

In addition to fuel systems, gun powder charges available as part of targets such as artillery carriers or reactive weapons are vulnerable to ignition risk.

1.6.2 Initiating Action of Fragments

When a high-velocity fragment hits a warhead that is part of a target (aircraft bombs, missile warheads, etc.), it may cause a detonation of the explosive, which leads to the complete destruction of the target. In some cases, the detonation of the target's warhead is the only way to damage it reliably. A typical example is the confrontation of shipborne air defense systems and anti-ship cruise missiles on close intercept boundaries (500–1000 m).

1.6.3 Effects of a Dense Flow of Fragments

The survivability of air targets is high. They can withstand a large number of direct hits of small caliber projectiles. The data collected during the Vietnamese campaign shows that the average number of hits for 20, 30, and 40 mm projectiles required to destroy an aircraft is, respectively, 6, 3, and 1 for tactical fighters, 18, 6, and 3 for heavy fighters, and 30, 10 and 5 for fighter bombers. Thus, damage to air targets at low fragment flow density is generally unlikely. With a single-component representation of an air target, its vulnerable area usually does not exceed 1.5 m^2.

The action of a dense flow of fragments – a flow containing several dozens of fragments per 1 m^2 of the cross-section of the flow – hitting the target with little timing delay, is more effective. The main characteristic of a dense flow is its specific energy E.

For preformed fragments $E = \Pi mV^2/2$, where Π is the density of a fragment flow (1/m^2). The condition of target destruction by a dense flow $E \geq E_{cr}$.

The action of dense flow on target leads primarily to the following effects: mechanical weakening or destruction of the airframe structure; air shock in closed spaces of the target; hydraulic shock in fuel tanks. *An air impact* is a shock pressure increase in closed spaces of the fuselage, keel, or wing under the action of a penetrating flow of fragments. *A hydraulic impact* is a shock pressure increase in fuel tanks under the influence of a penetrating fragment flow. Hydraulic impact leads to swelling and destruction of the walls of the tank.

1.7 Laws of Target Damage with Fragments

The probability of damaging a target with fragments is determined by the relative positions of the explosion and the target. Calculations can be made both in the coordinate system associated with the explosion and in the system associated with the target [1].

In the coordinate system associated with the explosion, the coordinate law G is the probability of damage to a target located on the ground surface at a point with defined coordinates. In this case, it is natural to accept flat polar coordinates r, ξ, i.e. to set $G = G(r, \xi)$. For a single-zone field, the coordinate law G is constant in angle and depends only on the radius r.

The probability of damage to the target P is the probability of hitting with at least one damaging fragment.

$$P = 1 - e^{-\langle n \rangle P_0}, \tag{1.36}$$

where $\langle n \rangle$ – the expected value of the number of fragments that hit the full (mid-view) target area S_0; P_0 – the probability of damage of the target with a single fragment.

With the average density of a fragment field in dynamics Π_d, we obviously have $\langle n \rangle = \Pi_d S_0$, hence:

$$G = P = 1 - e^{-\Pi_d S_0 P_0}. \tag{1.37}$$

S_0 multiplied by P_0, which has the dimension of the area, is a vulnerable target area S_v; from here

$$G = 1 - e^{-\Pi_d S_v}. \tag{1.38}$$

To determine the dynamic density of the fragment field, let us convert a static dispersion diagram into a dynamic one. The fragment field is assumed to be a single zone, i.e. it is located inside one angular zone with constant fragment density and velocity. The view of the angle zone for a motionless projectile (in statics) is shown in Figure 1.34, where V_0 is the initial velocity of fragments; φ_1 – is the front angle of the field in statics; φ_2 – the back angle of the field in statics; $\Delta\varphi = \varphi_2 - \varphi_1$. In practice, the angle of dispersion $\Delta\varphi$ is usually taken 80 or 90% of the fragments.

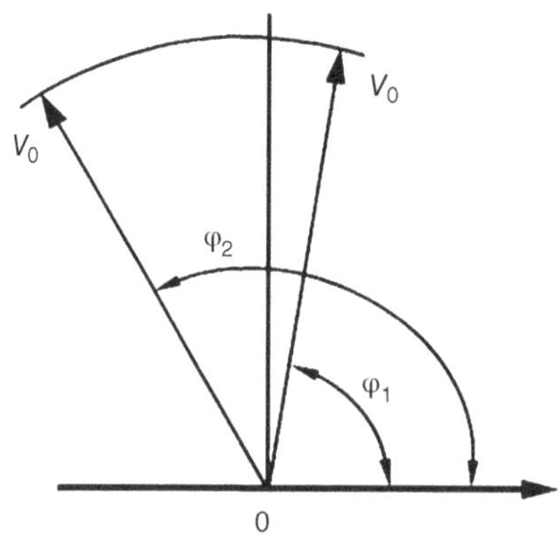

Figure 1.34 Angular dispersion zone for a motionless projectile. *Source:* From Balagansky et al. [2].

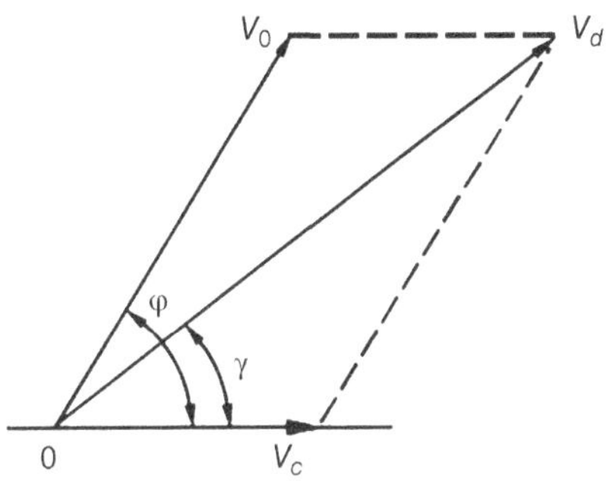

Figure 1.35 Vector summation of fragment velocity and own projectile velocity. *Source:* From Balagansky et al. [2].

If a fragment in statics moved with velocity V_0 at an angle φ to the direction of projectile movement, because of considering the projectile's own velocity V_c, i.e. vector summation of velocities V_0, V_c, (Figure 1.35), we obtain the following:

$$\gamma = \text{arctg}\frac{V_0 \sin \varphi}{V_c + V_0 \cos \varphi}; \tag{1.39}$$

$$V_d = \sqrt{V_0^2 + 2V_0V_c \cos \varphi + V_c^2}, \tag{1.40}$$

where γ is the angle of movement of the fragment in the dynamics; V_d – the resulting (dynamic) velocity of the fragment.

These relations determine the front and back dynamic angles of the field γ_1, γ_2, and the corresponding velocities V_{d1}, V_{d2}. The dynamic density of the field Π_d at different distances from the blast point is calculated by the formula

$$\Pi_d = \frac{N}{S_{sb}}, \tag{1.41}$$

where N is the number of fragments; S_{sb} is the area of the surface of a spherical belt.

Taking into account that $S_{sb} = 2\pi rh$, where $h = r(\cos\gamma_1 - \cos\gamma_2)$ (Figure 1.36), we get $S_{sb} = 2\pi r^2(\cos\gamma_1 - \cos\gamma_2)$ and, consequently, we obtain:

$$\Pi_d = \frac{N}{2\pi r^2(\cos\gamma_1 - \cos\gamma_2)}. \tag{1.42}$$

The formation of the damaged sectors on the ground at the ground explosion is shown in Figures 1.37 and 1.38. The fragment field comes from a point located in the ground plane. Boundaries of damaged sectors represent traces of cross-section of two conical surfaces by the ground surface passing through their common top. Angular boundaries of the damaged sector on the ground are determined by expressions:

$$\xi_1 = \arccos\frac{\cos\gamma_1}{\cos\theta_c}, \tag{1.43}$$

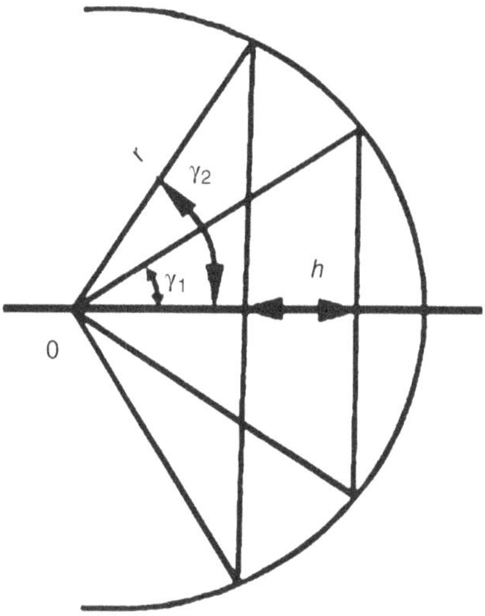

Figure 1.36 Fragment field on the sphere in dynamics. *Source:* From Balagansky et al. [2].

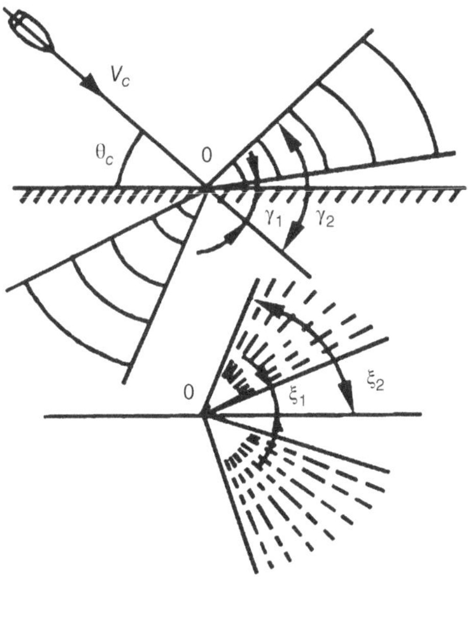

Figure 1.37 Dispersion of fragments after the projectile hits the ground. *Source:* From Balagansky et al. [2].

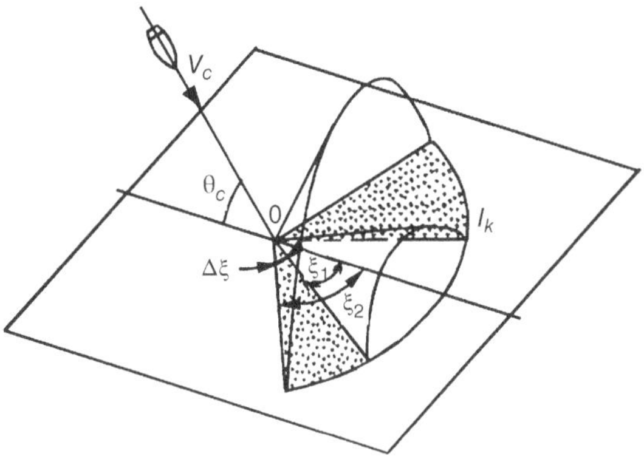

Figure 1.38 Formation of damaging sectors on the ground. *Source:* From Balagansky et al. [2].

$$\xi_2 = \arccos \frac{\cos \gamma_2}{\cos \theta_c}. \tag{1.44}$$

In the schemes of calculation with the implementation of the killing interval I_k, damaged sectors are limited to circle arcs with a radius equal to I_k. At small angles of incidence, typical, for example, for shooting from a tank, the angles on the ground and dynamic angles are approximately equal ($\xi_1 \cong \gamma_1$, $\xi_2 \cong \gamma_2$). At angles of incidence $\theta_c \geq \gamma_1$, the front boundary angle of the sector ξ_1 turns to zero (the front boundary cone of the field is entirely located underground). Finally, at an angle of incidence $\theta_c \geq \gamma_2$, the damaged area on the ground surface turns to zero (both boundary field cones are located underground; all the fragments go into the ground).

Summarizing all the above, for the fixed vulnerable area of the target S_v and fragments of the constant mass, we get a coordinate law of the damage in the sector $\xi_1 - \xi_2$ as follows:

$$G = 1 - e^{-\frac{NS_v}{2\pi r^2(\cos\gamma_1 - \cos\gamma_2)}} \quad \text{if} \quad r \leq I_k, \tag{1.45}$$
$$G = 0 \quad \text{if} \quad r > I_k.$$

1.8 Specified Zone of Target Damage with Fragmentation Munitions

1.8.1 The Area of the Specified Zone

The area of the specified zone of target S_{sp} is some nominal area, for which the explosion of a projectile inside the zone leads to the damage of the target with the probability equal to 1, while the probability is equal to 0 if the explosion occurs outside the specified zone. In general case,

$$S_{sp} = \iint_S G dS. \tag{1.46}$$

For a single-zone damage field in the polar coordinate system, the coordinate law by angle ξ is constant and depends only on the radius r, then

$$S_{sp} = 2\Delta\xi \int_0^{\infty} G(r) r dr. \tag{1.47}$$

For a fragmentation projectile with a single-zone field containing N identical fragments and with a constant vulnerable target area S_v, we will obtain

$$S_{sp} = 2\Delta\xi \int_0^{l_k} \left(1 - \exp - \left[\frac{NS_v}{2\pi r^2(\cos\gamma_1 - \cos\gamma_2)}\right]\right) rdr. \tag{1.48}$$

The value of the S_{sp} is a characteristic of the target, which depends on the damaging effect of the ammunition. It can be shown that the maximum efficiency is achieved when aiming at the center of gravity of the specified zone. In addition to the value, the configuration of the specified target area is very important.

The configuration of the specified zone on the ground (Figure 1.39) in case of a ground explosion of ammunition is similar to the configuration of damaged sectors on the ground (Figure 1.38), mirroring the line drawn on the ground through the point of explosion of a projectile perpendicular to the direction of

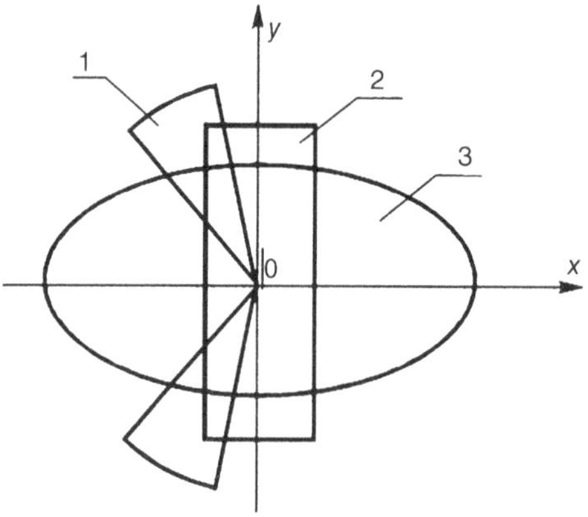

Figure 1.39 Illustration of the problem of concordance dispersion law and the coordinate law of damage: 1 – the real configuration of the specified zone; 2 – representation of the specified zone as a rectangle; 3 – dispersion ellipse. *Source:* From Balagansky et al. [2].

its symmetry axis. Naturally, the sizes of the specified zone are much smaller than the sizes of damaged sectors.

When calculating the effectiveness, it is common to replace the actual configuration of the specified zone with a rectangle, the area of which is equal to its value, and the dimensions of the sides of the rectangle for conventional artillery ammunition are chosen so that the longer side is perpendicular to the direction of the ammunition coming to the ground surface.

The concordance to which the law of dispersion and the coordinate law of damage are harmonized plays a decisive role in determining the effectiveness of ammunition. The problem of their concordance becomes especially critical in case of a low-angle fire (for example, in case of direct fire from a tank gun), when the dispersion ellipse is strongly elongated in the direction of firing (the ratio of the longitudinal axis of the dispersion ellipse to the transverse axis is 40–50), and the fragmentation field is located in the directions perpendicular to the firing direction (Figure 1.39). In the coordinate system associated with the explosion point, the dispersion ellipse and the specified damage zone have a small overlapping area in which the probability of target hitting is not equal to zero. From this, it is clear that in this kind of firing, it is not so much the value of the specified damage area that is important, but the shape of the damage area and its location relative to the ellipse of dispersion. It is this circumstance that determines the high efficiency of the firing of projectiles with axial flows of fragments, which primarily include shrapnel.

1.9 Methods for Optimizing the Parameters of Fragmentation Munitions

1.9.1 The Method of Bauman Moscow State Technical University (BMSTU)

Let us consider a simple method of optimization of the fragmentation ammunition proposed at BMSTU [1]. The following assumptions have been adopted to optimize the parameters of the fragmentation warhead in case of a ground explosion with use the criterion of the maximum of the specified area of the fragment damage S_{sp}:

1) a ground explosion of the projectile;
2) the size of the projectile is negligibly small compared to the size of the fragment field;
3) the blast effect of the projectile is negligible;

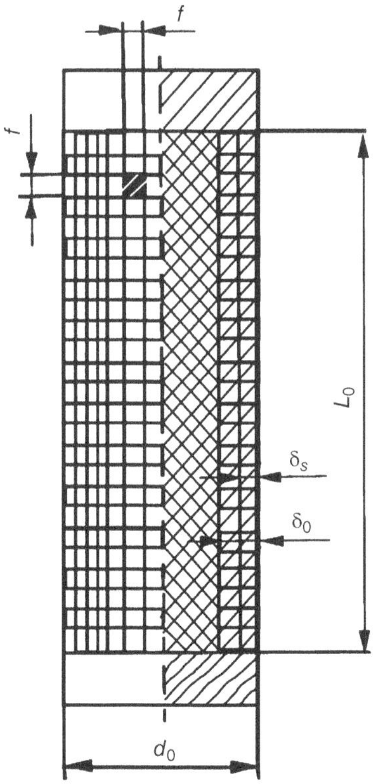

Figure 1.40 Equivalent cylindrical shell with HE charge. *Source:* From Balagansky et al. [2].

4) a warhead can be replaced by an equivalent cylindrical shell with a HE charge (Figure 1.40) with a total mass Q, charge mass C, and diameter d_0 equal for a warhead and an equivalent cylinder;
5) the fragmented shell is assembled from multiple layers of squarely packed fragments in a case;
6) the fragment field can be considered a one-zone field with a small angle of dispersion ($\Delta\varphi \leq 40°$) and a constant fragment velocity inside the angle of dispersion;
7) the target is one component, characterized by a constant vulnerable area S_v and steel equivalent h_{st}^{e}.

For the equivalent cylinder, the mass balance is

$Q = M + C + M_b,$

where M is the fragmented mass of metal; C – the mass of HE; M_b – the mass of bottoms.

The mass of the bottoms includes the mass of all components of the projectile that are not used for useful fragmentation (the mass of the bottom of the projectile and its bottom thick-walled part, the mass of the fuse, the mass of the stabilizer for the projectile for smoothbore systems, etc.).

Let us introduce the dimensionless parameters:

$$\alpha = \frac{C}{Q}; \quad \mu_b = \frac{M_b}{Q};$$

$$\xi = \frac{C}{M + C}.$$

The condition of the stable fragmentation $f \geq \delta_s$, from where the number of layers n_s and layer thickness δ_s

$$n_s = E\left(\frac{\delta_0}{\sqrt[3]{Y}} + 1\right); \quad \delta_s = \frac{\delta_0}{n_s}.$$

The E function means taking an integer part of an argument. The number of fragments at a known mass M is equal to

$$N = \frac{M\eta_\mu}{m},$$

where η_μ – the coefficient of the mass output of metal in controlled fragments. The physical sense of the parameters μ_b and η_μ is different. The relative mass of the bottoms μ_b is a characteristic of the projectile design, mass output into fragments, while η_μ is a physical and mechanical characteristic of the method of the predetermined fragmentation, properties of the metal and explosive.

Using also the previously introduced relations for fragment shape characteristics $\langle S \rangle$, Φ, A, we will write down 10 equations that determine the design-geometric parameters of the shell and fragments.

1) Local explosive filling ratio

$$\xi = \frac{\alpha}{1 - \mu_b}.$$

2) Wall thickness

$$\delta_0 = \frac{d_0}{2}\left(1 - \sqrt{\frac{\alpha\gamma_0}{\alpha\gamma_0 + (1 - \alpha - \mu_b)\rho_0}}\right).$$

3) Fragment volume

$$Y = \frac{m}{\gamma_0}.$$

4) Number of layers

$$n_s = E\left(\frac{\delta_0}{\sqrt[3]{Y}} + 1\right).$$

5) Layer thickness

$$\delta_s = \frac{\delta_0}{n_s}.$$

6) Square size

$$f = \sqrt{\frac{Y}{\delta_s}}.$$

7) Average midship of fragment

$$< S > = \frac{1}{2}\left(f^2 + 2f\delta_s\right).$$

8) Fragment shape parameter

$$\Phi = \frac{< S >}{Y^{\frac{2}{3}}}.$$

9) Ballistic coefficient of fragment

$$A = \frac{\rho_a < S > c_x}{2m}.$$

10) Number of fragments

$$N = \frac{Q(1-\alpha-\mu_b)}{m}\eta_\mu.$$

Using the previously introduced relations for the single-zone field, let us present them as the next subsystem of 13 equations linking kinematic and geometric characteristics of the fragment field.

1) Initial fragment velocity

$$V_0 = \varphi_0 \frac{D}{2}\sqrt{\frac{\xi}{2-\xi}}.$$

2) Front dynamic angle of the fragment field

$$\gamma_1 = \operatorname{arctg}\frac{V_0 \sin\varphi_1}{V_c + V_0\cos\varphi_1}.$$

3) Back dynamic angle of the fragment field

$$\gamma_2 = \operatorname{arctg}\frac{V_0 \sin\varphi_2}{V_c + V_0\cos\varphi_2}.$$

4) Dynamic velocity on the frontline of the field

$$V_{d1} = \sqrt{V_c^2 + 2V_cV_0 \cos \varphi_1 + V_0^2}.$$

5) Dynamic velocity on the backline of the field

$$V_{d2} = \sqrt{V_c^2 + 2V_cV_0 \cos \varphi_2 + V_0^2}.$$

6) Resulting velocity

$$V_d = \frac{1}{2}(V_{d1} + V_{d2}).$$

7) Kill velocity of a fragment (for a steel fragment at the ground level)

$$V_k = \frac{145h_{st}^e \Phi}{m^{\frac{1}{3}}},$$

here V_k (m/s); h_{st}^e (mm); m (g).

8) Kill interval of fragment

$$I_k = \frac{1}{A} \ln \frac{V_d}{V_k}.$$

9) Dynamic field density

$$\Pi_d = \frac{N}{2\pi r^2(\cos \gamma_1 - \cos \gamma_2)}.$$

10) The front angle of the damaging sector on the ground

$$\xi_1 = \arccos \frac{\cos \gamma_1}{\cos \theta_c}.$$

11) The back angle of the damaging sector on the ground

$$\xi_2 = \arccos \frac{\cos \gamma_2}{\cos \theta_c}.$$

12) Angle at the top of the damaging sector

$$\Delta\xi = \xi_2 - \xi_1.$$

13) Specified zone of target damage

$$S_{sp} = 2\Delta\xi \int_0^{I_k} \left[1 - \exp - \left(\frac{NS_v}{2\pi r^2(\cos \gamma_1 - \cos \gamma_2)}\right)\right] rdr.$$

These equations make up a system of 23 equations containing 25 connected variables α, ξ, V_0, γ_1,γ_2, V_{d1}, V_{d2}, V_d, ξ_1, ξ_2, $\Delta\xi$, δ_0, m, Y, n_s, δ_s, f, $<S>$, Φ, V_k, A, I_k, N, Π_d, S_{sp} and 15 constants: $Q, d_0, D, \rho_0, \gamma_0, \varphi_0, \varphi_1, \varphi_2, \mu_b, c_x, \eta_\mu, h^e_{st}, S_v, V_c, \theta_c$. Excluding 22 equations and 22 unknowns from the system, we obtain one equation linking three variables:

$$S_{sp} = f(\alpha, m).$$

Thus, the optimization problem is in finding the maximum of the function of two variables. The extremum of the function S_{sp} is to be found by numerical methods, for example, using the method of the steepest descent method, random search, etc. The range of the arguments depends on the ammunition type and target. For blast fragmentation projectiles of average caliber, it is possible to make a preliminary estimate $\alpha_{opt} = 0.1 - 0.3$, $m_{opt} \cong 0.25\left(h^e_{st}\right)^2$, were m (g), h^e_{st} (mm).

It is also necessary to consider the gyroscopic stability limitation on the projectile length if the projectile is stabilized by rotation.

$$L_0 = \frac{d_0}{\pi} \cdot \frac{Q(1-\mu_b)}{(d_0 - 2\delta_0)^2 \rho_0 + \left[d_0^2 - (d_0 - 2\delta_0)^2\right]\gamma_0}.$$

If the wall thickness of the body decreases, the condition of the shot strength may also be violated.

1.9.2 Warhead Optimization for the C-13 Unguided Aircraft Missile

Despite the simplicity of the above method, it was actually used to optimize the parameters of the C-13 blast fragmentation warhead of the unguided aircraft missile when assessing the possibility of effective damage of lightly armored targets like an armored personnel carrier.

The variant of the warhead optimization for the light-armored vehicle damage is shown in Figure 1.41. This calculation was performed by the OPTIM software developed at Novosibirsk State Technical University. In this case, the method of a random search for an extremum is used for optimization. The figure shows the search trajectory, position of the extremum point and gives the numerical values of the optimal shell filling ratio with HE and the optimal mass of the fragment.

Table 1.11 presents the main tactical and technical characteristics of the C-13 missile with a blast fragmentation warhead. Figure 1.42 shows the results of destruction action of heavy fragments on a 15-mm thick armored plate, Figure 1.43 shows the picture of an armored personnel carrier after being struck by a C-13.

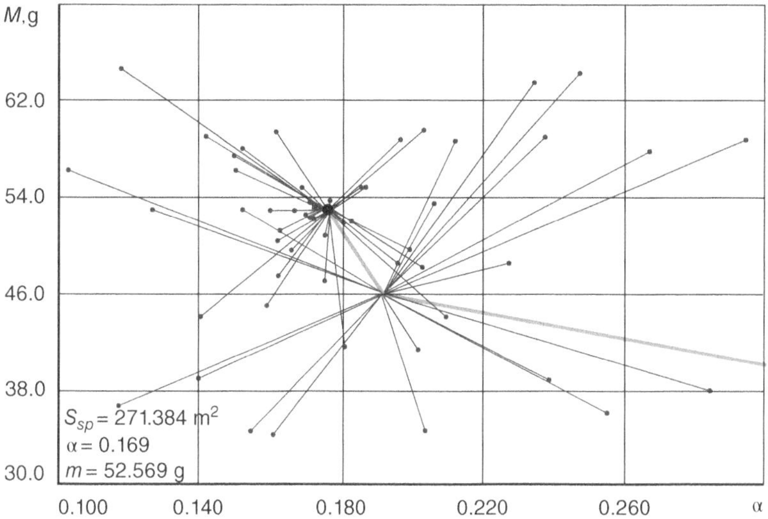

Figure 1.41 The variant of the warhead C-13 optimization for the light armored vehicle damage. *Source:* From Balagansky et al. [2].

Table 1.11 Main tactical and technical characteristics of the C-13 missile with a blast fragmentation warhead.

Caliber, mm	Length, mm	Mass of missile, kg	Mass of warhead, kg	Mass of HE, kg	Number of fragments, pcs.	Operating range, M	Missile's own speed, m/s
122	2898	69	33	7	450 with mass 25–35 g	1600–3000	530

Source: From Balagansky et al. [2].

1.10 Vulnerability Characteristics of Objects to the Effects of Fragments, Determination of Safe Distances

1.10.1 Methods of Efficiency Estimation

Previously, it was established that for fragmentation weapons and ammunition, a comprehensive characteristic of their target effects is the coordinate law of damage.

Figure 1.42 Results of heavy fragment action on the armored plate. *Source:* From Balagansky et al. [2].

Figure 1.43 Armored personnel carrier after a C-13 attack. *Source:* From Balagansky et al. [2].

In the previous sections, using the case of a fragmentation projectile with a one-zone fragment field in a ground blast as an example, it was shown how the coordinate law of damage can be obtained and the value of the specified zone of target damage assessed. From this example, it is clear that the vulnerability characteristics of targets must be known to assess the effectiveness of ammunition.

Vulnerability characteristics of the same target may vary depending on the required type of damage *A*, *B*, or *C*.

At present, methods of different levels are used to assess the effectiveness of the ammunition, depending on the target type, ammunition type and caliber, the required accuracy of calculation, and the form of representation of target vulnerability characteristics.

1) A complete statistical model of the ammunition effectiveness assessment for the target with detailed characteristics is the basis of the highest-level methods. Characteristics defining the vulnerability of the target to the action of fragments include a list of vulnerable aggregates, materials, and thicknesses of equivalent barriers, HE, gunpowder, and fuel tanks, aggregate filling factors with vulnerable elements, type of damage caused to the target, and damage hypotheses. Description of target geometry and armor thickness in the target coordinate system of flat quadrangles with specified coordinates of vertices, thickness, and material corresponding to physical characteristics of armor is specified. The internal equipment and the crew are displayed in the target coordinate system with elementary parallelepipeds and are in a certain connection reflecting the functional scheme of target vulnerability with respect to this type of damage.
2) The first-level method, which is a basic one for assessing the effectiveness of fragmentation munitions when operating on lightly vulnerable ground targets, is based on a simplified representation of the target as a spatial set of independent or functionally linked aggregates. The target model of lightly vulnerable targets without armor protection is similar to the model of a lightly armored vehicle's internal elements. Accounting for the actual location and mutual shielding of the aggregates is particularly important to assess the effectiveness of small-caliber ammunition, including cluster munitions, whose maximum effective range is often commensurate with target size. The vulnerability of personnel is assessed by damage to a single soldier.

The rapid estimation method corresponds to the target representation as a generalized parallelepiped and is used for optimization, preliminary analysis, and evaluation of medium and large calibers with axially symmetric wide-angle fragment fields of dispersion when operating on easily vulnerable targets. Sizes of the generalized parallelepiped are selected for the maximum approximation to the geometrical dimensions of the target within which its vulnerable aggregates are placed, and their vulnerable areas are evenly distributed ("smeared") along the corresponding edges of the parallelepiped.

When determining efficiency in the methods of the highest and the first levels, characteristics of the fragment field are specified in the form of the differential law of distribution in the elementary zones of the dispersion angle. In each elementary angular zone, the following are specified: the number of fractions and the number of fragments in each mass fraction; the average mass of a fragment and the shape parameter of the average fragment of each fraction; the average initial velocity of fragments, which is the same for all fractions.

In the express evaluation method, in addition to the differential law of distribution, it is allowed to use the integral law with a single law of fragmentation of the shell into fragments throughout the angle of dispersion. The number and sizes of angular zones are selected depending on the specific features of the ammunition and the level of the method used. The numerical values of the initial characteristics of the fragmentation are determined in accordance with the existing standards.

1.10.2 Characteristics of Target Vulnerability to Fragment Action

Critical values of specific energy (E_{cr}) of fragment required for damaging the vulnerable area of different targets (S_v) used for the evaluation of fragment efficiency are presented in Table 1.12.

Table 1.13 shows the critical values of specific impulse for the damage of manpower, lightly vulnerable, and lightly armored equipment.

Table 1.14 shows the values of steel equivalents for the same targets.

For a unified approach to the evaluation of the effectiveness of the damaging effect of ammunition, systems of initial data are being developed, which are obligatory to be used at all stages of development of ammunition. The consistency of the approach is provided by the unified initial data on vulnerability characteristics of typical elementary targets in relation to the damaging effect of the ammunition, unified methods for calculating efficiency indicators, and a unified set of initial characteristics of the ammunition.

Table 1.12 Characteristics of target vulnerability to fragmenting effect.

Parameters	UMP			PMP			LVE	MF	LAE
	Lying down	Sitting	Standing up	Lying down	Sitting	Standing up			
Vulnerable area, S_v, m^2	0.23	0.35	0.5	0.23	0.35	0.5	1.8	1.8	1.8
Specific energy E_{cr}, MJ/m^2	1.0	1.0	1.0	7.4	7.4	7.4	13.2	23.5	33.3

UMP, unprotected manpower; PMP, protected manpower (wearing bulletproof vests); LVE, lightly vulnerable equipment; MF, military fortifications (with a steel equivalent of 8–10 mm); LAE, lightly armored equipment (with armor approximately 16 mm thick).

Source: From Balagansky et al. [2].

Table 1.13 Average critical values of specific impulse i_{cr}, kPa s.

Unprotected manpower	5
Lightly vulnerable equipment	30–100
Lightly armored equipment	100–500

Source: From Balagansky et al. [2].

Table 1.14 Steel equivalents of targets, mm.

Unprotected manpower	1
Lightly vulnerable equipment	5
Lightly armored equipment	15

Source: From Balagansky et al. [2].

1.10.3 Determining Safe Distances

When calculating the ballistic coefficient and the damaging effect of the fragments on the targets, we assumed the averaging of the midship cross-section of the fragment S and the constancy of the drag coefficient c_x (see Section 1.5). These assumptions are unacceptable in determining the safe distances from the point of explosion of ammunition, for example when calculating the safety of storage facilities, and the safety of bomber aircraft in case of low-altitude bombing. The real midship cross-section (projections of the cross-section area of a fragment in the direction of its flight) is the value of the variable oscillating within $S_{min}, \ldots, S_{max}$. When assessing the safe distances to obtain a guaranteed result, it is advisable to consider the midship cross-section of a fragment equal to its minimum S_{min} value.

The assumption that the drag coefficient c_x is constant is also unacceptable for accurate estimates. For accurate calculations, it is necessary to consider the dependence of c_x of the fragment on its velocity V (or on the number of Mach $M = V/c_0$, where c_0 is the velocity of sound in air) [10]. In the subsonic region ($M < 1$), for a fragment of spherical shape, the c_x coefficient is approximately constant and equal to 0.5. In the region of $M \approx 1$, due to the change of flow modes and the occurrence of an attached (ballistic) shock wave, the air resistance increases sharply.

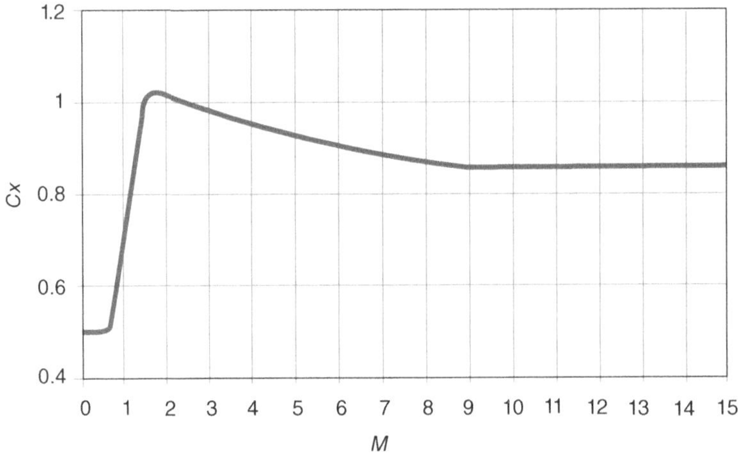

Figure 1.44 The dependence $c_x = f(V)$ for fragments of spherical shape. *Source:* From Bagiyan and Skrynnikov [10].

The dependence $c_x = f(V)$ for a fragment of spherical shape can be approximated by the following relations (Figure 1.44):

$$c_x = \begin{cases} 0.5 & \text{when } M \leq 0.5; \\ \dfrac{61}{80} + \dfrac{21}{80}\sin\left(\dfrac{\pi(M-0.5)}{1.1} - \dfrac{\pi}{2}\right) & \text{when } 0.5 < \mathrm{M} \leq 1.6; \\ 0.783 + 0.31e^{(-M/6.469)} & \text{when } 1.6 < \mathrm{M} \leq 9; \\ 0.86 & \text{when } M \succ 9. \end{cases}$$

1.11 Self-assessment Questions

1) Comment on the different classifications of fragmentation ammunition.
2) List the main methods to increase the range of blast fragmentation ammunition.
3) Why are the methods of controlled fragmentation of shells into fragments and preformed pieces used in fragmentation ammunition?
4) What are rod warheads and which targets are they used for?
5) List the main stages of the process of cylindrical shell fracture into fragments.
6) Calculate the initial velocity of fragments of the standard sample number 12, filled with phlegmatic octogen.
7) Calculate the angle of rotation of the shell θ at detonation velocity $D = 8000$ m/s and the initial shell velocity $V_0 = 1200$ m/s.

8) List the types of failure that are realized by crushing the shells into fragments.
9) How does the displacement of the point of initiation of detonation in HE charges influence the behavior of fracture of spherical shells?
10) What is the basic theorem of dimensional theory?
11) Which dimensions are independent?
12) List the main factors that influence the degree of fragmentation of shells from metal properties.
13) What is the scaling effect at fragmentation?
14) Why do we need standard fragmentation cylinders?
15) What sizes are the same for basic standard cylinders?
16) Comment the classification diagram in Figure 1.23.
17) List the main groups and grades of fragmentation steels.
18) Specify the fragmentation characteristics of manganese austenitic steels; when would it be appropriate to use them?
19) How is the law of fragment dispersion determined?
20) Name three main classes of fragment fields.
21) How are the characteristics of fragment fields determined experimentally?
22) What is the law of fragmentation of the shell into fragments and how is it determined experimentally?
23) In what coordinates are normalized histograms of fragment spectrums plotted and why?
24) List the main ways of semi-regular and regular crushing of shells into fragments and their advantages and disadvantages.
25) Give a description of the laws of fragment distribution by mass.
26) What is the median, mode, and average mass of a fragment of the spectrum?
27) What do you know about analytical views of the laws of fragment distribution by mass?
28) List the main characteristics of the shape of a fragment.
29) What is the ballistic coefficient, its dimension and physical sense?
30) Determine the killing interval of a fragment weighing 100 g flying at a velocity of 1700 m/s when damaging a target with the steel equivalent of 15 mm.
31) List three concepts of fragment damage of targets. Which one, in your opinion, most correctly reflects reality?
32) List the main kinds of damage caused by fragments on different targets.
33) Comment on the effect of the specified damage zone configuration on efficiency.
34) List the main assumptions adopted in the BMSTU optimization method.
35) What changes should be made to the BMSTU optimization method so that it can be used to optimize ammunitions with large fragment dispersion angles?
36) Give a brief description of the methods of different levels to assess the effectiveness of fragment action.

References

1 Odintsov, V.A. (2002). *Designs of Fragmentation Ammunition: Teaching Manual. Chapter 2.* Moscow: Bauman Moscow State Technical University [in Russian].
2 Balagansky, I.A. and Merzhievsky, L.A. (2004). *Effects of Means of Destruction and Ammunition.* Novosibirsk: Novosibirsk State Technical University [in Russian].
3 Odintsov, V.A. (2002). *Designs of Fragmentation Ammunition: Teaching Manual. Chapter 1.* Moscow: Bauman Moscow State Technical University [in Russian].
4 Odintsov, V.A. (1990). *Statistical Distribution at Fragmentation: Teaching Manual.* Moscow: Bauman Moscow State Technical University [in Russian].
5 Balagansky, I.A., Karanik, Y.A., Agureikin, V.A. et al. (2001). Fracture behavior of explosively loaded spherical molded steel shells. *Journal of Theoretical and Applied Fracture Mechanics* **36** (2): 165–173.
6 Sedov, L.I. (1944). *Similarity and Dimensional Methods in Mechanics.* Moscow, Sankt Peterburg: OGIZ [in Russian].
7 Odintsov, V.A. (1991). *Modeling of the Fragmentation Processes with the Help of the Unified Cylinders: Teaching Manual.* Moscow: Bauman Moscow State Technical University [in Russian].
8 Miropolsky, F.P., Sarkisian, R.S., Vishniakov, O.L. et al. (1996). *Aircraft Warheads and Their Investigation.* Moscow: Zhukovsky Academy [in Russian].
9 Andreev, S.G., Babkin, A.V., Baum, F.F. et al. (2002). *Physics of Explosion. V.2* (ed. L. P. Orlenko). Moscow: Fizmatlit [in Russian].
10 Bagiyan, A.A. and Skrynnikov, A.A. (2007). Ballistics of fragmentary damaging elements. In: *Proceedings of Zhukovsky Academy.* Moscow: Aircraft Armament [in Russian].

2

Ammunitions with Shaped Charges

2.1 Basic Concepts and Definitions. General Information

Ammunition with shaped charges (shaped charge ammunition) is designed to damage armored and lightly armored targets such as tanks, mobile artillery systems, infantry combat vehicles, armored personnel carriers, submarines. The effects of ammunition with shaped charges are based on the use of the cumulation phenomenon. Cumulation, i.e. the concentration of force, energy, or other physical quantity in a small volume, is an important phenomenon of nature.

This phenomenon is used both in technology and in everyday life, even without understanding the specifics of all its regularities and limitations. In the case of shaped charge ammunition, the cumulation phenomenon is to concentrate the energy of the explosion in a given direction, which leads to a significant local increase in the destructive effect.

Shaped charge ammunitions, the most common of which are HEAT (high-explosive anti-tank) ammunitions, are designed to penetrate armor. The explosive filling has a conical cavity in the front that is lined with a thin sheet of metal (typically copper). Upon detonation, this metal is converted to a thin jet with an extremely high velocity that can penetrate armor up to 10 times the caliber of the warhead. The casing of this projectile is usually thin as fragmentation properties are secondary considerations.

Shaped charge ammunition was widely used during the Second World War. Over the years, their designs have undergone significant changes. Modern ammunition with shaped charges is capable of penetrating homogeneous armor with a thickness of over 1000 mm. Nowadays, all armies have high-efficiency shaped charge munitions in service.

Damaging Effects of Weapons and Ammunition, First Edition. Igor A. Balagansky.

The following weapons with shaped charges can be distinguished:

- artillery projectiles and mines;
- grenades (hand and rocket-propelled);
- warheads of anti-tank-guided missiles;
- engineering mines and destruction charges;
- anti-tank aircraft bombs and cluster warheads;
- warheads of unguided aircraft missiles;
- warheads of guided missiles, and air bombs;
- warheads of torpedoes, mines, anti-submarine-guided missiles, and air bombs.

Let us consider the construction of some types of shaped charge ammunition.

2.1.1 Artillery Projectiles

Nowadays, almost exclusively smoothbore guns are used to fire artillery projectiles, because the rotation of projectiles when firing from rifled systems leads to a sharp decrease in efficiency of action. Let us look at some constructions of these projectiles.

Figure 2.1 shows the schematics of BK3M (100 mm caliber) and BK14M (125 mm caliber) projectiles for smoothbore artillery guns. The stabilizer blades are in the closed position and open after being fired. For projectiles with shaped charges, instantaneous fuses are used.

When the projectile meets the target, the fuse is triggered, the impulse from the fuse detonator is transmitted to the projectile detonator capsule located in the bottom part of the explosive charge at the top of the shaped liner. The explosion of the detonator capsule initiates the intermediate detonator and the explosive charge. When the detonation front passes along the liner, the pressure of several hundred thousand atmospheres acts on the metal liner, which leads to the collapse of the liner onto the axis of symmetry of the projectile and the formation of a metal jet moving at a velocity of 7–8 km/s, which has a strong damaging effect on armored targets.

2.1.2 Engineering Mines with Shaped Charges

Anti-tank mines with shaped charges are an effective tool for damaging tanks from below or sides. Bottom and side armors are much thinner than in the front of the tank, so the armor-piercing effect of the mines may be even lower than that of artillery projectiles, but it is important to have a high destruction effect within the armored space in the tank.

In such cases, ammunition with a shaped charge of a nonclassical scheme is used. The angle at the apex of the liner is much higher than the typical angle of

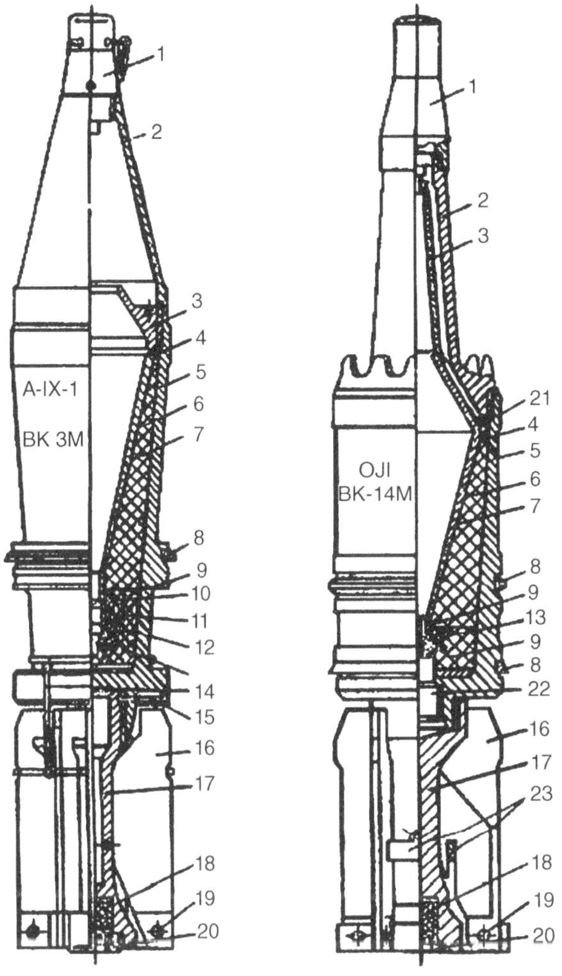

Figure 2.1 Anti-tank projectiles for smoothbore guns: 1 – GPV-2 fuse (BK3M) or piezo generator of the fuse device B15 (BK14M); 2 – head; 3 – protector (BK3) or contact cone (BK14); 4 – gasket; 5 – case; 6 - HE; 7 – liner; 8 – obturating belt; 9 – HE; 10 – gasket; 11 – plastic bushing; 12 – bottom capsule detonator K-1-T; 13 – plastic insert; 14 – wire fixer; 15 – stud; 16 – stabilizer blade; 17 – stabilizer body; 18 – tracer; 19 – blade axis; 20 – tracer nut; 21 – cuff; 22 – detonator of fuse; 23 – stabilizer ring with feet A. *Source:* From Balagansky and Merzhievsky [1].

such charges. Instead of the traditional jet, the charge forms a more massive element that moves at high velocity and has a highly damaging effect.

The American M21 mine (Figure 2.2) has a powerful shaped charge and a pin fuse with an overburden charge located coaxially with a shaped charge liner.

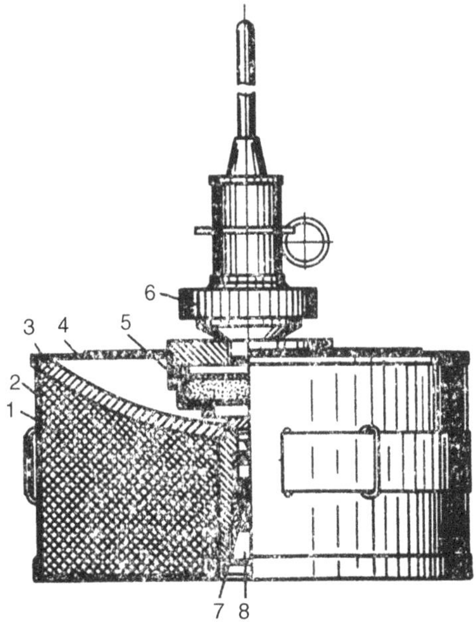

Figure 2.2 Anti-tank mine M21: 1 – explosive charge; 2 – case; 3 – liner; 4 – cover; 5 – powder overburden charge; 6 – pin fuse; 7 – pyrotechnical retarder; 8 – mine detonator. *Source:* From Balagansky and Merzhievsky [1].

When a tank hits the pin, it triggers the shock mechanism of the fuse, which ignites the powder of the overburden charge and pyrotechnic retarder of the mine detonator.

The gas pressure of burning gunpowder forces the lid together with the fuse and a camouflage layer of soil to be thrown from the liner, which provides normal conditions for the formation and action of massive compact damaging element onto the bottom of the tank.

The development of anti-bottom mines focused on significant improvements of their tactical and technical characteristics using noncontact fuses and the development of shaped charges specially designed to damage a tank from below distinguished by their increased anti-armor action. These include the French HPD mine and the Swedish FFV-028 mine. Both mines are equipped with shaped charges with the masses of explosives of 1.85 and 3.5 kg respectively, which ensure that, when penetrating a tank from below, several hundreds of fragments with high kinetic energy are brought into the tank's armored space.

Figure 2.3 shows the structure of an off-route stationary mine. Figure 2.4 demonstrates the installation and operation of this mine. The mine has a body with a shaped charge and is installed on the ground with the help of fixtures. The fixtures provide an opportunity to aim the mine in the required direction. The high-explosive charge is equipped with an initiating unit, which is connected to the output of the fuse device. Figure 2.4 shows a noncontact version of the fuse device mounted on the body of the mine. The signal to detonate the mine is given when a moving target comes into the activation zone of the fuse.

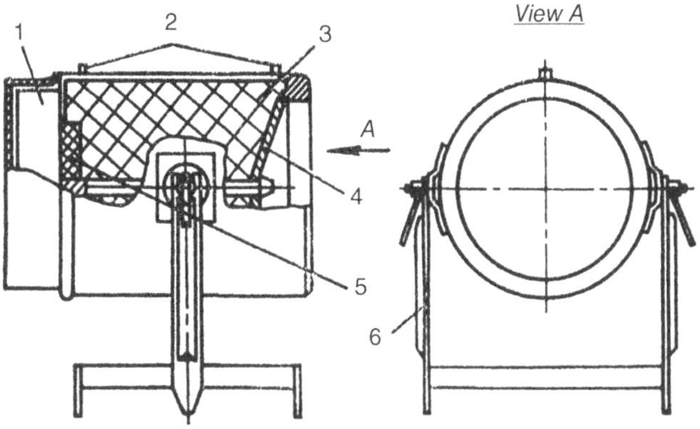

Figure 2.3 Structural scheme of an off-route anti-tank stationary mine: 1 – fuse unit; 2 – rear sight; 3 – explosive charge; 4 – metal liner; 5 – detonator; 6 – supports. *Source:* From Balagansky and Merzhievsky [1].

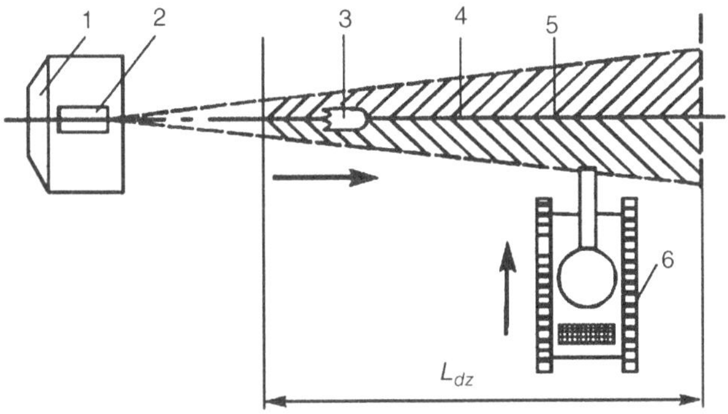

Figure 2.4 Deployment and action of an off-route mine: 1 – mine; 2 – rear sight; 3 – damaging element; 4 – zone triggering of a fuse device; 5 – flight direction of damaging element; 6 – an object of attack; L_{dz} – length of the damaging zone. *Source:* From Balagansky and Merzhievsky [1].

The advantages of such mines include constructive simplicity and low cost. The high velocity of the damaging element (about 2000 m/s) and the practical absence of the nonoperational zone near the mine allow achieving increased effectiveness of the mine at ranges up to 50 m for small-size and high-speed targets.

The effect of the damaging element, in comparison with traditional shaped charges, differs in the piercing of a hole of considerable diameter (up to 70% of the charge diameter) with the formation of massive rear fragments, which ensures high efficiency of target damage.

2.1.3 Anti-tank-guided Missiles (ATGM)

The creation of ATGM was a significant step forward in the development of anti-tank weapons. The anti-tank-guided missile systems damage targets at distances of more than 2 km, and anti-tank systems damage targets at distances of up to 10 km with a hit probability of up to 0.8–0.9 are currently being developed. At the same time, a high rate of fire and low vulnerability of anti-tank-guided missile systems are achieved due to the shorter time for shot preparation, change of firing position, projectile flight to the target, and the possibility of firing from closed positions. The low weight and dimensions of the whole complex facilitate maneuvering and camouflage on the battlefield. Modern ATGMs are capable of piercing homogeneous armor with a thickness of more than one meter, overcoming additional dynamic protection (explosive reactive armor).

Let us consider the features and functions of ATGM using as an example the classic Russian first-generation ATGM "Malyutka," the construction of which is shown in Figure 2.5.

ATGM "Malyutka" is aimed at the target manually and guided in-flight using wires. As the most widely produced anti-tank-guided missile, "Malyutka" is comparable with such weapons as the Kalashnikov assault rifle and RPG-7 grenade launcher. Currently, "Malyutka" is still in service with armies of dozens of countries and includes several modifications.

For the first time in the world, plastics were widely used in the construction of the missile.

The missile has no onboard power supply, but only one yes–no steering machine and a simple gyroscope with mechanical activation. The commands are transmitted by a thin cable with three copper wires of 0.12 mm diameter each in a textile

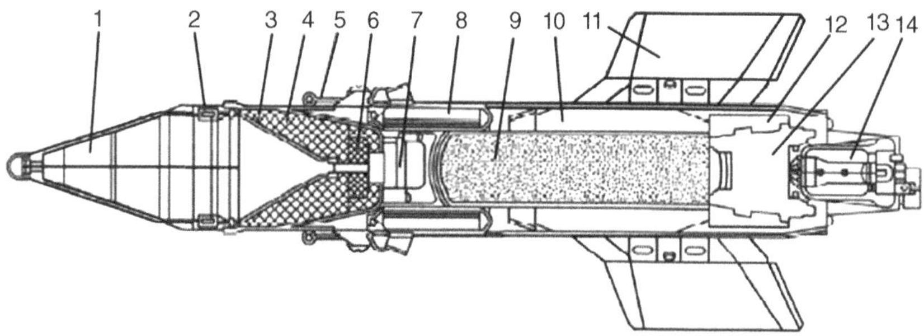

Figure 2.5 ATGM 9M14M of "Malyutka" complex: 1 – ballistic tip; 2 – piezoelectric element; 3 – liner; 4 – explosive charge; 5 – lock; 6 – lens; 7 – fuse; 8 – starting engine; 9 – march engine; 10 – spool with wire; 11 – stabilizer; 12 – on-board equipment; 13 – control system; 14 – gyroscope. *Source:* From Babkin et al. [2].

wrapper. The aerodynamic scheme of the missile is "tailless." Control is provided by changing the thrust vector of the main engine. To compensate eccentricity of the thrust of the engine, the missile is supposed to rotate around its axis with a speed of about 8.5 RPS. The missile's warhead has an A-IX-1 HE charge and a 9E219 piezoelectric fuse. In 1973, during the Arab-Israeli war, ATGM "Malyutka" damaged more than 800 Israeli tanks.

2.1.4 Anti-tank Bombs and Cluster Submunitions

Priority in the development of small aircraft bombs to destroy enemy tanks belongs to Soviet scientists and constructors. The first 2.5 kg PTAB 2.5 anti-tank bomb was developed in 1943 and was used against German tanks during the Battle of Kursk in World War II on 5 July 1943. It was used by IL-2 attack aircraft, and already on the first day, the Soviet pilots destroyed 30 enemy tanks.

As an example of a modern-shaped charge submunition for equipping cluster air bombs and cluster missile warheads, let us consider the PTAB – 1M caliber 1 kg submunition developed by Novosibirsk Institute of Applied Physics.

The PTAB-1M (see Figure 2.6) is developed for damaging armored vehicles: tanks, self-propelled artillery systems, infantry fighting vehicles. The submunition

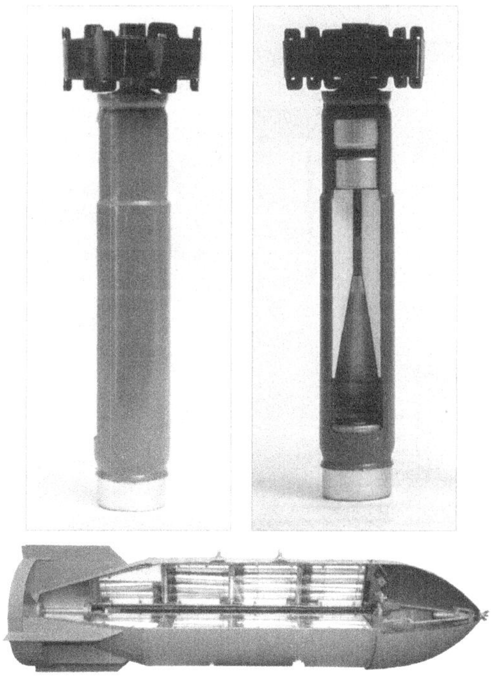

Figure 2.6 The PTAB-1M anti-tank submunition and the RBK-500 aircraft cluster bomb, equipped with PTAB-1M. *Source:* From Institute of Applied Physics. Advertising brochure [in Russian].

is used to equip cluster aviation ammunition: RBK-500, KMGU, warheads of guided, and unguided aircraft missiles.

The armor is penetrated by a shaped charge initiated by a piezoelectric fuse. PTAB-1M is equipped with a self-destruction mechanism.

The PTAB-1M with a diameter of 42 mm, length of 260 mm, and 940 g weight penetrates homogeneous armor with a thickness of at least 200 mm, which is quite enough to damage a tank from above. The RBK-500 cluster bomb has been developed for the PTAB-1M submunitions, which can be used from 300 to 25 000 m altitudes at the carrier velocity of 500–2300 km/h. The cluster diameter is 450 mm, the length is 1955 mm, the weight is 427 kg, and the number of submunitions in the cluster is 268.

2.2 Fundamentals of Cumulative Effects

2.2.1 The Phenomenon of Cumulation

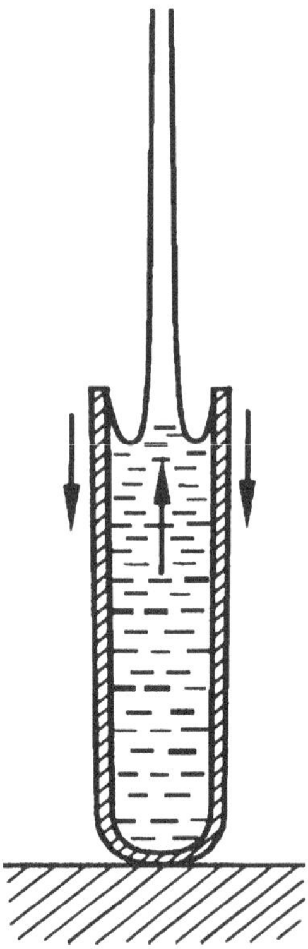

Figure 2.7 The experiment of G.I. Pokrovsky. *Source:* Lavrentiev et al. [4].

Let us illustrate this phenomenon with a simple experiment. Water is poured into a glass or metal test tube, and from a small height (10–20 cm), the tube falls vertically onto a table (Figure 2.7). Immediately after the tube hits the table, a thin stream of water over a meter high is knocked out of it. The effects of this kind are called cumulative effects.

Spherical cumulation is of great interest for practical use. In the simplest case, the problem of the collapse of a hemispherical cavity in a heavy liquid can be considered. In nature, this phenomenon can be observed during rain over a river, when nail-shaped water jets are formed on the surface of the water toward the raindrops.

Under certain conditions, as a result of an underwater explosion, an interesting phenomenon is observed, which was named "sultan" – water is thrown out over the free surface to a great height in the form of a narrow cone. It is noted that it is characteristic of a liquid medium and is not observed in underground explosions.

The tsunami problem is also directly related to the cumulation phenomenon. Tsunami waves arise from sudden movements of vast areas of the ocean bed during underwater earthquakes. As a rule, they make up a group of 2–3 waves, which in the open ocean are very long (100 km length) and flat (1 m height), so they are not dangerous. But when approaching the shore with decreasing depth, their height increases and can reach 30 m. When penetrating deep into the coastal territory, they cause destruction and human casualties.

Cumulative processes are unique: in contrast, the majority of natural phenomena that lead to smoothing out heterogeneity in the distribution of physical parameters, cumulation due to its inherent ability to concentrate the energy of large masses and volumes of matter into small areas leads to an increase in energy density in time and space.

The phenomenon of unlimited energy cumulation in converging detonation waves is of particular interest from the point of view of many applications, primarily for obtaining extreme states of matter, and E.I. Zababakhin made a great contribution to the study of this phenomenon [5]. In the case of converging waves, we come across the following picture: when a shock or detonation wave propagates, its surface decreases, and its specific energy and pressure at the front grow accordingly. At the center of the convergence, there is a special point where, if there were no inelastic effects, the pressure could become infinitely high.

Generally speaking, it is rather difficult to experimentally obtain strictly cylindrical or spherical convergent waves because of the necessity of simultaneous initiation of detonation strictly on the cylindrical or spherical surfaces, as well as because of the process instability at later stages.

In practice, the phenomenon of unlimited cumulation is used primarily in atomic bombs to compress a sphere of uranium-235 or plutonium-239 to transfer it to a supercritical state for initiating a nuclear reaction.

As an example of a device for the creation of ultrahigh pressures at unlimited cumulation, we present (Figure 2.8) a scheme of a unique hemispherical device – the so-called soft charge developed at the All-Russian Research Institute of Experimental Physics in Sarov [6]. An HE charge of a hemispherical shape is simultaneously initiated across its outer surface. The products of the explosion

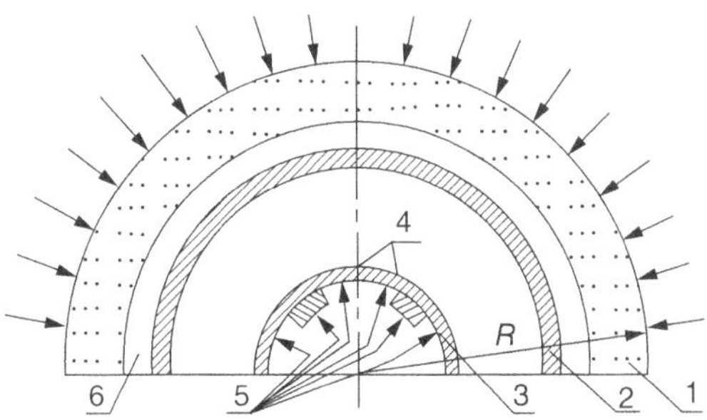

Figure 2.8 Hemispherical device "MZ": 1 – hemispherical charge of HE; 2 – steel shell; 3 – screen (iron); 4 – examined samples; 5 – electro contact gauge; 6 – air gap. *Source:* From Al'tshuler et al. [6].

of a spherically convergent detonation wave "softly" accelerate the steel hemispherical shell, the velocity of which at the convergence to the center (with a reduced radius) continuously increases. In existing systems, the shell velocity varies from 6.2 km/s, which corresponds to the pressure in iron of ~270 GPa, close to the pressures realized in planar acceleration systems, to 22.7 km/s, which corresponds to the pressure in iron of 1.8 TPa. The highest pressures in tantalum, reaching 2.5 TPa (25 million atm), were obtained using this system. Currently, pressures up to 3 TPa are available for studies of substances whose density exceeds 18 g/cm^3.

Another type of cumulation is related to the propagation of a flat shock wave in a layered barrier. Let us consider the movement of a wave with a front parallel to the layers in a system of alternating flat layers of light and heavy substances. If every heavy layer is made thinner than the previous one (the same for light layers), the impact may produce a higher velocity than the previous layer. The next layer will transmit an even higher speed and so on, i.e. the shock wave will accelerate. However, this will only happen if the energy loss on heating and residual kinetic energy of the layers is low.

An interesting effect of increasing the local effect of the explosion is observed in HE charges placed in a shell made of high-modulus ceramics with a sound velocity exceeding the velocity of detonation of HE [7]. The experiments were conducted with charges of various high explosives placed in a shell of silicon-carbide ceramics.

The scheme of experiments is shown in Figure 2.9. The longitudinal sound velocity in the shell was 11.1–11.2 km/s and significantly exceeded the detonation velocity of the HE charge. The measured pressure values in the detonation wave on the axis of charge in most cases markedly differed from the pressures of a

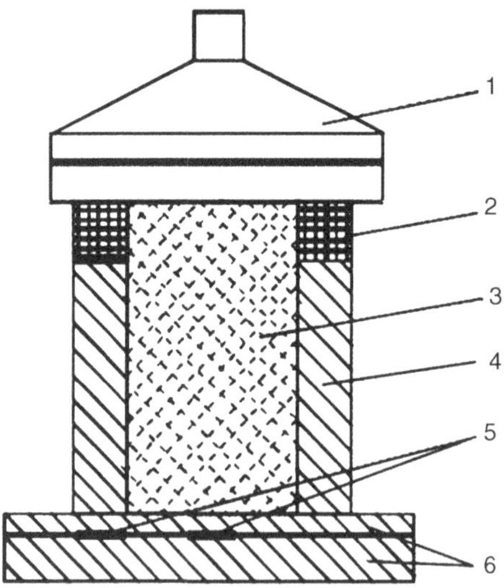

Figure 2.9 The experiments with ceramic shells: 1 – plane wave generator; 2 – ring of polystyrene foam; 3 – high explosive; 4 – ceramic shell; 5 – manganin gauges; 6 – copper plates. *Source:* From Balagansky et al. [7].

stationary detonation. In some experiments, a significant increase in pressure in the detonation wave was observed. In particular, for nitromethane, the growth of detonation pressure reached 56% in comparison with the pressure of stationary detonation. Subsequent analysis has shown that in such systems an important role is played by nonstationary detonation processes associated with the influence of the leading wave in ceramics on the detonation front.

Speaking of the phenomenon of cumulation, one cannot help but dwell on the so-called magnetic flux compression discovered in 1951 by A.D. Sakharov (Russia) and a little later by C.M. Fowler from the Los Alamos National Laboratory (USA). They suggested that the magnetic field should be rapidly compressed by a closed conductive shell. In those experiments, astonishingly high energy densities were achieved, hundreds of times higher than that of explosives and second only to nuclear warheads.

The effect of increasing the armor-piercing action of the projectile in the presence of a cavity in the explosive charge, the essence of which we will continue to analyze in detail, was discovered in the second half of the nineteenth century and was also called cumulative. Its use then was limited to some technical tasks in mining. A sharp increase in armor-piercing effect with the introduction of the metal liner was discovered a little later, and in 1914, the first patent for the use of the effect in the military was granted and armor-piercing ammunition based on the principle was developed. However, cumulative ammunition did not see any wide use until World War II. At the same time, the theory of this phenomenon was created.

2.2.2 The Cumulative Effect in Explosives Charges with Cavities

The theory of cumulation in HE charges with metallic liners of the cavity is most fully developed. The scheme illustrating the effect of a charge with a cavity is shown in Figure 2.10. Shaped charges are widely used both for military and industrial purposes. If a monolithic charge with axial symmetry leaves only a shallow conical dent on the surface of a steel block during an explosion, a charge with a conical cavity, which also contains a smaller amount of explosive, will knock out a crater five times deeper. When in the same charge the cavity is lined with metal, and the charge explodes at some distance from the barrier (1–3 times the charge diameter), a much deeper hole is produced (although its volume is smaller).

The action of charges without liners has been studied very poorly, as they are of no practical interest. Experiments show that after the detonation of the charge, the high-velocity flow of the explosion products is released from the cavity. Thus, the increase of efficiency of charge with cavity is based on the spatial concentration of the explosion products, which in a converging stream form a high-density gas jet moving at high velocity and density.

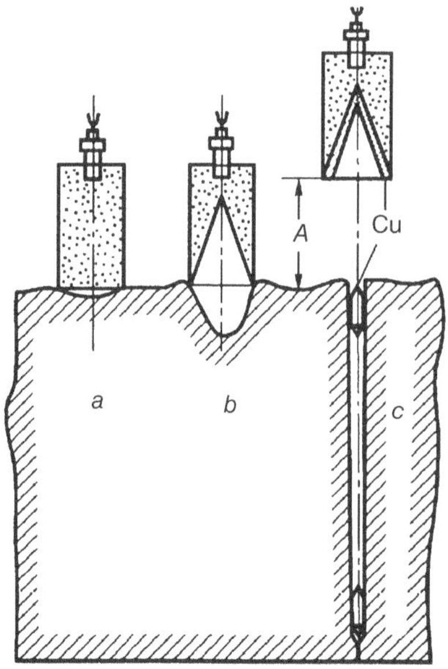

Figure 2.10 The action of HE charges on soft steel barrier: monolithic (a), hollow without liner (b), shaped charge with the lined cavity (c). *Source:* From Shell [8].

X-ray images of a shaped charge explosion with a copper liner (Figure 2.11) show that the formation of a metal jet is the most significant process in such charges.

However, in this case, the jet is formed from the liner material, which is accelerated by the explosion products toward the axis of the charge. Upon collision, a jet is formed from the inner part of the liner, which has a high velocity, and the outer part forms the so-called slug. The lined charge jet has a higher density, which does not decrease over a long time.

If the jet then hits the target, it turns around under the influence of braking, and the target material moves in a radial direction. This way, a crater is formed with a diameter significantly larger than the jet diameter. The depth of the crater depends on the shape and size of the charge, the composition of the HE, the geometry and properties of the liner, the distance to the target, and the material from which the liner is made.

It is common to indicate the damaging capacity of shaped charges of this kind by comparing the depth of the jet penetration with the diameter of the charge. Figure 2.12 shows the dependence of the penetration capacity of charges of the same type versus the distance A between the charge and the target (in this case the target is soft steel). It turns out that the depth of the crater achieved in this case increases with the distance A at first, and then slowly decreases. With an increase in the value of A, however, the spread in the values of the penetration depth also increases almost proportionally. The higher the energy content and detonation velocity of the explosive, the stronger the damaging ability of the charge. The

Figure 2.11 Impulse X-ray image of the shaped charge jet formation from a copper liner during HE charge explosion. *Source:* From Shell [8].

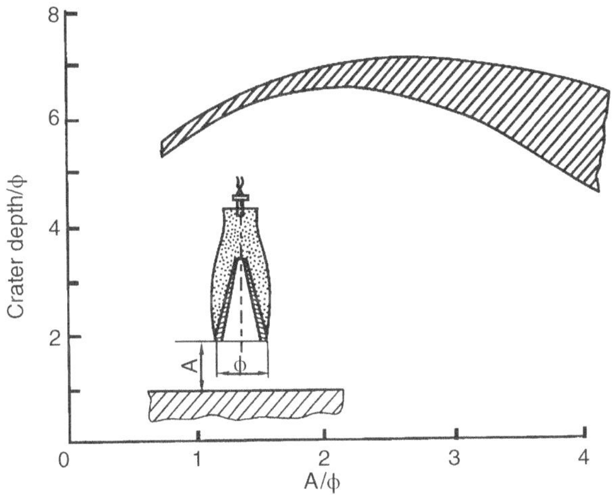

Figure 2.12 Penetration ability and scattering value of the jet penetration depth when soft steel is damaged. *Source:* From Shell [8].

volume of the crater turns out to be directly proportional to the energy content of the explosive.

The depth of the penetration, as well as its change with the distance A, significantly depends on the material from which the liner is produced. Heavy ductile metals of the copper group with a face-centered cubic crystal lattice and some alloys have a high penetration ability at large values of A.

They create continuous jets, the density of which is no more than 10% lower than the density of the liner material and which at a large elongation (about 10 times as long as the original length) do not break and keep high density. Other metals, such as iron or zinc, for example, at the initial stages also create continuous jets, which, in contrast, break much earlier. In shaped charges with liners of such metals, the maximum damaging effect is achieved at shorter distances to the target and has a smaller value than in the case of more ductile metals.

Brittle metals such as tungsten or titanium, as well as metals with high porosity produced by sintering, do not create continuous jets at all, they generate "discrete" jets, consisting of individual solid particles. In this case, the jet stretching does not lead to a decrease in its diameter, as in continuous jets, but to a decrease in the average density of the jet. The damaging ability of such jets in comparison with continuous jets is much lower. However, since discrete jets on many materials produce holes with a larger volume than solid ones, shaped charges of this kind are gaining importance in some applications.

The depth of penetration depends primarily on the density of the target material and to a much lesser extent on its strength. Table 2.1 compares the relative depths of penetration of a continuous jet formed by a standard charge explosion observed in different materials.

It should be noted that, strictly speaking, these numerical values are only applicable for a given kind of charge since the relative damage capacity depends on the design of the charge. Of two different designs, one would always be better, for example, at damaging steel, and the other more effective on a rock.

Table 2.1 Relative depth of penetration T in targets of various materials with a jet of standard shaped charge.

Material	Density ρ, g/cm^3	Relative depth T
Mild steel (standard)	7.8	1
Armor steel	7.8	0.8
Lead	11.3	1.3
Concrete	2.3	3.5
Perspex	1.2	2.8
Water	1.0	3.0
Polyethylene	0.92	5.0
Paraffin	0.9	7.2
Rubber	1.1	6.4

Source: From Shell [8].

The crater volume can be approximated as being directly proportional to the jet energy and inversely proportional to the strength of the target material.

2.2.3 Hydrodynamic Theory of Shaped Charges

The main statements of the theory of shaped charges with metal liners were published by Birkhoff [9] and Lavrentiev [10] for a stationary case using an ideal incompressible fluid model. They developed the hydrodynamic theory almost simultaneously and independently during the war but published the results much later. The main assumption of the theory is the representation of the liner material in the form of an ideal incompressible fluid.

2.2.3.1 Theory of Jets of Ideal Fluid

Let us first consider the picture of a flat-symmetric collision of two jets of an ideal incompressible fluid at an angle of 2α (Figure 2.13). Since the jets collide symmetrically, we can focus only on the upper part of the flow of the jets (Figure 2.14). Let us consider that the symmetry surface is smooth, i.e. there is no friction when the fluid moves along this surface.

Let's use u_0, M for the velocity and mass of the converging jet (mass of liquid flowing through the cross-section of the jet in a unit of time); u_J, M_J for the velocity and mass of the jet moving to the right along the $0x$ axis; and u_S, M_S for the velocity and mass of the jet moving to the left in the negative direction of the $0x$ axis. In this problem, the values of u_0, M and α are given. It is necessary to calculate the parameters of formed jets: u_J, u_S, M_J, M_S.

The laws of mass, impulse, and energy conservation are applicable to this process. According to the law of mass conservation:

$$M = M_J + M_S. \tag{2.1}$$

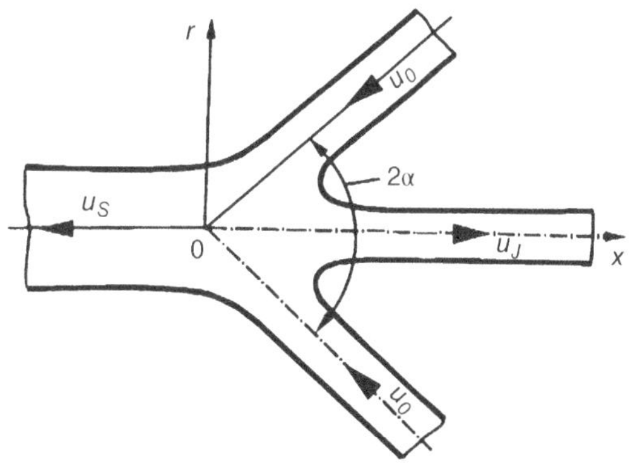

Figure 2.13 The scheme of a collision of two jets of incompressible fluid. *Source:* From Orlenko et al. [11].

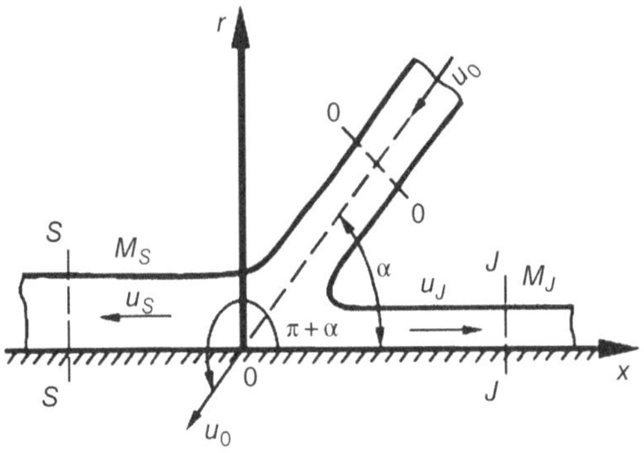

Figure 2.14 Scheme of single jet collision with a perfectly smooth surface. *Source:* From Orlenko et al. [11].

According to the law of conservation of momentum, the change in the amount of motion is equal to the impulse of the acting forces. As there are no forces acting on the fluid along the 0x axis, the change of the amount of motion is equal to zero.

$$(M_J u_J \cos 0^\circ + M_S u_S \cos \pi) - u_0 M \cos(\pi + \alpha) = 0 \tag{2.2}$$

or

$$M_J u_J - M_S u_S = -u_0 M \cos \alpha. \tag{2.3}$$

The projection of the equation of conservation of momentum on the 0r axis can be avoided because the jets move in the perpendicular direction to the 0r axis.

Since the jet collision process is stationary, the law of energy conservation can be written in the following form:

$$\frac{M_J u_J^2}{2} + \frac{M_S u_S^2}{2} = \frac{M u_0^2}{2}. \tag{2.4}$$

From the equations of momentum and energy conservation, we obtain that the absolute values of the velocities of all jets are equal:

$$u_0 = u_J = u_S. \tag{2.5}$$

Taking into account Eqs. (2.5), (2.1), and (2.3), we will obtain:

$$\begin{aligned} M_J + M_S &= M, \\ M_J - M_S &= -M \cos \alpha. \end{aligned} \tag{2.6}$$

From here, we define M_J and M_S:

$$M_J = \frac{M(1 - \cos \alpha)}{2} = M \sin^2 \frac{\alpha}{2}, \tag{2.7}$$

$$M_S = \frac{M(1 + \cos \alpha)}{2} = M \cos^2 \frac{\alpha}{2}. \tag{2.8}$$

2.2.3.2 Theory of Shaped Charge Jet Formation

Consider the transition from the theory of incompressible fluid jets to the theory of shaped charge jet formation. Let the liner of a shaped charge move with velocity V_0, perpendicular to the liner surface (Figure 2.15). Let us decompose this vector in two directions: along the liner surface and the $0x$ axis. As a result, we will get:

$$u_0 = \frac{V_0}{\text{tg}\,\alpha}; \quad V_K = \frac{V_0}{\sin\alpha}. \tag{2.9}$$

Now we will consider the process of jet motion relative to each of these components of the V_0 vector – the collapse velocity of the liner. The flow scheme shown in Figure 2.15 is divided into two parts (Figure 2.16). In Figure 2.16, on the left, we have the same flow scheme as in Figure 2.14. In this case, according to Eq. (2.5), two jets are formed whose velocities are equal, and, considering (2.9), are determined by the expression $u_0 = u_J = u_S = V_0/\text{tg}\,\alpha$. Figure 2.16 on the right shows the velocity $V_K = V_0/\sin\alpha$, which represents the velocity of the collision point of the shaped charge liner (points 0).

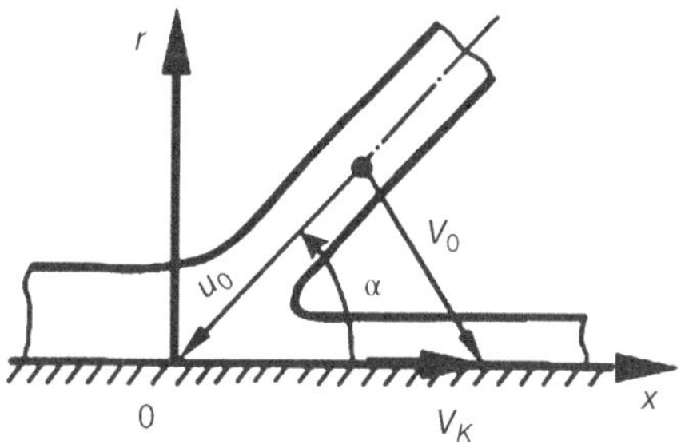

Figure 2.15 Scheme of shaped charge jet formation process in incompressible fluid. *Source:* From Orlenko et al. [11].

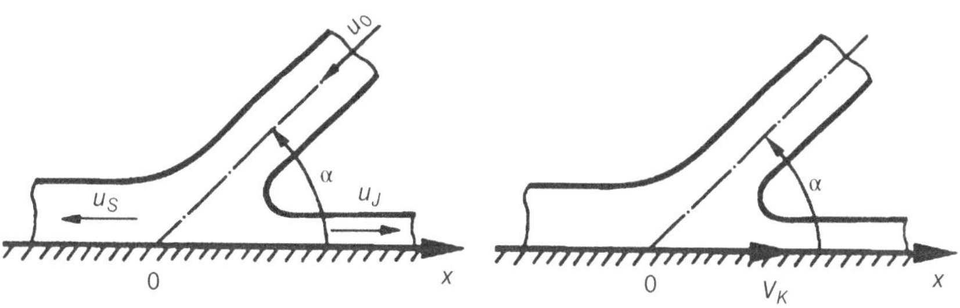

Figure 2.16 Scheme of shaped charge jet formation process decomposition. *Source:* From Orlenko et al. [11].

Connecting both flows, we get:

$$V_J = V_K + u_J = \frac{V_0}{\sin\alpha} + \frac{V_0}{\mathrm{tg}\,\alpha} = V_0\frac{1+\cos\alpha}{\sin\alpha} = V_0\frac{2\cos^2\frac{\alpha}{2}}{2\sin\frac{\alpha}{2}\cos\frac{\alpha}{2}} = \frac{V_0}{\mathrm{tg}\frac{\alpha}{2}}, \tag{2.10}$$

$$V_S = V_K - u_S = \frac{V_0}{\sin\alpha} - \frac{V_0}{\mathrm{tg}\,\alpha} = V_0\frac{1-\cos\alpha}{\sin\alpha} = V_0\frac{2\sin^2\frac{\alpha}{2}}{2\sin\frac{\alpha}{2}\cos\frac{\alpha}{2}} = V_0\mathrm{tg}\frac{\alpha}{2}. \tag{2.11}$$

Formula (2.10) determines the velocity of the shaped charge jet V_J, and formula (2.11) defines the slug velocity V_S. Accordingly, the mass M_J of the jet is determined by the formula (2.7), and the mass M_S of the slug is determined by the formula (2.8).

It should be noted that the collision angle α is not equal to the half of the angle of the original liner α_0. At the collision of the liner in the real charge, the sliding detonation front makes it rotate, so the angle of collision is always greater than the initial angle of the half of the angle of the original liner.

For conical liners, the collision angle increases from top to bottom, so the jet, in this case, has a gradient velocity along the length.

2.2.3.3 PER-theory

For the nonstationary case, when the elements of the conical liner move with different velocities, which is usually observed in practice (Figure 2.17), it is more convenient to use the PER-theory (Pugh E.M., Eichelberger R.J., Rostoker N.J., 1952). According to this theory, which is also based on the model of an ideal incompressible fluid, the shaped charge jet velocity is determined by the formula:

$$V_J = \frac{V_0}{\sin\alpha/2}\cos\left(\frac{\alpha_0-\alpha}{2} + \arcsin\frac{V_0}{2U}\right). \tag{2.12}$$

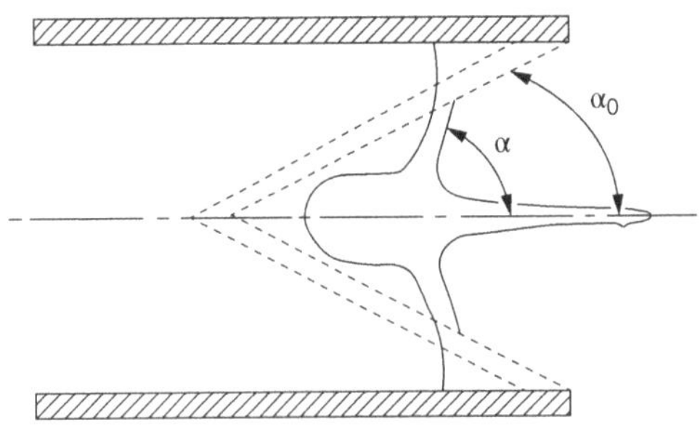

Figure 2.17 Scheme of conical liner collapse process. *Source:* From Walters and Zukas [12]. Reprinted with permission of Wiley.

Here, V_J is the velocity of the corresponding element of shaped charge jet; V_0 is the collapse velocity of the liner element; α_0 is half of the conical liner element angle; α-collision angle; $U = D/\cos\alpha$; where D is the detonation velocity of the HE charge. This equation describes the velocity of each element of the jet.

According to the PER-theory, the velocity of collision point can be determined by the following formula:

$$V_K = \frac{V_0}{\sin\alpha}\cos\left(\alpha - \alpha_0 - \arcsin\frac{V_0}{2U}\right). \tag{2.13}$$

2.2.4 Limitations of Hydrodynamic Theory

From the formulas of the hydrodynamic theory, it follows that the jet is formed at any angle α and any velocity V_0, and the theoretical limit of the jet velocity is equal to infinity when the collapse angle approaches zero. However, practice does not confirm these predictions. If the velocity $u_0 = \dfrac{V_0}{\text{tg}\,\alpha}$ is supersonic and $\alpha < \alpha_{cr}$, where α_{cr} is some critical angle of collision, the shaped charge jet is not formed.

On the other hand, if $\alpha > \alpha_l$, where α_l is the limit angle of collision, the jet is also not formed because the radial projection of the collision velocity of the liner is not sufficient to provide the deformation of the material required for the formation of the jet, because of the material's strength and internal viscosity that are not considered in hydrodynamic theory. Deformation of the liners with small values of the radial velocity of collision implemented at angles close to α_l leads to significantly different results in different conditions. In some cases, the explosion forms a compact element; in others, a charge with a similarly shaped liner can create a dispersed flow of fragments in the direction of the axis of the charge as a result of eversion and destruction of the liner after the explosion.

2.2.4.1 The "Reverse" Cumulation Mode

Let us consider separately the case when the angle of liner collapse $\alpha = \alpha_0 + \theta > \pi/2$, where θ is the angle of rotation of the liner in the process of throwing a sliding detonation wave [13]. Here, under certain conditions, the so-called reverse cumulation mode can be realized. A specific feature of this mode is the fact that the mass of the jet (the stream moving to the bottom of the liner) becomes greater than the mass of the slug (the stream moving to the top). In this case, the velocity of the jet in case of the application of powerful explosives may be about 5 km/s. This is less than in the normal mode, where the jet velocity reaches 8–10 km/s; nevertheless, the specific energy of the substance in the jet is 2–2.5 times higher than the specific energy of the HE. The boundary of the reverse cumulation mode at the further increase of the α_0 angle will be determined by

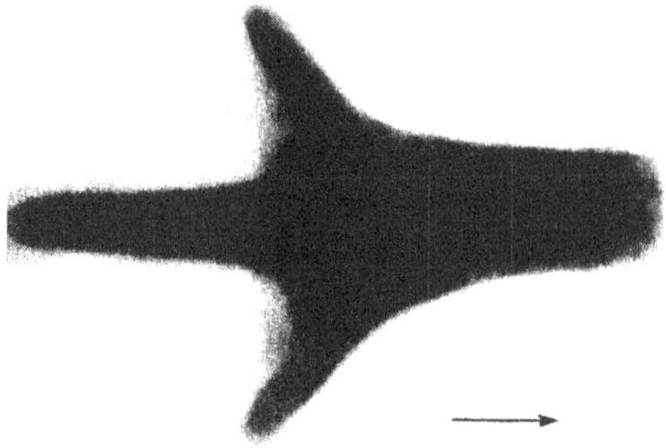

Figure 2.18 The "reverse" cumulation mode. *Source:* From Titov [13].

$$\alpha_0 = \frac{\pi}{2} - \frac{\theta}{2}.$$

Since the radial component of the liner deformation rate is small, it can be assumed that the reverse cumulation mode is not observed for all materials due to the negative effect of strength forces. For aluminum alloy and soft copper liners, the reverse cumulation mode has been experimentally observed.

In Figure 2.18, the impulse X-ray image of the deformation of such liner in case $\alpha > \pi/2$ is shown. The arrow shows the direction of movement.

2.2.5 Accounting for Compressibility of the Liner Material

Experiments show that at small collision angles, a shaped charge jet is not generated. To explain this effect, it is necessary to account for the compressibility of liner material in the process of its collapse.

Criteria of jet formation with an account of compressibility of liner material were defined in [14–16].

Let us consider the process of symmetric collapse of two plain jets of compressible fluid at an angle of 2α with the velocity u_0 in the coordinate system associated with the point of collision 0 (Figure 2.19). This problem is, in fact, analogous to the flow problem with a supersonic airflow at a blunt angle in aerodynamics (Prandtl–Meyer problem). The compressible supersonic flow at a velocity of u_0 is rotated by an angle α and slowed down. Such deceleration of a supersonic flow is carried out by the $0M$ front of an oblique shock wave. In this case, the flow before the shock front is not compressed. A rapid increase in pressure occurs behind the front of an oblique shock wave $0M$, and a jet in the direction of $0x$ is not formed.

According to the theory of oblique shock waves, there is a functional dependence $\beta = \beta(\alpha)$ between the angle of rotation of the supersonic flow α and the slope angle of the oblique shock wave front β for a given velocity u_0. When the angle α is increased, the connected front of the shock wave $0M$ can only exist up to an angle $\alpha < \alpha_{cr}$ (Figure 2.19). At angles $\alpha \geq \alpha_{cr}$, the front of the oblique shock wave $0M$ moves toward the incoming supersonic flow (Figure 2.20).

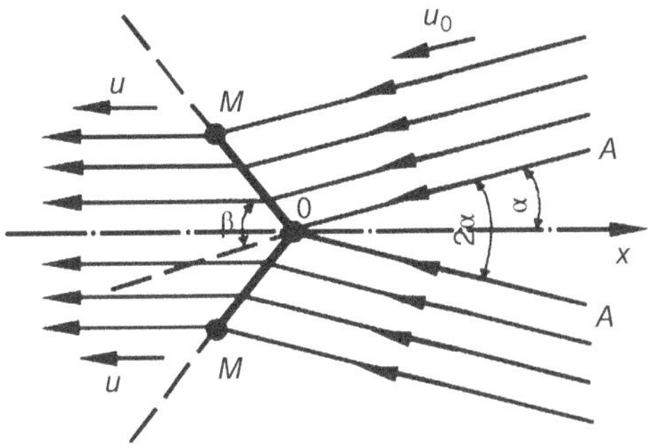

Figure 2.19 Scheme of the collision process of two compressible fluid jets. *Source:* From Orlenko et al. [11].

In this case, the pressure at the front of the shock wave MM' increases. The fluid behind the front of the shock wave MM' is shock-compressed, but since the surface of $AM0$ is free of shock compression, the fluid at section $0M$ begins to flow out to the right, and a jet is formed. Angle $\alpha = \alpha_{cr}$ is determined from the condition of the impossibility of a connected front of an oblique shock wave.

The analysis shows that the jet is not formed if $u_0 = \dfrac{V_0}{\text{tg}\,\alpha} \geq c_0$, is the bulk sound velocity in the liner material under normal conditions.

$$c_0 = \sqrt{\frac{K + (4/3)G}{\rho_0}}, \tag{2.14}$$

where K is the bulk compression module; G is the shear module; ρ_0 is the density of the liner material.

If $u_0 < c_0$, the jet should be formed both for incompressible and compressible fluid models at all angles of collision 2α.

A similar criterion of an absence of jet formation can be developed based on the following condition. In the coordinate system where the slug is fixed, the velocity of the collision point V_k should be supersonic, i.e. using the formulas (2.9) and (2.11), we get

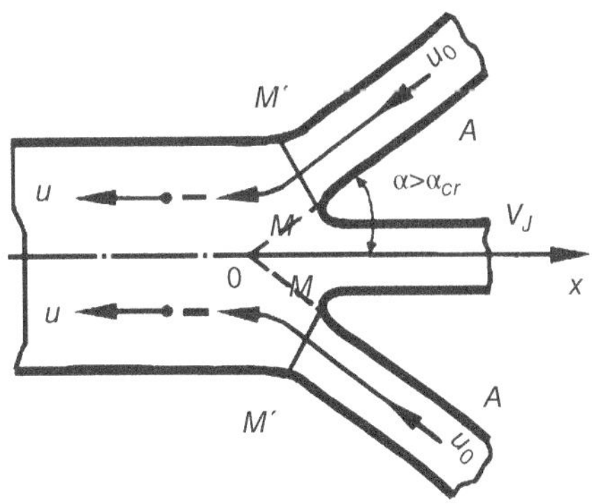

Figure 2.20 Scheme of jet formation in compressible fluid. *Source:* From Orlenko et al. [11].

$$V_K - V_S = \frac{V_0}{\sin\alpha} - \left(\frac{V_0}{\sin\alpha} - \frac{V_0}{\operatorname{tg}\alpha}\right) = \frac{V_0}{\operatorname{tg}\alpha} \geq c, \tag{2.15}$$

where c is the local sound velocity at the collision point 0.

In this case, the slug is fixed, and the collision point 0 moves at a velocity of $V_0/\operatorname{tg}\alpha \geq c$, i.e. no disturbance can overtake the point of collision 0; as a result, the jet is not formed. To form a jet, this condition must be satisfied:

$$\alpha \geq \operatorname{arctg}\frac{V_0}{c}. \tag{2.16}$$

This ratio is true for both plane and axially symmetrical liners.

The maximum jet velocity [15] that can be achieved with a liner made of a certain material is

$$V_J \leq 2.41 c_0. \tag{2.17}$$

This limit is correct for jets that can damage a target. Note that this limit does not have to be realized in a certain shaped charge.

Under appropriate conditions, gaseous jets with much higher velocities can be generated, but the density of these jets is so low that they can only be observed in a vacuum. The maximum velocity of these jets increases with decreasing atomic weight of the material and, for example, is equal to 90 km/s for beryllium [17]. But such gaseous jets do not produce any noticeable damaging effect.

Considering the general characteristics of cumulative processes and the basic relationships describing cumulative processes in shaped charges with a metal liner, we can say that they play an important role in nature and technical applications. Their research is an important scientific and technical problem.

A better understanding of the nature and properties of cumulative processes will allow, in particular, to increase the parameters of explosive and shock loading of materials, which is important in applications related to military purposes, increasing the effectiveness of explosive technologies, etc.

2.3 Explosion Loading of Shaped Charge Liners, Their Throwing, and Collapse

2.3.1 Calculation of Throw Velocity and Rotation Angle of a Shaped Charge Liner

2.3.1.1 The Planar Case

Let us first consider the case of planar symmetry as the simplest one. It corresponds to the collision of two flat plates originally placed at an angle of $2\alpha_0$. In this case, we need to consider only a half of the problem due to its symmetry. Then, the

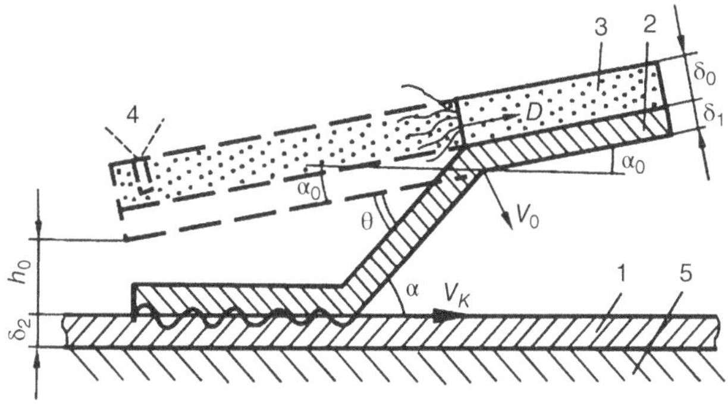

Figure 2.21 Scheme of plate throwing at explosion welding: 1 – fixed plate; 2 – throwing plate; 3 – explosive charge; 4 – detonator; 5 – massive base; h_0 – initial clearance between the plates; θ – the angle of the plate rotation in the process of throwing. *Source:* From Balagansky and Merzhievsky [1].

configuration of the resulting flows will be completely like the one shown in Figure 2.15. The acceleration of the plate to be thrown onto the symmetry plane is carried out by the means of an explosive charge placed on the outer surface of the plate. We need to determine the velocity of the plate V_0 and the collision angle $\alpha = \alpha_0 + \theta$, where θ is the angle of rotation of the plate in the process of throwing by a sliding detonation wave. This scheme is typical for explosion welding problems (Figure 2.21). The physics of processes occurring in explosion welding is similar to that of shaped charges processes. The difference is mainly due to lower intensity of the shock-wave processes occurring in welding.

The throwing velocity of the upper plate depends on the ratio of the mass of explosives to the mass of the thrown plate r [18]. If the areas are equal, we have

$$r = \frac{m_{HE}}{m_{pl}} = \frac{\delta_0 \rho_0}{\delta_1 \rho_1}, \tag{2.18}$$

where δ_0, ρ_0 – thickness and density of the HE charge; δ_1, ρ_1 – thickness and density of the thrown plate; V_k – velocity of the collision point. At the initial-parallel position of the plates $V_k = D$.

In general

$$V_k = D \sin(\alpha - \alpha_0)/\sin \alpha, \tag{2.19}$$

where D is the detonation velocity of the HE charge; α – collision angle; α_0 – the initial angle between plates.

The collision angle is related to the velocity of throwing V_0:

$$\alpha = \alpha_0 + \theta = \alpha_0 + 2 \arcsin\left[\frac{V_0}{2D}\right]. \tag{2.20}$$

The throwing velocity V_0 can be estimated by the following formulas:

$$V_0 = D\frac{\sqrt{1+(32/27)r}-1}{\sqrt{1+(32/27)r}+1} \text{ for } k = 3, \tag{2.21}$$

where k is the polytropic parameter of detonation products:

$$V_0 = 1.2D\frac{\sqrt{1+(32/27)r}-1}{\sqrt{1+(32/27)r}+1} \text{ for mixed explosives.} \tag{2.22}$$

You can use formulas that are also based on the Gurney throwing model and the laws of impulse and energy conservation:

$$V_0 = \frac{r}{r+2}D\sqrt{\frac{k}{k^2-1}}; \tag{2.23}$$

$$V_0 = Dr\sqrt{\frac{k}{(k^2-1)(r^2+5r+4)}}. \tag{2.24}$$

It should be noted that the exact value of the throwing velocity can only be obtained by direct measurement. Contact, optical, and X-ray pulse methods are widely used to measure the throwing velocity and other kinematic parameters.

Some decrease in the throwing velocity of the plate takes place because from the outer open surface of the charge, a portion of explosive is dispersed without taking part in the reaction (Yu.B. Chariton layer). It is suggested to introduce a certain equivalent layer constant for each HE that does not participate in the reaction to account for these losses. The other quantity of explosive reacts completely. Based on experimental data, it was shown that the thickness of this equivalent layer can be close to half the value of the critical diameter of this HE d_{cr}.

Thus, when calculating the velocity and collision angles, it is necessary to take the incomplete detonation of HE into account. For this, the r_e parameter is introduced into the formula for calculating the throwing velocity instead of the r parameter.

$$r_e = \frac{[\delta_0 - (d_{cr}/2)]\rho_0}{\delta_1\rho_1} = r - \frac{d_{cr}\rho_0}{2\delta_1\rho_1}. \tag{2.25}$$

In the presence of a massive shell on top of the plate to be thrown, the throwing velocity increases significantly with the same amount of HE.

2.3.1.2 The Case of Axial Symmetry

Unfortunately, there are no such simple dependencies for calculating the throwing characteristics and collapse of the shaped charge liner. This is due to at least two reasons. First, the distribution of mass over the length of a conical liner is a variable; thus, throwing velocity and angle of collapse will vary along the length of the charge. Second, the thickness of each element of the liner collapsing on the axis of symmetry increases, which should also be taken into account in the calculations.

2.3.2 Investigation of a Shaped Charge with a High-modulus Ceramic Tube

Explosion loading of axially symmetrical liners, their throwing, and collapse for clarity will be demonstrated using a shaped charge with a tube made of high-modulus ceramics as an example [19]. The importance of this problem relates to the goal of increasing the shaped charge jet velocity, which is necessary for many applications. Since the maximum velocity of a jet depends on the sound velocity of the liner material, it is advisable to use high-modulus ceramic materials, where sound velocity significantly exceeds the detonation velocity in the explosive charge.

2.3.2.1 Experiments

Figure 2.22 shows the scheme of the experimental assembly. Aluminum oxide tubes with internal diameter of 15 mm, length of 100 mm, and thickness $\delta = 0.5$ mm; 1.0 mm; 1.5 mm were obtained. The external diameter depended on the thickness of the tubes. The assemblies were filled with cast HE TNT/RDX 50/50. The weight of each HE charge was equal to 200 g. The outer diameter of the charge was 36 mm, the length of the charge was 100 mm. The charges were installed at a distance of 50 mm from the target. U7A high carbon steel disks with a diameter of 100 mm and a thickness of 19 mm were used as targets.

Figures 2.23 and 2.24 show the front and rear sides of steel witness specimens and the effects of shaped charge jets on them.

The peculiar feature of the targets affected by the ceramic particle flow was the unexpectedly large dimensions (diameter and depth) of the resulting caverns. On the front side of the samples, the diameter of the caverns was 20 25 mm. The depth of the caverns increases with the thickness of the ceramic tube. For the charge with a ceramic tube thickness of 1.5 mm, the jet penetrated through the sample and created a cavern in the base plate of the target. In this case, the total depth of penetration was 30 mm. The sample itself was fractured to fragments formed by radial cracks.

Metallographic analysis performed with scanning electron microscopy methods showed that the temperature in the target material near the crack walls was abnormally high. In particular, the formation of spherical pores indicates that the material had reached its boiling point. It is known that the boiling point of pure iron is about 2750 °C, but with the addition of alloying elements typical of steel, the boiling point could be somewhat lower. Nevertheless, the data presented suggest that the growth of cracks formed on the surface of the hollow surface occurred at extremely high speeds in conditions close to adiabatic.

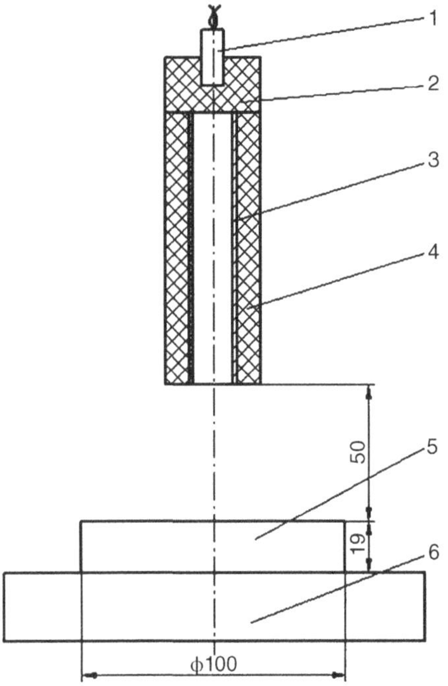

Figure 2.22 Scheme of the experimental assembly: 1 – electric detonator; 2 – initiating charge; 3 – main charge; 4 – ceramic tube; 5 – target; 6 – base. *Source:* From Balagansky et al. [19]. Reprinted with permission of Pleiades Publishing.

Figure 2.23 Effects of shaped charge jets on front sides of the witness specimens for ceramic tubes of different thicknesses: 1 – δ = 0.5 mm; 2 – δ = 1.0 mm, 3 – δ = 1.5 mm. *Source:* From Balagansky et al. [19]. Reprinted with permission of Pleiades Publishing.

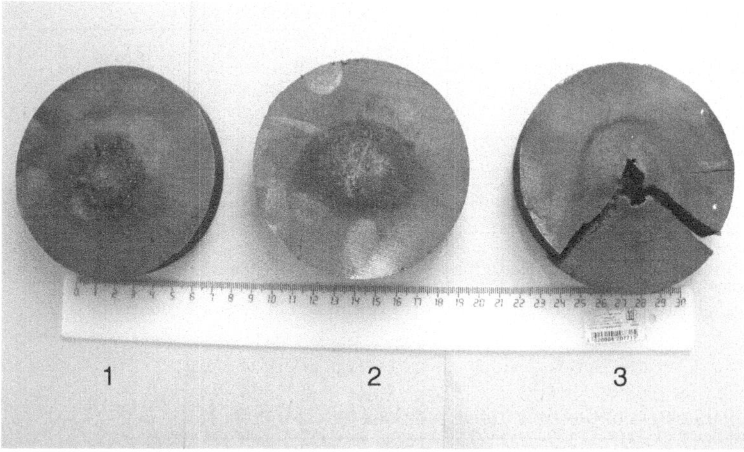

Figure 2.24 Effects of shaped charge jets on rear sides of the witness specimens for ceramic tubes of different thicknesses: 1 – δ = 0.5 mm; 2 – δ = 1.0 mm, 3 – δ = 1.5 mm. *Source:* From Balagansky et al. [19]. Reprinted with permission of Pleiades Publishing.

The boiling of the target material indicates that the impact velocity of the particles exceeded the threshold velocity. The threshold velocity is characterized by a "thermal explosion." Melting and evaporation of the substance in the rarefaction wave occur in a certain volume around the impact point. This behavior is determined by the thermophysical properties of the material at extremely high pressures.

2.3.2.2 Numerical Modeling

To explain the results of the experiments on the impact of ceramic jets on steel targets, a numerical simulation of the process of explosive loading of three corundum tubes with different wall thicknesses was performed. The geometry of the problems completely corresponded to experimental settings. For numerical modeling, the ANSYS AUTODYN software was used.

The modeling was performed in a two-dimensional axisymmetric posting on a fixed Eulerian mesh. The lower boundary of the calculated area was the symmetry axis. The spatial resolution was 20 mesh cells per mm. The detonation was initiated from the left end of the explosive charge. The parameters of the models used in calculations in AUTODYN terminology are presented in Table 2.2. Parameters for the corundum necessary for modeling the process of explosion collapse are presented in Table 2.3.

At the stationary stage of the jet formation, all three variants of the simulation with tubes of different thicknesses are characterized by similar pressures at the collision point and the velocity of the shaped charge jet.

Table 2.2 The parameters of the models used in calculations.

Material name	Equation of state	Strength	Failure
Comp B	JWL	None	None
Al_2O_3	Shock	von Mises	Hydro (P_{min})
Steel 1006	Shock	Johnson Cook	None
Al 2024	Shock	None	None

Source: From Balagansky et al. [19]. Reprinted with permission of Pleiades Publishing.

Table 2.3 Parameters of the constitutive relations for corundum.

EOS	Shock $D = a + \lambda u$
Reference density	3.92 g/cm^3
Gruneisen coefficient	0.5
Parameter a	0.871 cm/μs
Parameter λ	0.713
Strength	von Mises
Shear modulus	1.0 Mbar
Yield stress	0.08 Mbar
Failure	Hydro P_{min}
P_{min}	−0.0075 Mbar

Source: From Balagansky et al. [19]. Reprinted with permission of Pleiades Publishing.

Thus, the simulation results are discussed in detail below using the 1.5 mm setup as an example.

The geometry of the assembly at the initial state is shown in Figure 2.25. It should be noted that the explosive charge is surrounded by a shell. Figures 2.26–2.28 show the fields of material flows and pressure contours at specific moments of time.

The results show that the jet is a flow of ceramic particles. The pressure at the point of contact at the stationary stage is approximately 0.9–1.0 Mbar, the maximum velocity of the leading part of the jet is approximately 23 km/s, and the velocity of the main jet is 14 km/s.

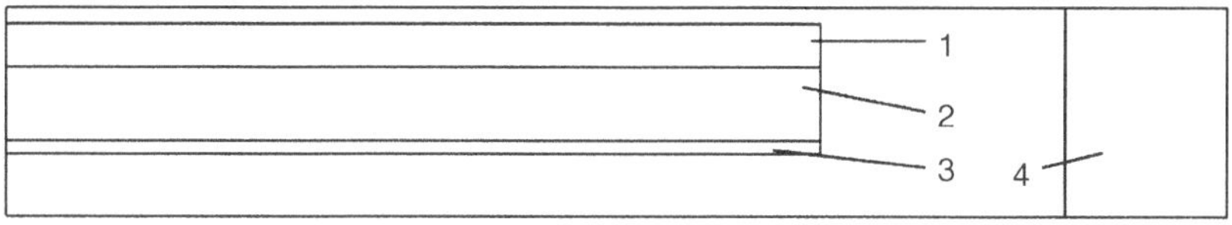

Figure 2.25 Initial setup of the simulation: 1, duralumin shell; 2, HE charge; 3, ceramic tube; and 4, steel target. *Source:* From Balagansky et al. [19]. Reprinted with permission of Pleiades Publishing.

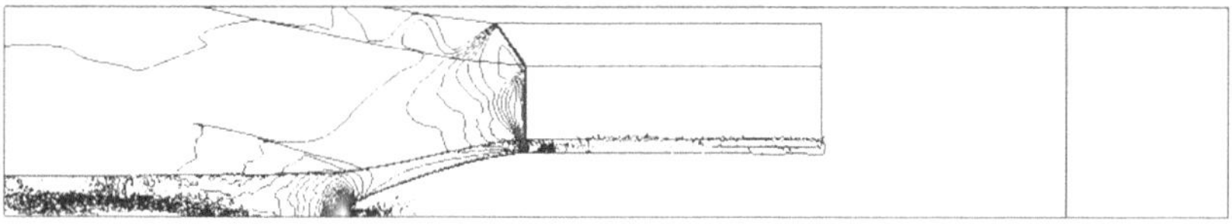

Figure 2.26 Flow fields and pressure contours at the stationary stage at time $t = 8\ \mu s$. *Source:* From Balagansky et al. [19]. Reprinted with permission of Pleiades Publishing.

Figure 2.27 Flow fields and pressure contours at time $t = 10.25\ \mu s$. *Source:* From Balagansky et al. [19]. Reprinted with permission of Pleiades Publishing.

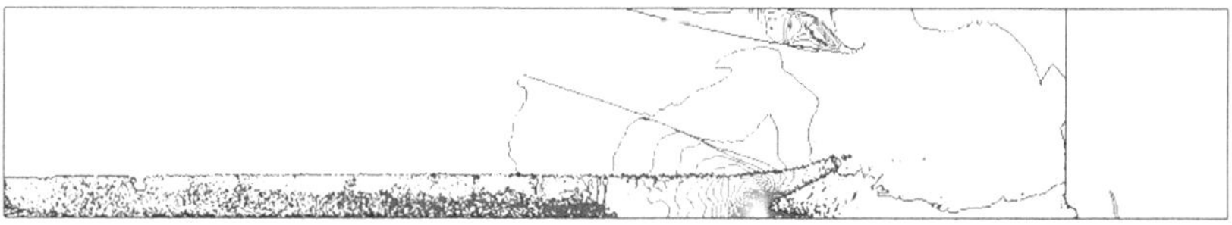

Figure 2.28 The beginning of the penetration of the jet into the target at time $t = 14.25\ \mu s$. *Source:* From Balagansky et al. [19]. Reprinted with permission of Pleiades Publishing.

The results of numerical modeling match the results of structural investigations. The simulation results indirectly confirm experimental data showing extremely high values of temperature and pressure in the area of the jet-to-target interaction.

2.4 Formation, Tension of Metal Jets, and Their Penetration into Targets

2.4.1 Movement and Breaking of Shaped Charge Jets

Metal jets formed by the explosive collapse of shaped charge liners are moving as elongated axisymmetric bodies with a high velocity. Distribution of axial velocity of different parts of the jet formed in the process of explosive collapse is characterized by the ΔV_J difference from head to tail elements, with the head part having a velocity comparable to the orbital speed (7–8 km/s), while the tail elements, as a rule, move with the velocity of about 2 km/s.

The behavior of the axial velocity distribution along the jet length sets the value of the initial strain rate $\dot{\varepsilon}_{x0,}$ the local value of which is determined by the ratio of the axial velocity difference ΔV_J to the initial length l_0 of the jet element, which is considered to be equal to the length of the corresponding section of the shaped charge liner. The value of the initial velocity of deformation $\dot{\boldsymbol{\varepsilon}}_{\boldsymbol{x0}}$ changes along the length of the jet. The majority of axisymmetric jets are characterized by values $\dot{\varepsilon}_{x0} \sim (10^4 - 10^5)\ \mathrm{s}^{-1}$. Under the action of the velocity gradient, the jet in free flight is stretched in the axial direction with a simultaneous reduction of its transverse size. At the same time, most jets at the initial stage of their existence are characterized by uniform stretching along the entire length without localization of deformation and with preservation of close to cylindrical or weakly conical shape (see Figure 2.11). Then, stretching is gradually localized in areas of multiple narrow "necks" formed in the jet.

As a result, the jet disintegrates into separate gradient-free elements that do not change their length later, although the distance between these elements continuously increases. It is important to note that after breaking of the jet, its elements begin to fly away in a radial direction, which leads to a decrease in the piercing action.

The behavior of jets breaking into separate elements is different and depends on liner material properties and on geometrical and kinematic characteristics of the jet itself (initial radius of elements R_{J0}, initial rate of deformation $\dot{\boldsymbol{\varepsilon}}_{\boldsymbol{x0}}$). For example, high-gradient copper jets break into individual elements in a surprisingly regular way. This type of failure (plastic) is characterized by the formation of geometrically similar individual elements with a developed neck, the radius of which is close to zero. Failures in jets of nickel, niobium, pure aluminum happen similarly.

Based on dimension analysis, the defining parameter responsible for stretching and breaking of the jet was selected by L.A. Merzhievsky [1]:

$$\frac{Y_0}{\rho_0 \dot{\varepsilon}_{x0} R_{J0}^2},$$

where Y_0 is the yield strength of the jet material. Physically, this parameter characterizes the ratio between plastic and inertial forces.

Fracture of jets made of other materials can demonstrate different behavior. Jets of lead and tungsten are destroyed volumically. The break of jets generated from steel liners in most cases occurs by a "quasi-brittle" break without a significant neck.

The common property of all materials in jets is their abnormally high plasticity compared to static conditions. High-gradient copper jets after breaking have a total length of elements approximately 10 times greater than the initial length of the jet. Some parts of the jet have even greater elongation before breaking. For example, for some parts of niobium jets, the value of the maximum elongation factor can reach $n_b = 26$. It is due to this abnormally high plasticity of materials, stretching jets can reach significant lengths before interacting with targets, and this fact explains the high penetration ability of shaped charges.

2.4.2 Penetration of Shaped Charge Jets into Barriers

In the process of jet penetration into a barrier (steel, aluminum, concrete, etc.), the pressure generated on the interface between the materials of the jet and the barrier can be one to two orders of magnitude higher than the strength properties of the barrier. This applies to those parts of the jet, which have a velocity of $V_J > 4$ km/s. In this case, the strength of the jet and the barrier can be neglected.

Let us consider an approximate theory of shaped charge jet element penetration into a homogeneous barrier. Let us denote the length of the element as l, its initial velocity – V_J, density – ρ_J, velocity of penetration into the barrier – u_x, initial density of the barrier – ρ_T.

Under the action of high pressure at the interface between the jet and the barrier (Figure 2.29), the jet "spills out" and its material flows in the direction opposite to its velocity. The material of the barrier also "de-parts" from the high-pressure zone, and part of this material is carried out together with the jet to the free surface of the barrier (a), and the other part moves out of the zone of the resulting hole (b).

In the coordinate system where the boundary between the jet and the barrier is fixed, the velocity of the jet element is $V_J - u_x$, and the velocity of the barrier is u_x. The time of "spreading" (near point X) of the jet element and the formation of a hole with a depth of L

$$t_L = \frac{L}{u_x} = \frac{l}{V_J - u_x}. \tag{2.26}$$

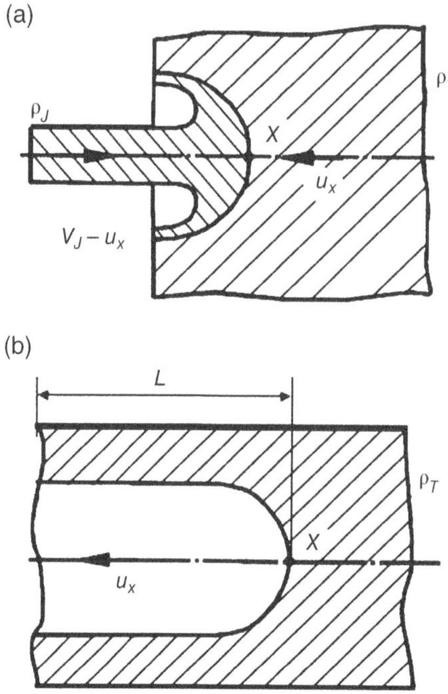

Figure 2.29 Scheme of a jet element penetration in the coordinate system connected with the boundary of penetration. *Source:* From Orlenko et al. [11].

From here, we get the formula for determining the depth of penetration:

$$L = l\frac{u_x}{V_J - u_x}. \tag{2.27}$$

Based on this, approximate formulas for different models of both incompressible and compressible fluid can be constructed, and in the first approximation, the effect of strength on the penetration process can be taken into account.

Let us consider the case when the jet and the barrier materials are ideal incompressible fluids. As the penetration process shown in Figure 2.29 is stationary, Bernoulli's equation is valid for it.

$$\frac{\rho_J (V_J - u_x)^2}{2} + p_J = \frac{\rho_T u_x^2}{2} + p_T = p_x, \tag{2.28}$$

where p_J, ρ_J are the initial pressure and density of the jet; p_T, ρ_T are the initial pressure and density of the barrier; p_x is the pressure at the point X on the boundary between the jet and the barrier, where the jet and barrier velocities are equal to zero. Since $p_x >> p_J$ and $p_x >> p_T$, at $p_J = p_T$ from Eq. (2.28), we obtain

$$\rho_J (V_J - u_x)^2 = \rho_T u_x^2. \tag{2.29}$$

From here

$$\frac{u_x}{V_J - u_x} = \sqrt{\frac{\rho_J}{\rho_T}} = \mu. \tag{2.30}$$

The pressure and velocity of penetration of an ideal incompressible jet into an ideal incompressible barrier are:

$$p_x = p_T + \frac{\rho_T u_x^2}{2}; \quad u_x = \frac{\mu V_J}{1 + \mu}. \tag{2.31}$$

Based on (2.27) and (2.31), we obtain a formula for determining the depth of penetration:

$$L = l\sqrt{\frac{\rho_J}{\rho_T}} = l\mu. \tag{2.32}$$

According to this formula, the depth of penetration depends only on the jet length and density, and the density of the barrier, and does not depend on the jet velocity V_J, compressibility of the jet and barrier materials, or their strength.

Formulas for calculating the depth of penetration into the barrier taking into account the strength and compressibility of the jet and barrier material can be obtained as well, but we will not do it here.

2.5 The Influence of Design Parameters and Manufacturing Technology of Shaped Charges on the Penetration Effect

The choice of design parameters and technology of producing a shaped charge, whether it is a military charge or a perforator for opening oil and gas wells is determined by specific requirements, such as:

- type and material of the barrier to be pierced;
- conditions under which the shaped charge is applied;
- depth and diameter of the pierced hole;
- potential spalling and other behind-armor damage.

2.5.1 Shaped Charge Liner

The most important element of the shaped charge that determines its piercing ability and behind-armor effect is the liner. In shaped charges, single- and multilayer liners are used. A multilayer liner usually consists of two or more layers, with the former, often called bilayer liners, most commonly used. Liners can be combined, consisting of different components with different materials, thicknesses, or geometric shapes.

To increase the penetration effect of shaped charges, the main directions of improving the liners are associated with improving the mechanical, technological, and operational properties of the liner material, optimizing the geometric shape and size of the liner, and developing various options for multilayer and combined liners.

Single-layer metal liners of materials with high density and plasticity at high strain rates are widely used. The most common materials are copper, technical iron, and low carbon steel, since they are the most readily available materials forming high-density solid monolithic jets with high penetration ability.

The influence of the atomic crystal structure of the metal, such as its crystal lattice, on formation of shaped charge jets has already been noted. In particular, it has been determined that the best results are observed in liners made of metals with cubic crystal lattices (copper, iron, aluminum), and the worst in metals with hexagonal lattices (cadmium, cobalt, magnesium). There is also a correlation between the production technology, structure, and mechanical properties of liners and the penetration ability of shaped charges.

The increase in the penetration capacity of military-shaped charges is associated with the use of tungsten and tungsten-based alloys, as well as depleted uranium liners. There are also known solutions to increase the penetration ability of shaped charges by using liners made of tantalum and hard alloys based on it, or copper alloys with the addition of noble metals (silver, gold, platinum).

2.5.2 High-explosive Charge and Case

2.5.2.1 HE Charge

The penetrating effect of the shaped charge is influenced by detonation and energy characteristics of the explosive used, determined by its density ρ_0, detonation velocity D, and several other parameters. In general, the use of more powerful explosives in a shaped charge at the same velocity gradient along the jet allows to increase the thickness of the liner, which provides an increase in the release of metal from the collapsed liner into the jet, the ultimate stretch of the jet before breaking and the corresponding increase in the penetration effect at extended distances from the charge to the barrier.

When producing shaped charges by pressing, phlegmatized RDX and HMX are most often used, while alloys of TNT with RDX and HMX are used for casting. In recent years, thermoplastic and plastisol explosive compositions have been widely used, which improve the quality of manufacturing shaped charges.

2.5.2.2 The Shape of HE Charge

The penetration action is also affected by the shape of the HE charge. The most common form is a cylinder, but the desire to reduce the mass of HE without reducing the penetration action requires optimization of its external shape, which consists of defining and preserving the "active" mass of the HE charge. This can be achieved by transitioning from the cylindrical form to the cylindrical-conical one, and if there was a "lens" (detonation front control unit) in the original charge, it is desirable not to reduce its diameter to preserve the penetration action.

2.5.2.3 HE Charge Case

Protecting the explosive charge from damage and harmful effects of the atmosphere, the case is the most important structural element of improving its efficiency. In a decisive way, it influences the detonation ability of the charge and the blast energy release by the liner. Another important function of the case is related to the possibility to control the front geometry of the detonation wave propagated in the explosive charge. It is established that at a certain combination of properties of the explosive charge and case material, interesting physical phenomena can be observed. For example, it is shown in several works that when detonation is propagated in a charge encased in a beryllium or high-modulus ceramic shell having a sound velocity higher than the detonation velocity of HE, the detonation velocity and pressure at the detonation front can change.

2.5.3 Detonation Front Control

The main problems solved with the help of the control system of detonation front shape are directed to increasing the velocity of the collapse of shaped charge liner and synchronization of the collapse process along the length of the shaped charge liner. The main ways to solve these problems can be:

- use of profiled inserts ("lenses") made of inert or exothermally reactive materials embedded in HE charges;
- addition of air or vacuum channels or cavities;
- varying detonation characteristics along the length or diameter of the charge;
- application of nontraditional initiation schemes and regimes of detonation propagation;
- transition from single to multipoint detonation initiation systems;
- and various other approaches.

From the practical point of view, the first approach – the use of lens units – is the most realistic. The lens unit serves as a detonation wave generator – it converts the initially divergent spherical detonation front into a toroidal front (Figure 2.30). With point initiation of a detonation in a shaped charge without a lens, the laws

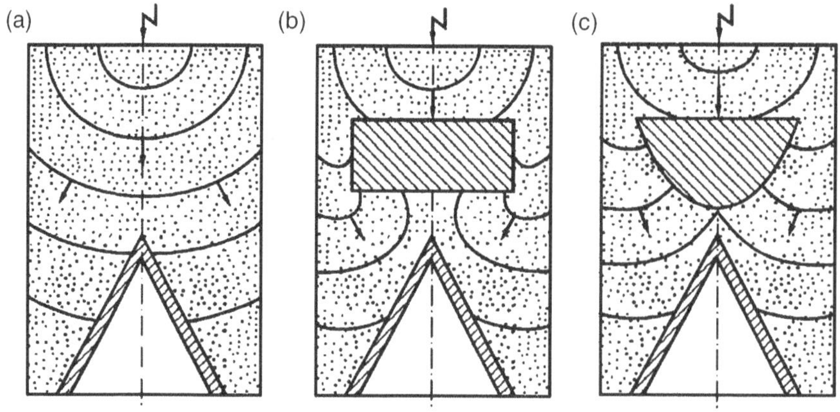

Figure 2.30 Influence of geometrical shape of the lens on the shape of detonation front: (a) charge without lens; (b, c) charges with lenses of different shapes. *Source:* From Orlenko et al. [11].

of the detonation front propagation are subject to the laws of geometric optics: the detonation wave is interpreted as a regular spherical surface moving at a constant velocity.

The presence of a lens of an inert material in the HE charge leads to the distortion of the detonation front shape.

Formation of a certain profile of the detonation front is determined by certain requirements to the geometric shape and size of the lens, its material, the method of installation of the lens in the charge, and the distance from the end plane of the lens to the top of the shaped charge liner. In this case, the stability and efficiency of the shaped charge action are determined: firstly, by the correctness and symmetry of the shape of the convergent detonation front and, secondly, by the mismatch (both by displacement and skewness) of the liner and detonation front.

The geometrical parameters of the lens units are determined by the kind of HE, the material and shape of the lens, as well as the selected equipment method. The shapes of the lenses used in practice are very diverse. Typically, the lenses are made of plastics, press material AG-4B, or porous polymers like rigid polystyrene foam or elastic microcellular polyurethane foam with a density of 0.6–0.8 g/cm^3.

2.5.4 Shaped Charge Manufacturing Technology

The efficiency of a shaped charge is significantly influenced by the technology of its production. It is known from the military and civil use of shaped charges that the level of penetration of hard steel barriers achieved by such charges manufactured using conventional technologies and using traditional liner materials does not exceed six diameters of the charge. At the same time, increasing the accuracy of shaped charge manufacturing, meeting all the requirements for the quality of liner material and HE charge, the use of advanced HE and shaped charge explosive filling methods, development of initiation and control units for detonation front can increase this limit to 8–10 diameters. Finally, for perfectly manufactured shaped charges, the depth of penetration of the steel barrier is predicted to be 12 calibers.

Dependence of the depth of penetration on the accuracy of manufacturing of individual parts and assembly of the structure is due to the fact that technological errors, the appearance of which is inevitable in the production conditions, directly affect the symmetry of the liner collapse process and, accordingly, the symmetry of formation and movement of the shaped charge jet.

The analysis of the influence of technological errors on the penetrating effect of shaped charges allows us to identify two main mechanisms: the curvature of the jet axis and its early break-up in the axial direction. The curved jet has a lower penetration ability since a significant part of its kinetic energy is unproductively spent on the expansion of the formed cavern due to the effect of "smearing" of jet elements on the walls of the cavern or even failure to hit it. The early breakup of the jet under axial tension also leads to a decrease in the penetration ability of the shaped charge, primarily due to a decrease in the total length of the jet.

Ignoring external factors associated with the presence of gas-dynamic effects on jet elements in the process of their formation and movement, we can identify the primary technological factors that determine the behavior of the jet and affect its penetration ability.

1) Variation in wall thickness of the shaped charge liner and case.
2) Variation in form and density of HE charges.
3) The grain size of the liner material.
4) Misalignment of the liner with the charge.
5) Initiating impulse displacement, as well as misalignment of the lens unit.
6) Clearances and misalignments between individual parts of the shaped charge unit introduced during the assembly.

Obtaining quantitative estimates of the influence of these factors on the penetrating effect of the shaped charge is difficult because their effects vary between the charges of different designs and even between different parts of the same unit.

Nonideal detonation processes in a shaped charge. However, technological imperfection of the charges is not the only factor that can reduce their penetration ability. Even with high quality of manufacturing, a dramatic decrease in penetration can sometimes be observed, especially at low ambient temperatures. In this case, the processes related to the reduction of penetration are associated with the nonideal detonation processes, especially at weak initiation (Figure 2.31).

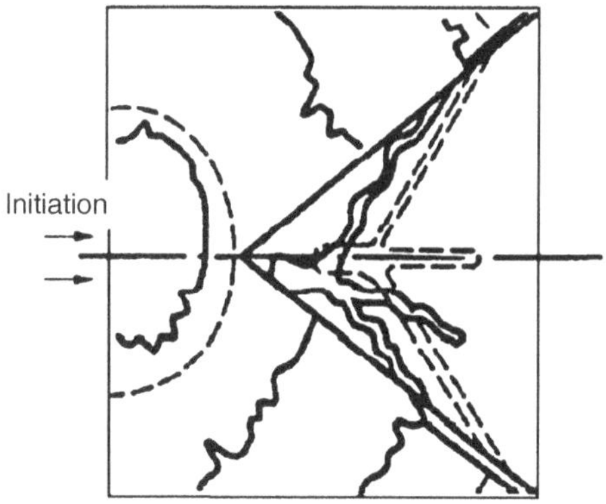

Figure 2.31 Asymmetry of the detonation front. *Source:* From Orlenko et al. [11].

The factors contributing to the nonideal nature of detonation processes are:

- protracted transition of the initiating shock wave into the detonation wave;
- acquired longitudinal–transverse instability of the detonation front;
- preservation of congenital and generated features of the flow structure behind the detonation front at considerable distances from the place of their origin;
- imperfection of the transfer process – the extraction of energy from the explosion products by the liner with imperfect energy release behind the detonation front.

All these factors are especially manifested in the case of weak initiation when the initiation shock wave changes into a detonation wave through a complex transformation in time and space.

2.5.4.1 Reasons for Longitudinal–Transverse Instability of Detonation Wave Propagation

Depending on the degree of ideality (nonideality) of the detonation process, different reasons for the longitudinal–transverse instability may be predominant. The main ones are the following:

- frontal instability during the transition of an initial shock wave into a detonation wave;
- structural and rheological heterogeneity of the HE charge, and consequently of the energy release;
- amplitude and time characteristics of detonation initiation;
- appearance of HE desensitization zones;
- rapid changes in detonation front direction under conditions close to the detonation propagation limit.

A combination of several reasons can be responsible for the instability as well. Certain systems with so-called inherent natural instability are well-studied, such as the system of nitromethane–acetone, for which the detonation front always pulses, which determines the only possibility of its propagation. It should be especially emphasized that these instability effects are observed with weak initiation.

2.5.4.2 Longitudinal–Transverse Instability of Initiating Shock Waves

When plane-symmetric and divergent initiating shock waves are evolving near critical conditions of detonation initiation, longitudinal–transverse instability of the front might appear. The size of these emerging inhomogeneities can be over an order of magnitude larger than the inhomogeneities of stationary detonation fronts in liquid explosives. Figure 2.32 shows the charge assembly and a photograph obtained using ring and linear streak diaphragms at the ends of the charges.

The "SI" arrows in the picture mark the initiation surfaces, including the whole butt of the HE tablet; a spot with a diameter of several millimeters at the axis of the charge; a narrow-ring surface on the periphery of the charge with a conical cavity.

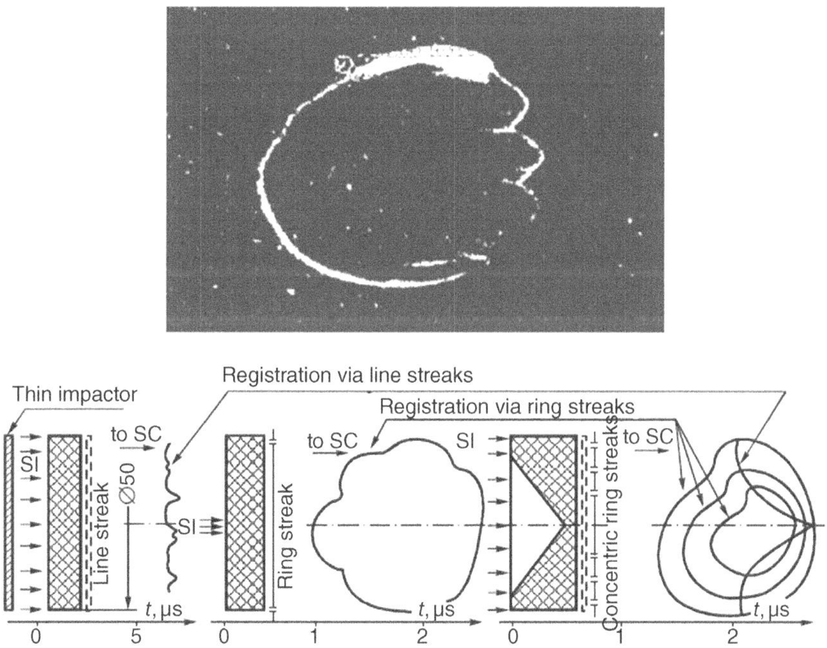

Figure 2.32 Typical frontal irregularities; SC-streak camera. *Source:* From Orlenko et al. [11].

The appearance of irregularities indicates the existence of local depressions at the front of the initiation shock wave, the probability of which is especially high in the case of pressing tablets with a large diameter-to-height ratio.

As the distance traveled by the initiation shock wave increases, the irregularities are smoothed. However, at the depths exceeding the occurrence of a stationary detonation, there is a high probability of local inhomogeneous destructions with the topography characteristic of an initiation shock wave in the zone of its transition to detonation.

2.6 Influence of the Operational Conditions of Ammunitions with Shaped Charges on Their Damaging Effects

The effectiveness of the damaging effect of shaped charge ammunition is largely determined by the conditions in which they operate and the design and properties of the targets they damage. Let us consider how the distance between the charge and the target and the rotation of the ammunition influence its destructive action.

2.6.1 Standoff Distance

As already mentioned, the depth of penetration of the barrier by the shaped charge jet depends on the distance between the charge and the barrier. The distance from the shaped charge to the barrier at which the maximum depth of penetration is reached is called the "standoff distance."

Standoff distance is determined by the design of the charge, the accuracy of its manufacture, and the characteristics of the barrier, particularly its density. Standoff distance increases with the increasing angle of shaped charge liner, power of the explosive charge, density of the barrier material, and accuracy of the shaped charge manufacture.

Let us consider the reasons that affect the change of the standoff distance. As a shaped charge jet moves, it is elongated, and when it reaches a certain elongation, it breaks down into separate elements. Due to the imperfection of the charge manufacturing technology, the shaped charge jet is not ideal and does not move strictly along the axis but dissipates at a solid angle equal to 0.5–1.5° for modern shaped charges. After the breakdown of the jet into elements, the latter receive a transverse impulse because of the asymmetry of the jet, which, in turn, leads to an increase in angle dispersion.

A corresponding contribution to this process is made by the aerodynamic forces acting on the jet elements, which after the disintegration rotate relative to the center of mass, which can be clearly seen in X-rays. During the angular dispersion, the jet interacts with the walls of the formed crater losing its piercing action a part of the jet is simply "smeared" onto the crater walls. For ideal shaped charges made without any deviations in size, such "smearing" should be absent. The higher the manufacturing tolerances are (variation in wall thickness, the inhomogeneity of the HE charge density, etc.), the stronger this effect is manifested. In this case, the number of misaligned elements of the jet penetrating into the barrier increases, which leads to a decrease in the standoff distance.

The more precisely the shaped charge is manufactured, the greater its standoff distance and correspondingly the value of target penetration will be. If for conventional shaped charges with conical copper liners, the standoff distance is (1–4) *d*, then for precision charges it reaches (6–8) *d*.

The change of the standoff distance is largely determined by the fragmentation process of shaped charge jet. For high-velocity high-gradient jets, the proportionality of time of existence of a monolithic jet (from the moment of formation and up to its breaking into elements) to its initial diameter is established experimentally. In this case, the head parts of the jet having a smaller initial diameter are disintegrated earlier than the tail parts. However, the head parts of the jet before disintegration can move further from the end of the charge. Based on the analysis of the jet disintegration process depending on the lining parameters and properties of the HE, it is possible to explain the reasons for changing the standoff distance. When the angle of the conical liner decreases, the mass of the liner transferred to the jet and, accordingly, the jet diameter and the time of its disintegration decrease. In this case, the depth of the crater increases, and the jet formed by such tall liners is disintegrated at lower distances from the end of the charge compared with liners with broader angles.

2.6.2 The Effect of Rotation on the Shaped Charge Effect

It is known that the penetrating effect of rotating shaped charge ammunition is much lower than that of nonrotating ammunition. As the angular velocity of rotation of the ammunition increases, the negative effect of rotational motion increases. This is explained by the fact that due to the law of angular momentum conservation, elements of the jet will have an angular velocity determined by the angular momentum of the corresponding element of the liner relative to the axis of rotation under its collapse.

The maximum angular velocity of the jet is determined by the expression:

$$\omega_J = \omega_0 \left(\frac{R_0}{R_S} \right)^2,$$

where ω_0 – angular velocity of rotation of the projectile; R_0 – radius of the base of the shaped charge liner; R_S – radius of the base of the slug after its full collapse. If we consider that the projectile rotation velocity when leaving the channel of the rifled gun barrel is about 20 000 rpm, it is not difficult to imagine what centrifugal forces will be produced when the liner is collapsed. In this case, even small errors in the production of shaped charges will lead to a violation of symmetry in the process of jet formation, which is the main cause of the jet splitting into two or more parts, its faster fragmentation, as well as other types of disintegration. In the further movement of the jet under the influence of centrifugal forces, the velocity of its fragments in the radial direction increases, which leads to a loss of penetration.

2.7 Formation and Effect of Explosively Formed Projectiles

The effective action of the shaped charges described above appears only at short distances to the target (about 5–7 diameters of the charge). To ensure long-range penetration action, special shaped charges with very shallow metal liners – so-called explosively formed projectiles (EFP) – are used.

One type of long-range elements produced by the explosion of such a shaped charge is the projectile that combines the slug and jet parts (separated in classic shaped charges) into a single whole. As a result, the initial velocity of such a projectile is lower than the average velocity of a typical jet and reaches 3.5–5.0 km/s. However, the mass of this explosively formed projectile is much larger than the mass of a typical jet and can make up to 95% of the initial mass of the liner. It is possible for the EFP to retain its initial shape formed during the explosion of the charge along its whole trajectory.

As a rule, compact projectiles are formed from low-profile conical and hemispherical liners, for which the ratio of height to diameter of the base of its internal (facing the barrier) surface $h/d < 0.3$. In turn, the projectiles can be lengthened, disintegrating into several fragments, and compact, capable of keeping their integrity when moving at long distances. The former usually form at $0.2 < h/d < 0.3$, and the latter at $h/d < 0.2$.

Depending on the constructive parameters of the charge, several mechanisms of the formation of a compact projectile can be realized. The primary ones are "eversion," "leakage," and "retraction." However, much more often, the behaviors combining the basic ones are observed. Figure 2.33 shows the primary and combined mechanisms of the formation of a compact projectile.

It should be noted that the formation of the most efficient (in terms of increasing the penetration), elongated EFPs with the relative elongation larger than l_E/d_E is usually achieved with the help of deformation mechanisms of eversion or retraction. Here, l_E and d_E are the length and diameter of the EFP, respectively. In this case, segmented shallow liners are usually used, the central part of which is perpendicular to the axis of symmetry, which under explosive loading allows to

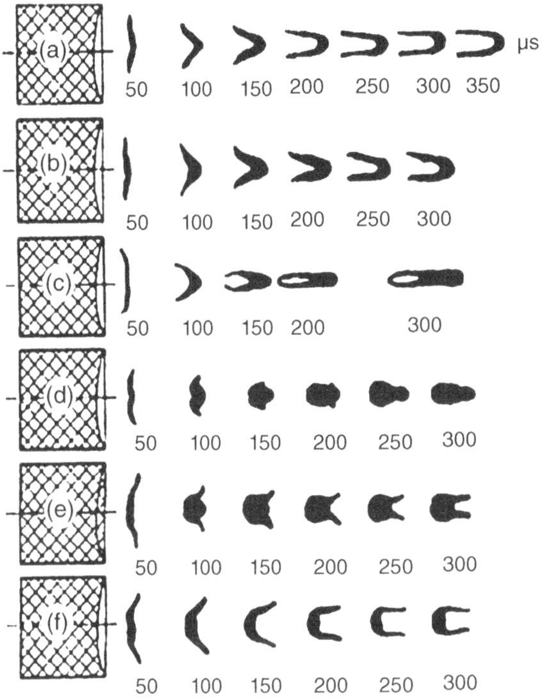

Figure 2.33 Basic (a, d, f) and combined (b, c, i) mechanisms of compact EFP formation: (a) eversion; (b, c) eversion and leakage; (d) leakage; (i) leakage and retraction; (f) retraction. *Source:* From Orlenko et al. [11].

provide it with only the axial component of velocity and exclude the manifestation of negative effects inherent to the leakage.

The most detailed analysis of the behavior of liners under explosive loading and their subsequent inertial deformation can be achieved with numerical simulation. In this case, very high requirements are applied to the models describing the behavior of materials, since even small errors in the modeling of materials will lead to significant deviations in the shape of the resulting compact EPF.

Let us consider the process of deformation of a segmented shallow metallic liner located at the end of cylindrical HE charge (Figure 2.34).

At the initial stage of throwing under the action of detonation products of the charge, each element of the liner obtains a certain velocity, directed along the normal of the surface of the liner. Due to its concave shape, the liner is compressed and turned out because the height of the active mass of shaped charge decreases from the center to the periphery of the liner. Thus, the impulse of pressure forces transmitted by detonation products to various elements of the liner is decreased accordingly.

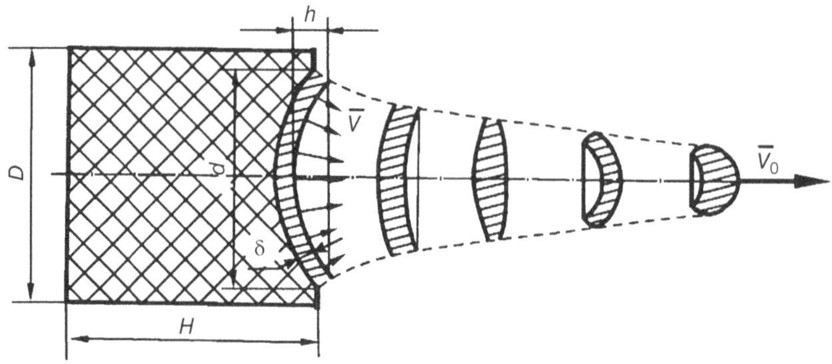

Figure 2.34 The process of forming a compact projectile. *Source:* From Miropolsky et al. [20].

Due to the action of viscous friction forces and forces of resistance to elastic-plastic deformation of the liner material, the velocity of its particles is gradually leveling.

There is a certain range of values of relative geometrical proportions of the metal liner (Figure 2.34), within which a compact element is not disintegrated. These values are determined experimentally and fall within the intervals of $2 \leq h/\delta \leq 5$, $3 \leq d/\delta \leq 30$. In this case, the ratio of the charge height to diameter H/D should not be less than 0.5.

Ammunition that forms a compact projectile during an explosion has a damaging range of up to 100 m or more. Therefore, issues related to the ballistics of the projectile are important.

Numerous experiments show that with accuracy sufficient for practical applications, when estimating the loss of EFP velocity in ballistic calculations, the value of the drag coefficient of such projectiles can be considered equal to one.

Having a large mass and high velocity, the projectiles knock out a significant number of secondary fragments of large mass from the armor, which can effectively damage the vital elements of the target. At distances of up to tens of meters, the projectiles are capable of breaking through armor barriers with a thickness equal to (0.4–0.6) of the original liner diameter.

2.8 The Effect of Ammunition with Shaped Charges on the Armor of Modern Tanks

2.8.1 Characteristics of Modern Tank Armor

Shaped charge ammunition produced quite a sharp jump in the damage capacity compared to the protective capabilities of monolithic steel armor. They created a problem that could not be solved by traditional methods because of the unacceptable increase in the weight of the tank. This circumstance gave an impetus to the development of a new generation of tanks with combined frontal armor, which included fillers such as fiberglass and ceramics. This produced abnormally high protective properties compared to the predictions the hydrodynamic theory. The nature of this anomaly lies mainly in the active destructive effect of the filler material surrounding the cavern on the jets. It is caused both by the release of the energy stored by the filler during the jet passage and by releasing the internal energy of the filler itself when glass and ceramics are used as fillers.

It should be noted that the combined frontal armor of Soviet tanks in the 1970s did not provide protection from the whole range of available shaped charge ammunition, but only from the most common ammunition of that time, such as the shaped charge ammunition for 105-mm tank and anti-tank guns and shaped grenades. The competition between tank armor and anti-tank ammunition of this type lasted until the early 1980s. Further improvements of fillers provided protection against the shaped charge ammunition of 122-mm rifled guns.

At the same time, in the arsenals of all countries, there was an accumulation of anti-tank-guided missiles with shaped charge warheads whose armor penetration capability was 1.2–1.5 times higher than that of the shaped charge ammunition and, accordingly, exceeded the protective capabilities of the tank armor.

By the end of the 1970s, Soviet engineers have developed practical anti-shaped charge armor based on the principle of counter-explosion (the so-called explosive reactive armor), work on which began in the 1950s [21]. The emergence of explosive reactive armor developed by Dr. M. Held during the Arab-Israeli War of 1982 (for M48 and M60 tanks in service with the Israeli army), intensified the adoption of the technology by the Soviet military. Thanks to fully prepared technical, design, and technological solutions, Soviet Union's main battle tanks were equipped with explosive reactive armor (ERA) in record time of just one year. Installation of ERA on the T-64A, T-72A, T-80B tanks, which already had quite powerful armor, almost instantly depreciated the arsenals of anti-tank-guided weapons of potential opponents of that time period.

The use of ERA has initiated worldwide intensification of new developments in the field of anti-tank weapons, starting a completely new stage in the projectile vs. armor competition. Now it was possible to "attack" the incoming projectile just as it attacked passive armor before. It was such a leap in anti-shaped charge ammunition resistance that the subsequent upgrades of TOW-1 anti-tank-guided missiles and even TOW-2 missiles of larger caliber were unable to neutralize it. The M829 antitank projectile was also unable to solve the problem.

Only the technological breakthrough in the field of tandem-shaped charge ammunition (TOW-2A, HOT-2T, PARS-3, and others) capable of penetrating 1000–1250 mm of steel armor behind the mounted ERA has created new challenges for armor development. Tandem ammunition includes two shaped coaxial charges that operate one after the other. Armor penetration of these charges is thus combined. Along with the intensified development of ammunition for penetrating tank frontal armor, unconventional methods are being considered as well, including the explosively formed projectiles designed to damage tanks through their tops, sides, and bottoms.

The difficulty of addressing this new challenge lies in the fact that increasing the equivalent thickness of a side armor plate by one unit leads to an increase in the mass of the vehicle several times greater than the equivalent frontal armor enhancement.

2.8.2 Interaction of Shaped Charge Jets with Explosive Reactive Armor

Explosive reactive armor is now used for the protection of armored vehicles (tanks, infantry fighting vehicles, etc.), which consists of high explosives in a metal shell. In 1957–1961, B.V. Voitsekhovsky, V.L. Istomin, S.V. Zhuravlev, and others experimentally demonstrated high efficiency of the simplest device (Figure 2.35a) consisting of two metal plates (1, 2) (made of aluminum or steel) and a layer of HE between them (3) against metal shaped charge jets. This effect was called "dynamic protection," and the two metal plates and an explosive layer between them are called dynamic protection elements.

There are various modifications of the dynamic protection (explosive reactive armor) of armored vehicles against shaped charges. Let us consider the classification, design, and the mode operation of dynamic protection.

Four types of dynamic protection are distinguished in terms of location in relation to the main armor protection of the tank.

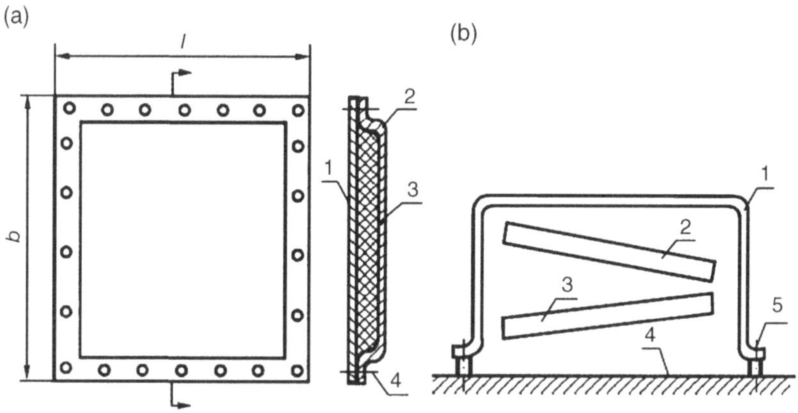

Figure 2.35 Design scheme of a dynamic protection element (a) and a hinged dynamic protection device (b). *Source:* From Orlenko et al. [11].

2.8.2.1 External Dynamic Protection

In this case, the dynamic protection, i.e. its elements 2, 3, and case 1 in which these elements are installed with the fixings 5 are placed on the external surface of the tank's armor plate 4 (Figure 2.35b). The advantage of this type of dynamic protection is in its relatively small mass and high efficiency. The external dynamic protection element does not exceed the weight of a 20-mm armor plate, providing protection 5–15 times better than passive armor, depending on the angle of the shaped charge jet. The disadvantage of external protection is its vulnerability to the action of bullets, fragments, shock waves, and detonation products that can easily detach the external protection from the tank's primary armor.

2.8.2.2 Built-in Dynamic Protection

In this case, dynamic protection is covered with 10–30 mm armored sheets protecting it from bullets, fragments, shock waves, and products of detonation. The disadvantage of such dynamic protection is that it has approximately twice the weight of external dynamic protection.

2.8.2.3 Dynamic Protection Embedded into the Armor

This is a promising type of dynamic protection, but it requires special manufacturing of armor. This type of dynamic protection can be combined with external or built-in dynamic protection.

Layered dynamic protection designed to combat tandem-shaped charges.

Classification of dynamic protection can be based on the shape of its elements: the elements of dynamic protection can be flat plates with explosives and their combinations, or shaped into cylinders, balls, or crosses where metal shells or cells of armor with explosives and inert materials are combined. In practice, flat elements have become widespread because of their efficiency, ease of manufacture, and operation.

Dynamic protection of different designs and sizes is characterized by different times of active action on a shaped charge jet. According to this criterion, dynamic protection can be classified as short- (up to 150–200 μs) and long-lived (up to 1–2 ms).

Let us look at the interaction of dynamic protection with a shaped charge jet (Figure 2.36). Shaped charge jet moving at velocity V_J pierces the upper plate 1, hits the explosive material 2, and initiates it. In a short period of time, the plates accelerate to the velocities V_1 and V_2. These velocities, for example, for plates of the same mass can be calculated using formulas for throwing the plates:

$$V_1 = V_2 = \frac{D}{\sqrt{8}}\sqrt{\frac{3\beta}{3+\beta}}, \tag{2.33}$$

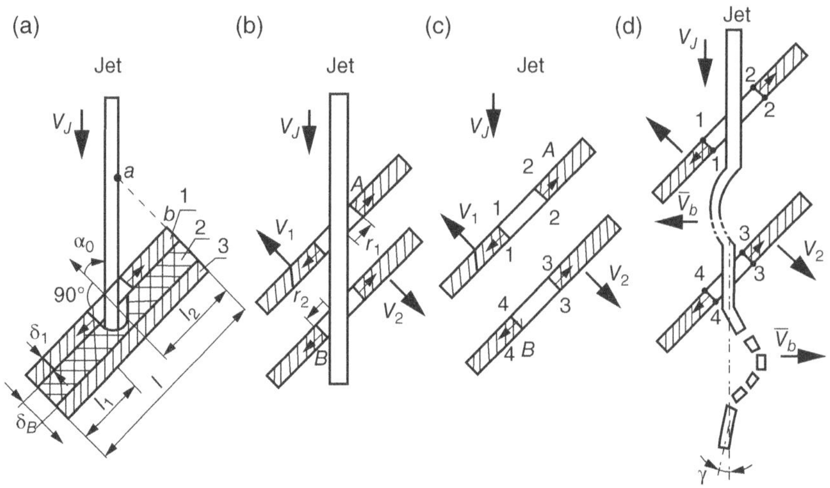

Figure 2.36 Scheme of interaction between a shaped charge jet and an element of dynamic protection. (a) Initiation of HE; (b) acceleration of plates; (c) interaction of the of the jet with plates; (d) jet breakage. *Source:* From Orlenko et al. [11].

where D is the detonation velocity of HE; $\beta = m/M$, M is the mass of one plate; m is half the mass of HE located between plates of the same thickness.

After piercing the top plate 1 and HE 2, the jet pierces the bottom plate 3. Widening holes begin to form in the plates that have the current radius of r_1 and r_2, but because the plates are moving at velocities V_1 and V_2 at an angle of α_0 to the jet, at some point in time the edge of A hole in the top plate, and then the edge of B hole in the bottom plate hit the side of the jet at a velocity of $V_s = V_J \sin \alpha_0$ (without taking into account the rate of expansion of the hole in the plate at the time of impact). The kinetic energy of the triggered part of the jet is converted into the energy of the motion of half-holes of the upper and lower plates (cross-sections 2-2 and 4-4 in the figure). As a result, the edges of the holes in the plates depart from the jet under the action of the impulse, which arises from the interaction of the triggered part of the jet and the edges of the holes in the plates. Under this action, a part of the jet obtains a side velocity V_s, which can reach hundreds of meters per second. As a result, part of the jet is bent in the direction perpendicular to the jet and then torn into pieces. In this case, the undamaged part of the jet rotates at an angle γ relative to the geometric axis of the jet.

After the first interaction of the plates with the jet, the edge of 2-2 or 4-4 moves away from the jet at some velocity, but then, as the half-hole expands, the expansion velocity decreases and may drop to zero. As a result, the edge of the hole in the plate catches up with the jet and hits it again at a velocity of V_s, and then this process is repeated many times. The efficiency of dynamic protection depends not only on parameters of its elements but also on the distance of elements from the main obstacle as some time is required for the undamaged part of the jet to move and get the maximum damage from the side impulses.

2.9 Methods for Evaluating the Effectiveness of Ammunition with Shaped Charges

When penetrating the armor with shaped charge projectiles, people, equipment, and devices located within the tank may be directly damaged by a shaped charge jet, fragments formed from the back surface of the armor, as well as by a shock wave. For classic-shaped charge munitions, the leading role is played by the mechanical factor of the jet, amplified by the shock wave. The damaging effect of the jet, the zone, and angle of secondary fragments, the pressure in the shock wave decreases with increasing thickness of the armor, and the angle of the projectile to the armor, counted from the normal.

To increase the behind-armor effect, it is necessary to maximize the armor-piercing action of ammunition, ensuring not only armor penetration but also maximum damage inside the target. The most vulnerable elements inside the tank are ammunition. When a jet hits the cartridges with gunpowder charges or fuel tanks, they ignite. If the jet hits a projectile filled with explosives, an explosion may occur.

Note that the tank is most vulnerable from the top and bottom. The thickness of armor in these places is the smallest. The largest thickness of the armor of the tank is in its frontal part.

Since there is currently no satisfactory model to describe the complex process of shaped charge effects on different types of targets, the effects of shaped charge ammunition are being evaluated experimentally. Experimental methods of determining the effects of shaped charge ammunition are to explode it in a static or dynamic mode, recording the results of its piercing and behind-armor effect on real military equipment or its mock-ups. In experiments, the ammunition is positioned relative to the target according to the specified conditions of its contact with the target, the most important of which are the angle of contact, the angle of view, the distance of the explosion point from the intended point of hitting the target surface, taking into account the detonation time of the fuse. Explosions are made at different points around the target at different angles of contact and in different directions (viewpoints).

Not only the piercing effect of the shaped charge jet is considered, but also its behind-armor effect: the volume of the residual part of the jet, its piercing, ignition, and initiation effect; characteristics of the secondary fragment flow. Based on the analysis of the blast results, a conditional damage law can be formulated, and the average number of hits required can be determined in accordance with the specified type of damage.

2.10 Self-assessment Questions

1) What is the principle of energy cumulation?
2) What is the approximate velocity of the head and tail of the shaped charge jet?
3) What are the main differences between a standard shaped charge and an explosively formed projectile?
4) What are the approximate throwing velocity and mass of compact projectile?
5) Name and explain the main cumulative processes in nature and technology.
6) What are the purposes of conical cavity liners in shaped charge ammunition?
7) Which section of the shaped charge liner is used to form the jet? Which forms the slug?
8) How is the maximum depth of the crater influenced by the distance between the shaped charge and the target?
9) How does the crater volume change?
10) Why is the strength of materials ignored in the hydrodynamic theory of shaped charge?
11) What are the similarities and the differences between the impact of jets and the formation of shaped charge jets when the liner collapses?
12) Which angle is larger: the angle of the shaped charge conical liner or the angle of collapse and why?
13) Why is the velocity of the jet head higher than that of the tail in shaped charges with conical liners?
14) Why no jets are formed at very small and very large collapsing angles of the liner?
15) What is the criterion of the absence of jet formation?
16) How can one estimate the maximum possible jet velocity for a given liner material?
17) Can this velocity really be achieved in a given shaped charge?
18) How many times does the copper jet increase in length and why?
19) What does the depth of target penetration in hydrodynamic theory depend on?
20) How does the accuracy of manufacture affect the penetration action of a shaped charge and why?
21) Why are lenses used in shaped charge ammunition?
22) Why is it necessary to use a powerful initiating impulse when initiating a cumulative charge?
23) What is the standoff distance of a shaped charge?
24) Why are smoothbore cannons usually used to fire shaped charge ammunition?

25) What are the main advantages of explosively formed projectiles over classical shaped charges?
26) What is combined tank armor made of?
27) What is "explosive reactive armor"?
28) How can a conditional damage law be obtained?
29) List the primary damaging factors of the shaped charge ammunition on armored vehicles.

References

1 Balagansky, I.A. and Merzhievsky, L.A. (2004). *Effects of Means of Destruction and Ammunition*. Novosibirsk: Novosibirsk State Technical University [in Russian].

2 Babkin, A.V., Veldanov, V.A., Graznov, E.F. et al. (2008). Chapter 6 – Cumulative weapons and ammunition. In: *Means of Destruction and Ammunition: Textbook* (ed. V.V. Selivanov). Moscow: Bauman Moscow State Technical University [in Russian].

3 Institute of Applied Physics. Advertising brochure [in Russian].

4 Lavrentiev, M.A. and Shabat, B.V. (1977). *Hydrodynamics Problems and Their Mathematical Models*. Moscow: Nauka [In Russian].

5 Zababakhin, E.I. and Zababakhin, I.E. (1988). *The Phenomena of Unlimited Cumulation*. Moscow: Nauka [in Russian].

6 Al'tshuler, L.V., Trunin, R.F., Krupnikov, K.K. et al. (1996). Explosive laboratory devices for shock wave compression studies. *Adv. Phys. Sci.* **166**: 575–581. [In Russian].

7 Balagansky, I.A., Razorenov, S.V., and Utkin, A.V. (1993). Detonation parameters of condensed high explosive charges with long ceramic elements. In: *Proceedings of the 10th International Detonation Symposium, Boston, USA (12–16 July 1993)*. Arlington: Office of Naval Research.

8 Shell, R. (1967). Detonation physics. In: *Kurzzeitphysik*, vol. **2** (ed. K. Vollrath and G. Thomer), 276–349. Wien: Springer.

9 Birghoff, G., Mac Daugall, D.P., Puch, E.M. et. al. (1948). Explosives with lined cavities. *J. Appl. Phys.* **19**: 563–582.

10 Lavrentiev, M.A. (1948). Fundamentals of the theory of cumulative charges and their armor-piercing action. *News Artill. Acad.* **56**: 46–91. [in Russian].

11 Orlenko, L.P. (ed.) (2002). *Physics of Explosion, V.2*. Moscow: Fizmatlit [in Russian].

12 Walters, W.P. and Zukas, J.A. (1989). *Fundamentals of Shaped Charges*. New York: Wiley.

13 Titov, V.M. (1979). Possible modes of hydrodynamic cumulation upon collapse of the liner. *Proc. USSR Acad. Sci.* [in Russian].

14 Walsh, G.M., Shreffler, R.G., and Willig, F.G. (1953). Limiting conditions for jet formation in high-velocity collisions. *J. Appl. Phys.* **24**: 349–359.

15 Chou, P.C., Carleone, J., and Karpp, R.R. (1976). Criteria for jet formation from impinging shells and plates. *J. Appl. Phys.* **47**: 2975–2982.

16 Kinelovskii, S.A. and Trishin, Y.A. (1980). Physical aspects of the hollow-charge effect. *Combust. Explos. Shock Waves* **16**: 504–515.

17 Koski, W.S., Lucy, F.A., Shreffler, R.G. et al. (1952). Fast jets from collapsing cylinders. *J. Appl. Phys.* **23**: 1300–1305.

18 Deribas, A.A. (1980). *Physics of Hardening and Explosion Welding*. Novosibirsk: Nauka [in Russian].

19 Balagansky, I.A., Merzhievskii, L.A., Ul'yanitskii, V.Y. et al. (2018). Generation of hypervelocity particle flows by explosive compression of ceramic tubes. *Combust. Explos. Shock Waves* **54**: 119–124.

20 Miropolsky, F.P., Sarkisian, R.S., Vishniakov, O.L. et al. (1996). *Aircraft Warheads and Their Investigation*. Moscow: Zhukovsky Academy [in Russian].

21 Rototaev, D. and Grigoryan, V. (1999). *Projectile – Armor: Which is Stronger?* 32. Moscow: Military Parade [in Russian].

3

High-explosive Ammunitions

3.1 Basic Concepts and Definitions. General Information

High-explosive ammunition is designed to achieve target damage through the effects of an explosion, producing overpressure resulting from the detonation of high explosives.

Thus, the blast effect is understood to be the destructive effect of the explosive charges themselves, which are equipped with weapons and ammunition during their explosion in different environments (air, water, soil). A distinction is made between local and general blast effects. The detonation of an explosive charge generates strongly heated and compressed to very-high-pressure gaseous explosion products, which, expanding, displace the environment and destroy the objects located in it (local explosion action). In addition, as a result of the impact of explosion products on the environment, a shock wave occurs at the front of which there is a sharp spike in pressure and other environmental parameters. Distributing in the environment with supersonic velocity and involving in motion the environment itself, the shock wave can cause damage to objects located at a significant distance from the explosion. This is a common blast effect. Thus, the blast effect of ammunition is due to the joint destructive effect of explosion products and shock waves.

As the definition makes clear, all weapons and ammunition filled with explosives have some degree of blast effect. This is why blast ammunition is usually understood to be large-caliber ammunition containing a significant amount of explosives, or such weapon for which, regardless of caliber, the blast effect is the main one.

A relatively new kind of blast ammunition is a thermobaric weapon, which forms a cloud of a fuel–air mixture when activated and then detonates and creates a field of damage. Thermobaric blast ammunition is comparable in its damaging effect to low-power nuclear weapons. For the first time in combat conditions, such weapon was used by the United States during the war in Vietnam in 1969. These weapons were called "vacuum bombs" by journalists.

Fuel–air explosives are one of the most well-known types of thermobaric weapons. Considering the principle of operation of the fuel–air explosives, the analogy with accidental explosions at the chemical industry plants, at the warehouses of flammable and volatile substances, explosions of the dwelling house during the domestic gas leakage is drawn. In such cases, the explosion has a similar behavior – the formation of an aerosol mixture of flammable substances with air oxygen, which then ignites accidentally or with the help of a detonator. In this case, under certain conditions in such a fuel–air mixture detonation process in the form of an explosion causing significant destruction may occur. To imagine the possible destruction in case of explosions of fuel–air mixtures, let us give some examples.

On 16 December 1976, in the port of Los Angeles, there was an explosion of the Liberian tanker "Sansinena." The ship was being filled with ballast in very low wind. On the deck, displaced fuel vapors from the hold formed a cloud, which was ignited by an unknown source. The flame penetrated the hold and caused a detonation. The tanker deck and deck superstructures were thrown up by about 250 m. 6 people were killed, 3 were missing, 58 wounded. Damage amounted to 21.6 million dollars.

On 3 June 1989, at 22:14 Moscow time at the 1710th kilometer of the Ulu-Telyak-Osh section of Kuibyshevskaya railway near Ufa there was a catastrophic explosion of a fuel–air mixture cloud. Two passenger trains "Novosibirsk–Adler" and "Adler–Novosibirsk" were approaching this place, where the railroad line crosses a wooded depression. When both trains reached the depression, there was a strong explosion. A flame immediately covered the railway carriages and some of them were derailed by an explosive wave. From the ground near the burning trains, a huge flaming torch was raising to the sky: turned out, a liquified gas pipeline passing under the railway bed about 500 m away was burning. When the cloud exploded, the front of the flame was 1.5–2.0 km wide. The immediate cause of the accident was a gas leak in one of the pipes.

The main damaging factor of fuel–air explosives is the overpressure at the shock wave front, which reaches the value of 20 atm and increases to 50 atm when the wave is reflected from the ground and various objects. In the detonation zone, temperatures can reach up to 2500–3000 °C for several tens of microseconds.

The overpressure and impulse of the shock wave of the fuel–air explosives, starting from a certain distance from the place of the blast, exceed the corresponding values at the front of the shock wave of conventional blast ammunition. Outside of the cloud, a shock wave with extended areas of compression and rarefaction propagates at a velocity of up to 1500 m/s, and at 100 m, the overpressure at the front of the shock wave can reach 1 atm.

Since a fuel–air mixture is capable of penetrating areas inaccessible to conventional bomb fragments and is formed according to the terrain profile, no fortification structures can protect against them when conditions for detonation are created. In addition, infiltrating buildings through the ventilation shafts or open windows, fuel–air mixtures are in more favorable conditions for the development of detonation, causing the destruction of supporting structures from the inside. Such properties of the fuel–air mixture ammunition allow to consider this weapon as a means of destroying undefended and weakly defended personnel, military equipment in open areas, fortifications and defensive structures, clearing minefields, preparation of temporary areas for the landing of helicopters, destruction of buildings, etc.

First, we will consider the design and effect of classical blast weapons filled with condensed high explosives.

3.1.1 Artillery Projectiles

The separation of projectiles into the blast and blast-fragmentation ones is based on their constructive differences and is to a large-degree arbitrary. Blast projectiles are intended for destruction of non-concrete fortifications, wire, mine, and other barriers, damage of hidden personnel; for removal of ground cover from long-term fortifications. If necessary, high-explosive projectiles may be used for firing at open personnel and enemy tanks.

A fuse of a blast projectile is usually set to slower action so that the projectile is immersed into the ground at the optimal depth before detonation. When the fuse is set for instantaneous action, the blast ammunition has a significant fragmentation and some ignition effect.

It is reasonable to have blast projectiles only for large caliber guns. At present, blast projectiles are available in ammunition of 180 mm caliber and above (Figure 3.1).

F-572, F-625D, F-644, F-643 projectiles are of long-range shape, while the F-625 projectile is shaped for short ranges. Constructive characteristics of the projectiles are given in Table 3.1. The projectile bodies are made of 45X and 45X 2 chromium steel. The projectiles are filled with TNT.

3.1.2 Artillery Mines

The blast mines are intended for hinged firing to destroy the trenches equipped with the top cover, trenches, timber, and stone defensive structures, for removing

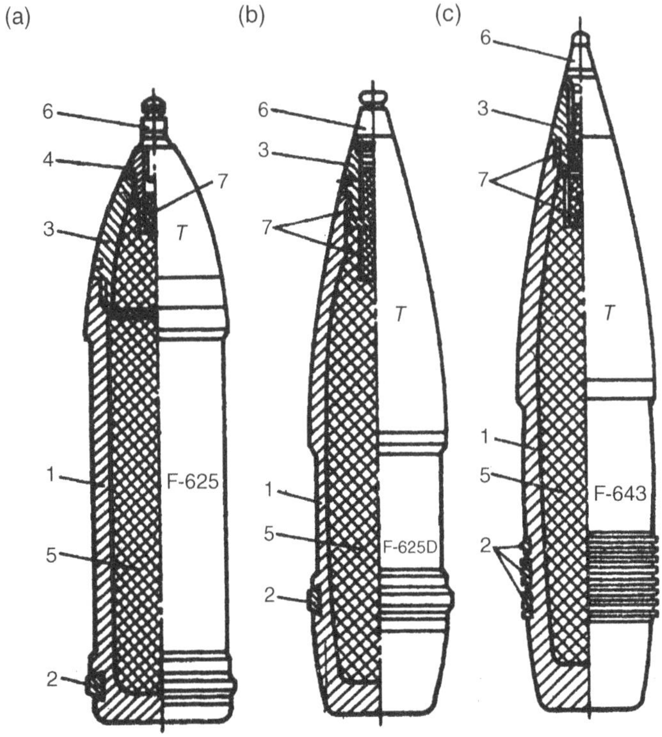

Figure 3.1 High-explosive projectiles: (a) 203-mm short-range projectile F-625; (b) 203-mm long-range projectile F-625D; (c) 210-mm projectile F-643; 1 – projectile body; 2 – leading belt; 3 – screw head; 4 – firing cup; 5 – explosive charge: 6 – head detonator; 7 – additional detonator. *Source*: From Balagansky et al. [1].

the protective embankment from the durable reinforced concrete structures and destroying their combat cover, as well as for destroying and suppressing the enemy's firing means and equipment, located openly or in shelters. In addition, blast mines can be used to destroy and suppress enemy personnel located openly or in shelters, especially on the reverse slopes of the terrain.

The blast mines are available in 160- and 240-mm calibers and are used to fire from the 240 mm M-240 mortar (at a range of 800–9700 m), the 160 mm M-160 mortar (at a range of 750–8040 m), and the 160 mm mortar of 1943 model (at a range of 620–5100 m).

Constructive characteristics of some samples of blast mines are given in Table 3.2, and their design is presented in Figure 3.2.

Steel mine's chamber is filled with TNT, and steel cast iron mine's chamber is filled with A-80 amatol. To ensure reliable detonation of the explosive charge, pieces made of pressed PETN are placed in its socket.

Table 3.1 Construction specifications of Russian high-explosive projectiles.

Ordnance	Projectile index	The total length L, cal	Weight of projectile q, kg	Wall thickness δ, cal	Weight of explosive charge ω, kg	Filling ratio of HE α, %
180-mm C-23 gun	F-572	5.38	88	0.15	10.7	12.1
203-mm howitzer B-4	F-625	4.63	100	0.08	23.4	23.4
Same here	F-625D	4.65	100	0.13	15.8	15.8
210-mm gun BR-17	F-643	5.19	133	0.13	18.8	14.4
210-mm C-33 gun	F-642	—	134	—	—	—
210-mm gun B-3	F-644	5.20	133	0.16	15.6	11.7

Source: From Balagansky et al. [1].

Table 3.2 Construction specifications of blast mines.

Name	Mine index	Length with fuse L, cal	Mass of the mine q, kg	Thickness of the shell δ, cal	Mass of HE ω, kg	Filling ratio of HE α, %
240-mm steel mine	F-864	6.40	130.70	0.062	31.95	23.3
160-mm steel mine for M 160	F-853C	7.00	41.14	—	9.00	21.8
160-mm steel cast iron mine for M-160	F-853A	7.00	41.14	0.103	7.72	18.8
Same here	F-853AУ	7.15	41.14	0.103	7.72	18.8
160-mm steel mine for M-160	F-853У	7.05	41.14	0.094	9.00	21.8
160-mm steel cast iron mine for M-160	F-853	6.85	40.87	—	7.45	18.2
160-mm steel cast iron mine for a mortar of 1943	F-852	6.85	40.87	—	7.45	18.2

Source: From Balagansky et al. [1].

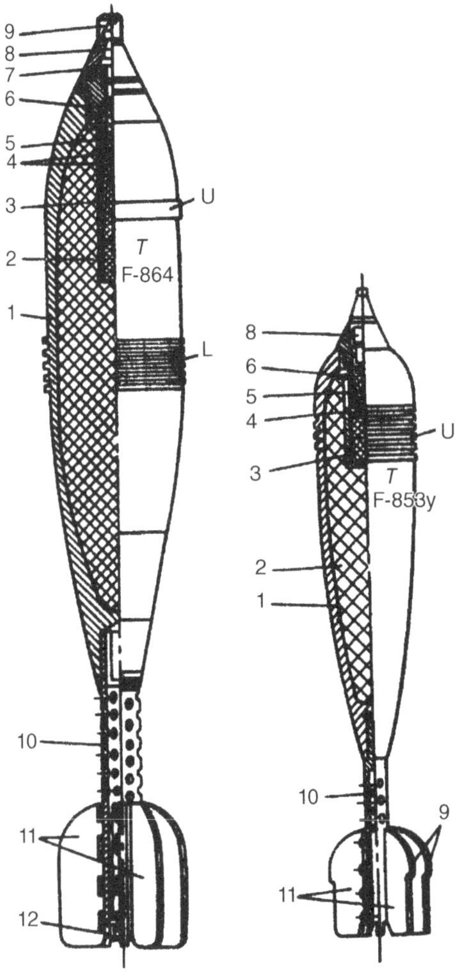

Figure 3.2 Blast mines of calibers 240 and 160 mm: 1 – body; 2 – explosive charge; 3 – additional detonator; 4 – gaskets; 5 – detonator; 6 – firing cup; 7 – large safety cap; 8 – fuse; 9 – small safety cap; 10 – stabilizer; 11 – stabilizer feathers; 12 – plug; U, L – upper and lower centering thickening. *Source*: From Balagansky et al. [1].

3.1.3 Aviation Bombs

In the postwar years, several types of blast bombs of 100, 250, 500, 1500, 3000, 5000, and 9000 kg calibers were added to the Soviet Army's arsenal. The blast bombs, accepted for use in the late 1940s to early 1950s, were mainly intended for attacking large ships. Only FAB-1500 was considered acceptable for attacks on industrial objects, dams, and underground structures.

The construction of a typical blast aviation bomb is shown in Figure 3.3. The nomenclature of Soviet blast bombs is given in Figure 3.4.

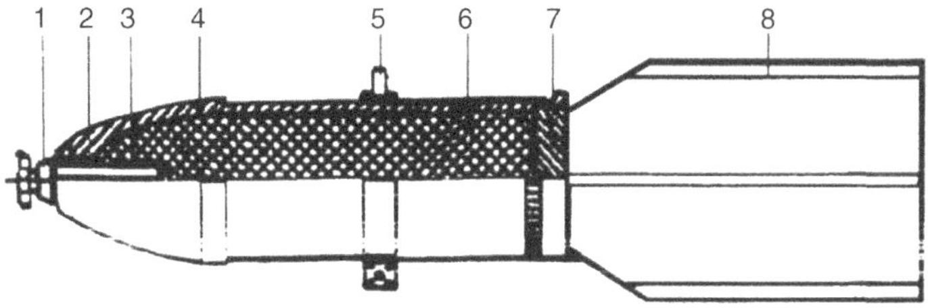

Figure 3.3 Blast bomb: 1 – fuse; 2 – transition bush; 3 – detonator; 4 – shell; 5 – bugel; 6 – explosive; 7 – bottom; 8 – stabilizer. *Source*: From Balagansky et al. [1].

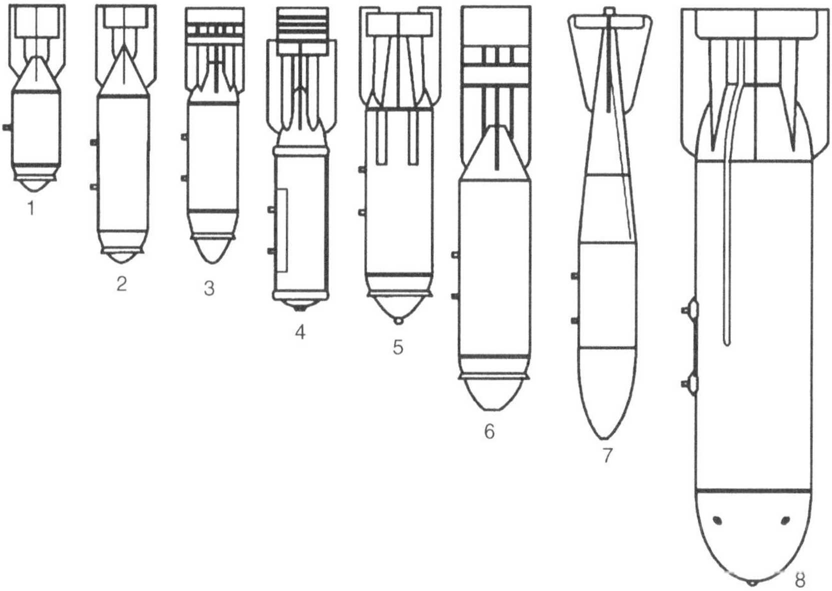

Figure 3.4 Blast bombs: 1 – FAB-100; 2, 3 – FAB-250; 4 – OFAB-250; 5–7 – FAB-500; 8 – FAB-5000. *Source:* From https://ru.wikipedia.org/wiki/.

A typical FAB-1500 bomb has 18-mm thick walls and contains 675 kg of explosive. FAB-3000M-46 and FAB-3000M-54 each contained 1400 and 1387 kg of TNT, while FAB-9000M-54 contained 4297 kg of TNT.

The data of free-fall blast bombs and radiuses of damage zones are given in Tables 3.3 and 3.4, respectively. In Table 3.4, the zone of heavy damage should be understood as brickwork damage, the zone of medium damage – damage to wooden walls and parts of buildings, the zone of light damage – glass breakage, shifting tiles, etc.

Table 3.3 Free-fall bombs from 1940 to 1950s.

Bomb index	Weight of bomb, kg	Weight of warhead, kg	Weight of explosive, kg	Length, mm	Diameter, mm
FAB-100	100	70	—	964	267
FAB-250	250	230	99	1589	285
FAB-500	500	450	213	2142	392
FAB-1500	1400	1200	675	3000	580
FAB-5000	4900	4200	2207	3107	642

Source: Data from https://ru.wikipedia.org/wiki/.

Table 3.4 Radiuses of damage zones.

Name of bombs	Distance, m		
	Zone of heavy damage	Zone of medium damage	Zone of light damage
FAB-50	12	25	50
FAB-100	18	35	70
FAB-250	28	56	112
FAB-500	40	80	160
FAB-1500	56	112	224

Source: Data from https://ru.wikipedia.org/wiki/.

3.1.4 Volumetric Explosion Ammunition

The first samples of volumetric explosion bombs (fuel–air explosives or thermobaric weapons), as already noted, were developed in the United States in the late 1960s. They were quite small in size and capacity (up to 10 gallons) (Figure 3.5). After the drop, at a relatively low altitude (30–50 m), the brake parachute was opened, which provided the bomb with stabilization and velocity most favorable for the sequence of detonation operations (charge explosion and opening of the bomb shell), spraying of the fuel mixture, scattering and subsequent explosion of detonators. A 1.5-m-long probe pulls out of the nose of the bomb, which touches the ground and triggers the explosion process.

When a bomb containing 10 gallons (approximately 38 l) of ethylene oxide detonates, a fuel–air mixture cloud with a radius of 7.5–8.5 m and a height of up to 3 m is formed. After 125 ms, this cloud is exploded by several detonators. The resulting shock wave has a pressure of 2.1 MPa at the front. For comparison: to create this pressure at a distance of 8 m from a TNT charge, about 200–250 kg of TNT is required. At a distance of 3–4 radii or 22–34 m, the overpressure in the shock wave quickly decreases to about 0.1 MPa. For damaging an aircraft by a shock wave, the pressure of 0.07–0.09 MPa is required. Therefore, such a bomb is capable of completely disabling an aircraft or helicopter at an airfield parking lot within a radius of 30–40 m from the explosion site.

Ethylene oxide, propylene oxide, methane, propyl nitrate, MAPP (a mixture of methyl acetylene, propadiene, and propane) have been tested and found to be suitable for use as fuel for fuel–air explosives bombs.

Attempts to create larger caliber ammunition failed at the time due to technical difficulties. A workaround, cluster bombs, was found. There were several 32.6 kg fuel–air explosive bombs in one cassette. These bombs were distributed over a certain area, thus increasing the size of the cloud.

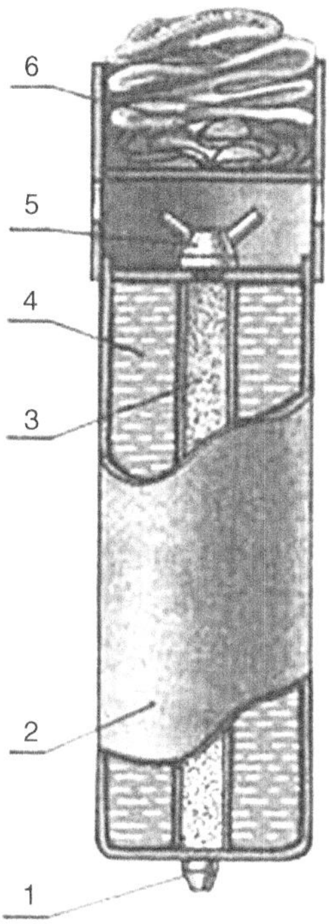

Figure 3.5 Volumetric explosion ammunition: 1 – fuse; 2 – cylindrical case; 3 – central explosive charge; 4 – liquid fuel; 5 – explosion initiation device; 6 – brake parachute. *Source*: From http://army.armor.kiev.ua/hist/obomvzryv.shtml.

According to the modern classification, such volumetric explosion ammunition is called *two-stroke*, as its operation is divided into two stages: cloud formation and cloud explosion. Two-stroke ammunition was a qualitative jump in the development of blast ammunition; however, they are not devoid of a num ber of shortcomings, such as a strong influence of weather conditions on their effectiveness. All this has led to the creation of a weapon with a single-stroke operation scheme, the principal difference being the existence of a single process of fuel–air mixture formation, initiation, and development of an explosive transformation reaction in it. The most widespread among *single-stroke* volumetric explosion munitions are the so-called thermobaric ammunition.

The principal design scheme of thermobaric ammunition is shown in Figure 3.6. It includes a container with thermobaric mixture 1, intermediate detonator 2, and capsule detonator 3.

The emergence of thermobaric ammunition is connected to the development of special compositions – thermobaric mixtures. As a rule, thermobaric mixtures are paste-like mixtures based on liquid fuel with large negative oxygen balance (for example isopropyl nitrate $C_3H_7NO_3$), in which a powder of high-brisance explosive substance (for example hexogen) and fine powder of combustible metal (for example aluminum) are introduced. Such mixtures can detonate; however, this requires a sufficiently powerful intermediate detonator, and the parameters of detonation are relatively small (detonation velocity of about 5000–6000 m/s). As a result of the explosive transformation reaction, gaseous products containing non-reacted combustible components are formed behind the detonation front, and metal particles are heated and begin to react with oxidizing components of the explosion products (CO_2 and H_2O).

After the detonation reaches the surface and the container shell is destroyed, the gaseous products begin to expand into the air while being subjected to intense deceleration. Burning metal particles outrun the gaseous products and reach the air where the intensity of their combustion increases significantly. The energy released in this process is used to supply the air shock front parameters and to form a long high-temperature area with an increased thermal effect (see Figure 3.6). At a later stage, due to the development of turbulent diffusion, the explosion products are mixed with air and the non-reacted combustible components are burned. In this case, energy is released in the "tail" of the shock wave compression phase and leads to its longer duration.

Several times volumetric explosion ammunition was used in various wars of 1980–1990. On 6 August 1982, during the war in Lebanon, an Israeli aircraft dropped such a bomb (American-made) on an eight-floor apartment building. The explosion occurred close to the building at the level of 1–2 floors. The building

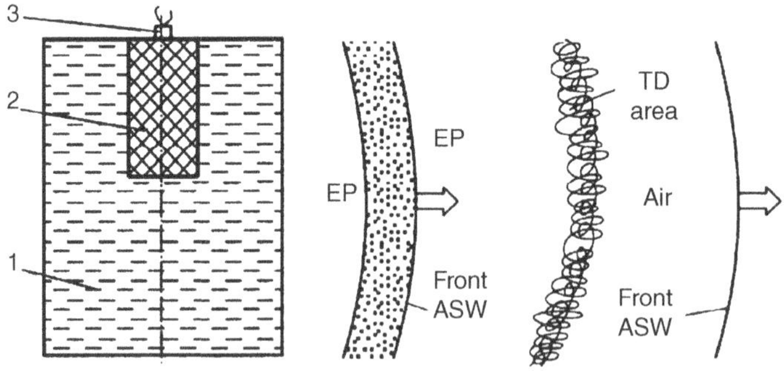

Figure 3.6 Principal scheme of thermobaric ammunition: 1 – thermobaric mixture; 2 – intermediate detonator; 3 – capsule detonator; TD – turbulent diffusion; EP – explosion products. *Source*: From Babkin et al. [2].

was destroyed. About 300 people were killed (mostly not in the building, but near the explosion location).

In August 1999, an ODAB-500 large-caliber volumetric bomb was dropped on the Dagestani village Tando, where a significant number of Chechen fighters had accumulated. The militants suffered huge losses. In the following days, the mere appearance of a single Su-25 attack aircraft over any settlement forced the militants to hastily leave the village.

The Russian army is armed with several types of volumetric explosion ammunition.

The aircraft bomb ODAB-500PMV has a diameter of 50 cm, a length of 238 cm, the scope of the stabilizer 68.5 cm, total weight of 525 kg, with a 193 kg charge. It is used with airplanes and helicopters. After separation from the carrier at an altitude of 30–50 m, the brake parachute located in the tail of the bomb opens, and the radio altimeter turns on. At an altitude of 7–9 m, a charge of an ordinary explosive is detonated. In this case, the thin-walled case of the bomb is destroyed, and liquid fuel is sublimated. After 100–140 ms, an initiating detonator is detonated in a capsule attached to a parachute and the thermobaric mixture is exploded.

300-mm 9M55S missile for the "Smerch" multiple launch rocket system.

Missile type	Reactive
System	"Smerch" multiple launch rocket system
Missile caliber	300 mm
Missile length	760 cm
Missile weight	800 kg
Warhead weight	280 kg
Maximum range	70 km
Minimum distance of fire	20 km
Warhead type	Monoblock thermobaric

The missile structurally consists of a head block with the flight control system, a warhead, and a propulsion system with a solid-propellant rocket engine. Although the "Smerch" system is capable of firing the full allowance of ammunition (12 missiles) in 20 seconds, the volumetric explosion missiles are fired either by a single missile or at time intervals that allow the previous missile to explode before the next one approaches the explosion zone. The missile is divided into three parts as it approaches the target at the downward end of the trajectory: head, combat, and propulsion. At the altitude of 60–70 m, the braking parachute is opened, and the radio altimeter is activated. Then everything happens in the same way as with the aircraft bomb.

The Russian Army also has other volumetric blast weapons, such as S-8DM and S-13D unguided aircraft missiles, and so on.

3.2 Parameters of an Air Shock Wave During the Explosion of High Explosives

3.2.1 Physical Phenomena Accompanying the Explosion of a Charge in the Air

Formed as a result of the explosion of the charge of condensed explosives, strongly heated and compressed to very-high-pressure explosion products (EP) upon expansion produce a sharp impact on the environment. At the same time, the pressure at the interface of EP-air decreases and becomes equal to 500–900 atm, and deep into EP from their external surface with the velocity equal to the local velocity of sound, the rarefaction wave (RW) starts to propagate (Figure 3.7a,b). The explosion of 1 l of condensed HE produces gases at a pressure of about 100 kbar, which after expansion to atmospheric pressure and cooling to room temperature will take a volume of about 1 m^3. For typical condensed explosives, at the initial moment of interaction between EP and air, the EP

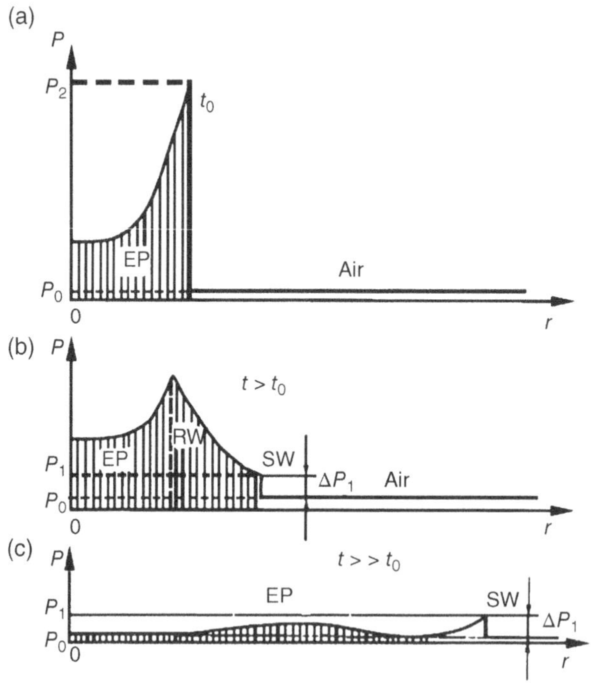

Figure 3.7 The pressure distribution in explosion products (EP) and the air at different times (a–c). *Source*: From Miropolsky et al. [3].

expansion velocity can reach several kilometers per second. As the EP continues to expand, their velocity decreases as they compress and engage more and more air masses. The layer of compressed air explosion products forms an air shock wave (ASW), at the front of which a pressure spike of $\Delta p_1 = p_1 - p_0$ is formed (see Figure 3.7c).

Over time, both the pressure inside the EP and the overpressure at the front of the ASW Δp_1 decrease. The velocity of EP decreases to zero, and the pressure becomes less than the pressure in the environment. As a result, the explosion products change the direction of movement and begin to move in the opposite direction – toward the center of the explosion. This is the stage in the evolution of the explosion process when the SW is detached from the EP and continues to propagate independently as an independent gas-dynamic complex.

At the same time, a zone with lower atmospheric pressure is formed between the ASW and the EP boundary (Figure 3.8). The distance at which the ASW is detached from the EP is usually called the critical distance r_{cr}. The air masses adjoining the EP will also be involved in the movement toward the center of the explosion. The air compressed by the SW is moving after it.

Let us define in D the velocity of SW (the velocity of point A in Figure 3.8), and in u_1 – the mass velocity of air at the front of SW. Obviously, the mass velocity of the layer of air compressed by a shock wave will be the slower, the lower pressure in this layer is, i.e. the further this layer is away from the front of SW. The difference in velocity of the wave front (point A) and the boundary of the zone with pressure below atmospheric (point B) leads to the fact that over time (with increasing distance, passing the front of the SW), the width of the compression zone increases.

In this case, the overpressure at the front of SW Δp_1 decreases, because as SW is moved away from the point of explosion, the area of the wave surface increases, and, consequently, the mass of air involved in the movement increases.

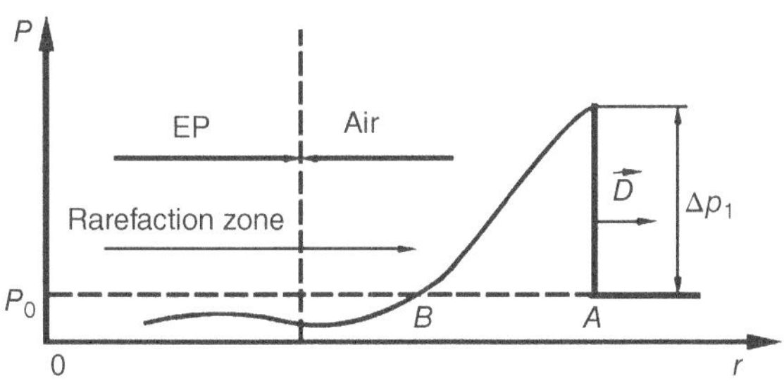

Figure 3.8 Separation of the shock wave from the explosion products. *Source*: From Miropolsky et al. [3].

Besides, pressure decrease is related to the irreversible nature of air compression in SW, accompanied by a transformation of its kinetic energy into heat. A free-running shock wave, leaving behind heated air, eventually completely loses its energy and at long distances from the point of explosion turns into a normal sound wave.

The numerical simulation of the process of detonation of an explosive charge in the air provides a detailed picture of the flow field. For example, it was found that there is a secondary SW in the EP, which is formed immediately after the main air SW at the tail of the rarefaction wave (RW). Secondary SW is moving against the EP flow. At the same time, its intensity is continuously growing. After the collapse in the center (see *r*–*t* – diagram of the spherical charge explosion process in Figure 3.9), the reflected secondary SW will catch up with the main SW after a certain time.

In Figures 3.10, 3.11, the profiles of mass velocity and density of EP and air at different moments of time were obtained as a result of numerical calculation of a spherical charge explosion of pressed TNT with density $\rho_{HE} = 1590$ kg/m^3 and radius $r_0 = 0.1$ m are given as examples. The EP-air boundary is marked in these figures with a dotted line.

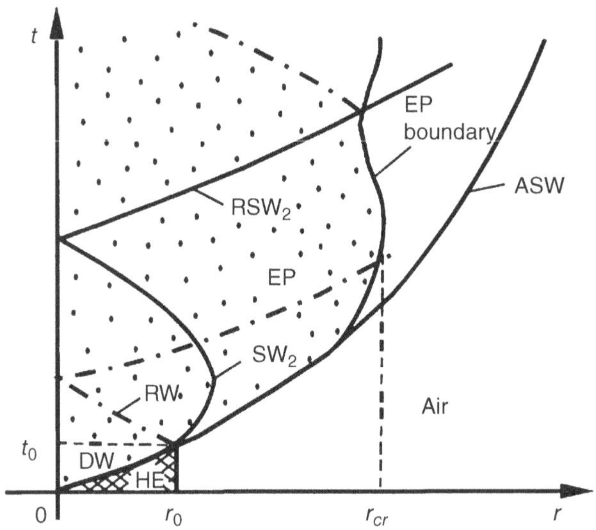

Figure 3.9 Diagram of the explosion process in phase plane (*r*–*t*): ASW – air shock wave; DW – detonation wave; SW_2 – secondary shock wave; RSW_2 – reflected secondary shock wave; RW – rarefaction wave. *Source*: From Miropolsky et al. [3].

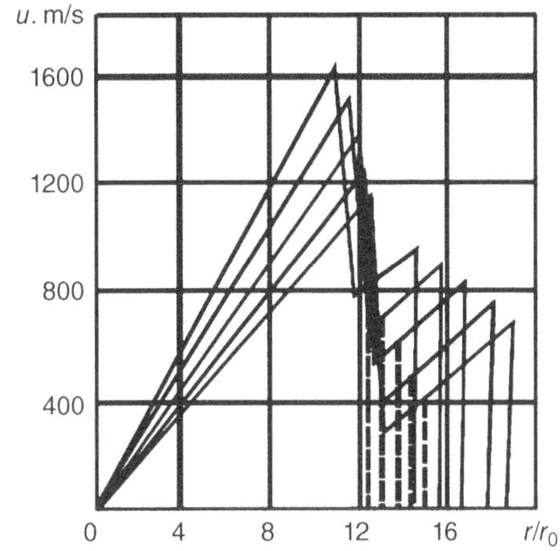

Figure 3.10 Distribution of mass velocities in air and detonation products under detonation of TNT charge with radius r_0 = 0.1 m and density ρ_{HE} = 1590 kg/m^3 in different moments of time. *Source*: From Miropolsky et al. [3].

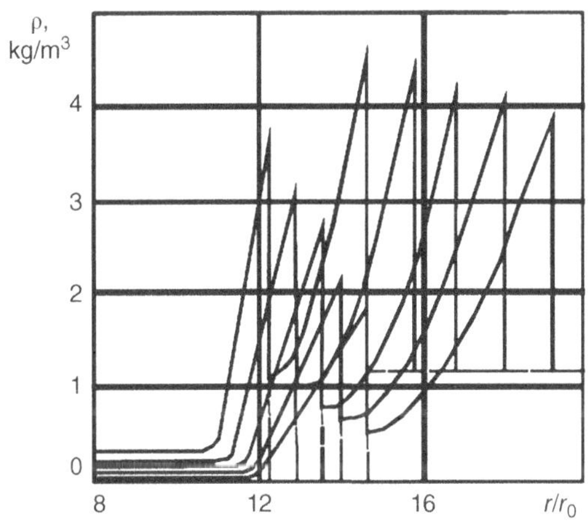

Figure 3.11 Distribution of densities in air and detonation products under detonation of TNT charge with radius r_0 = 0.1 m and density ρ_{HE} = 1590 kg/m^3 in different moments of time. *Source*: From Miropolsky et al. [3].

3.2.2 Air Shock Wave (ASW) Parameters

Using the basic relations of the elementary theory of shock waves in gases, it is easy to obtain formulas allowing us to determine parameters at the front of shock wave depending on the value of overpressure Δp_1.

If in the Hugoniot adiabatic equation for an ideal gas

$$\frac{v_1}{v_0} = \frac{(k+1)p_0 + (k-1)p_1}{(k+1)p_1 + (k-1)p_0}, \tag{3.1}$$

where k is the adiabatic index, we represent p_1 through overpressure Δp_1 ($p_1 = p_0 + \Delta p_1$), then, bearing in mind that $\rho = 1/v$, we get

$$\frac{\rho_0}{\rho_1} = \frac{v_1}{v_0} = \frac{2k + (k-1)\Delta p_1/p_0}{2k + (k+1)\Delta p_1/p_0}, \tag{3.2}$$

where ρ_1 – density, and v_1 – specific air volume at the SW front.

Expressing Δp_1 and p_0 in atmospheres and assuming for air $k = 1.4$, $p_0 = 1$ atm, we will have

$$\frac{\rho_1}{\rho_0} = \frac{7 + 6\Delta p_1}{7 + \Delta p_1}. \tag{3.3}$$

From the formula (3.3), it follows that for strong SW ($\Delta p_1 \gg 1$ atm) air density at the front of the wave increases no more than six times.

Using a known formula for shock wave velocity

$$D = v_0\sqrt{\frac{p_1 - p_0}{v_0 - v_1}}. \tag{3.4}$$

and the expression for v_1/v_0 from (3.2), we get

$$D = \sqrt{\frac{v_0 \Delta p_1}{1 - v_1/v_0}} = \sqrt{\frac{v_0}{2}[2kp_0 + (k+1)\Delta p_1]}. \tag{3.5}$$

Let $p_0 = 1$ atm, $\rho_0 = 1.225$ kg/m^3, $k = 1.4$. Then, since $v_0 = 1/\rho_0$, we have

$$c_0 = \sqrt{kp_0 v_0} = 340 \text{ m/s}. \tag{3.6}$$

By substituting these values in (3.5) and keeping in mind (3.6), we will have

$$D = c_0\sqrt{1 + \frac{k+1}{2k}\cdot\frac{\Delta p_1}{p_0}} = 340\sqrt{1 + 0.86\Delta p_1}, \tag{3.7}$$

where Δp_1 is expressed in atmospheres, velocity D – in meters per second.

The mass velocity of air flow behind the shock wave front u_1 can be determined by the formula

$$u_1 = \sqrt{(p_1 - p_0)(v_0 - v_1)}. \tag{3.8}$$

Taking into account expression (3.4), this formula can be presented in the form of

$$u_1 = \frac{\Delta p_1 v_0}{D} = \frac{\Delta p_1}{\rho_0 D}. \tag{3.9}$$

Substituting in (3.9), the value $\rho_0 = 1.225$ kg/m^3 and expressing Δp_1 in atmospheres (1 atm = $1.013 \cdot 10^5$ Pa), we obtain, taking into account (3.7),

$$u_1 = \frac{8.3 \cdot 10^4 \Delta p_1}{D} = 243 \frac{\Delta p_1}{\sqrt{1 + 0.86 \Delta p_1}} \text{ (m/s)}. \tag{3.10}$$

The temperature of the air at the SW front can be found by using the ideal gas equation

$$T_1 = \frac{p_1 v_1}{p_0 v_0} T_0 = \frac{p_0 + \Delta p_1}{p_0} \cdot \frac{\rho_0}{\rho_1} T_0. \tag{3.11}$$

By substituting in (3.11), $p_0 = 1$ atm, $T_0 = 288$ K and having in mind (3.3), we get

$$T_1 = 288 \frac{(1 + \Delta p_1)(7 + \Delta p_1)}{7 + 6\Delta p_1}, \tag{3.12}$$

where Δp_1 is expressed in atmospheres.

Finally, the velocity of sound at the front of SW we will determine by the formula

$$c_1 = \sqrt{kRT_1}, \tag{3.13}$$

which at $k = 1.4$ and value of the gas constant for air $R = 287$ m^2/(s$^2 \cdot$ K) taking into account (3.12) gets the following form

$$c_1 = 20\sqrt{T_1} = 340 \sqrt{\frac{(1 + \Delta p_1)(7 + \Delta p_1)}{7 + 6\Delta p_1}} \text{ (m/s)}. \tag{3.14}$$

The resulting formulas for SW parameters in the air are applicable only for overpressure Δp_1 not exceeding 10 atm. The reason is that at high pressures (and temperatures), the phenomena of dissociation of molecules and ionization of air atoms take place. The increase in the number of elementary particles, in this case, leads to a change in the equation of air state. One of the ways to account for physicochemical transformations in gas is to introduce the effective adiabatic index k_{ef}. For strong SW ($\Delta p_1 \gg 100$ atm), it is possible to consider that $k_{ef} = 1.1, ..., 1.2$.

Table 3.5 presents data obtained using formulas (3.1)–(3.14) at different values $k = k_{ef}$. For example, at $\Delta p_1 = 1000$ atm, the air density behind the SW front reaches 19.754 kg/m^3, which is 16.1 times higher than the initial density ρ_0. If we do not take into account the change of k, the maximum air density in SW will not exceed $6\rho_0$.

3.2.3 Overpressure, Specific Impulse, and Time of Action of the Air Shock Wave

As can be seen from the previous section, to determine the parameters of the ASW and to estimate the destructive action of the explosion, it is necessary, in particular, to know the value of the overpressure at the front of the wave Δp_1 [4].

Table 3.5 Parameters of shock waves in the air.

Δp_1, atm	D, m/s	u_1, m/s	ρ_1, kg/m^3	c_1, m/s	T_1, K	k_{ef}
0.01	340	2.4	1.234	340	289	1.4
0.1	354	23.4	1.311	344	296	1.4
1.00	464	179	1.991	377	354	1.4
10.0	1 053	788	4.828	567	804	1.39
100	3 237	2564	9.816	1146	3 630	1.26
1000	10 443	7948	19.754	2408	17 878	1.13

Source: From Miropolsky et al. [3].

It is obvious that the pressure Δp_1 depends on the mass of the charge HE m, the specific heat of the explosion Q_s, distance to the point of the explosion r, and initial atmospheric pressure p_0. Dependence

$$\Delta p_1 = f(m, p_0, Q_s, r)$$

is determined experimentally. The methods of similarity theory make it possible to significantly reduce the required amount of experimental research. Let us write out formulas of dimensions of all considered values:

$$\begin{aligned} &\dim \Delta p_1 = \dim p_0 = \frac{M}{LT^2}, \\ &\dim Q_s = \frac{L^2}{T^2}, \quad \dim r = L, \quad \dim m = M. \end{aligned} \tag{3.15}$$

It is convenient to take m, r, p_0 as parameters with independent dimensions.

$$\Pi = \frac{\Delta p_1}{p_0}, \quad \Pi_1 = \frac{mQ_s}{p_0 r^3}. \tag{3.16}$$

Thus, the general functional relationship between these values:

$$\frac{\Delta p_1}{p_0} = f_1\left(\frac{mQ_s}{p_0 r^3}\right) = f_2\left(\sqrt[3]{\frac{mQ_s}{p_0 r^3}}\right). \tag{3.17}$$

3.2.3.1 Overpressure

The dependence (3.18) proposed by M.A. Sadovsky on the processing of experimental data related to the detonation of TNT charges without a metal shell ($Q_s = 4.1$ MJ/kg) under normal atmospheric conditions ($p_0 = 1$ atm) at an altitude excluding the influence of the reflective action of the Earth's surface on the Δp_1 value has found wide practical application.

$$\Delta p_1 = 0.84\frac{\sqrt[3]{m}}{r} + 2.7\frac{\sqrt[3]{m^2}}{r^2} + 7.0\frac{m}{r^3}. \tag{3.18}$$

Processing of these TNT explosions on the Earth's surface has shown that in this case the dependence can be approximated by the following formula:

$$\Delta p_1 = 0.95\frac{\sqrt[3]{m}}{r} + 3.9\frac{\sqrt[3]{m^2}}{r^2} + 13.0\frac{m}{r^3}. \tag{3.19}$$

When an explosion is carried out on the ground surface, it can be approximated that all the energy of the explosion is distributed not in a spherical volume, as is the case with an explosion in the air, but in a hemisphere. Therefore, a ground explosion is equivalent to an air explosion of a double charge. In this case, the calculated formula can be obtained from the formula (3.18) by substituting the mass of charge $2m$:

$$\Delta p_1 = 1.06\frac{\sqrt[3]{m}}{r} + 4.3\frac{\sqrt[3]{m^2}}{r^2} + 14.0\frac{m}{r^3}. \tag{3.20}$$

The difference between the coefficients in the formulas (3.19) and (3.20) is explained by the fact that the dependence (3.20) corresponds to an explosion on an absolutely rigid surface, and the formula (3.19) indirectly takes into account the energy losses of the explosion for the formation of a crater in the ground.

In formulas (3.18)–(3.20), the mass of the HE charge is in kilograms, distance in meters, and the numerical value Δp_1 is expressed in atmospheres. Using the above formulas, it is possible to determine the values of SW parameters at the explosion of charges not only of TNT but also of other explosives, including in conditions different from normal ones. It is not difficult to show that for explosives with a specific heat of explosion Q_s' at an altitude where the pressure is equal p_0', the formula (3.18), for example, can be presented in the form of

$$\Delta p_1 = 0.84\frac{\sqrt[3]{k_T m}}{r}\left(\frac{p_0'}{p_0}\right)^{2/3} + 2.7\frac{\sqrt[3]{(k_T m)^2}}{r^2}\left(\frac{p_0'}{p_0}\right)^{1/3} + 7.0\frac{k_T m}{r^3}, \tag{3.21}$$

where $k_T = Q_s'/Q_s$ – energy TNT coefficient of equivalence of HE (Q_s – specific heat of TNT explosion). The value $m_T = k_T \cdot m$ is called the TNT equivalent of the explosive charge. Specific heats of the explosion of various explosives are given in the Appendix.

Later experimental research has shown that M.A. Sadovsky's formulas give a good approximation for overpressures $\Delta p_1 \leq 10$ atm and for HE charges weighing more than 2 kg in TNT equivalent. In the close zone of the explosion, other approximations should be applied.

3.2.3.2 Time of Action of the Shock Wave

Using the theory of similarity, you can also get a functional ratio for the time of action of SW τ^+ (Figure 3.12). Since dim $\tau^+ = T$, we get

$$\frac{\tau^+}{\sqrt{\frac{m}{p_0 r}}} = \varphi\left(\sqrt[3]{\frac{mQ_s}{p_0 r^3}}\right). \tag{3.22}$$

Experimental studies show that

$$\varphi(x) = \frac{a}{x}. \tag{3.23}$$

M.A. Sadovsky proposed a dependence for the time of SW action under air explosion of TNT charge in normal atmospheric conditions

$$\tau^+ = 1.3\sqrt[6]{m}\sqrt{r}\cdot 10^{-3}. \tag{3.24}$$

Here, too, the mass of charge m is substituted in kilograms, distance r in meters, and the numerical value τ^+ is expressed in seconds.

For a ground explosion, doubling the mass of the charge, we get

$$\tau^+ = 1.46\ \sqrt[6]{m}\sqrt{r}\cdot 10^{-3}. \tag{3.25}$$

3.2.3.3 Specific Impulse

In approximate calculations on the estimation of the destructive action of an explosion, the specific impulse of SW I_1 is often used, which is defined as the area under the overpressure curve $\Delta p(t)$:

$$I_1 = \int_0^{\tau^+} \Delta p(t)dt. \tag{3.26}$$

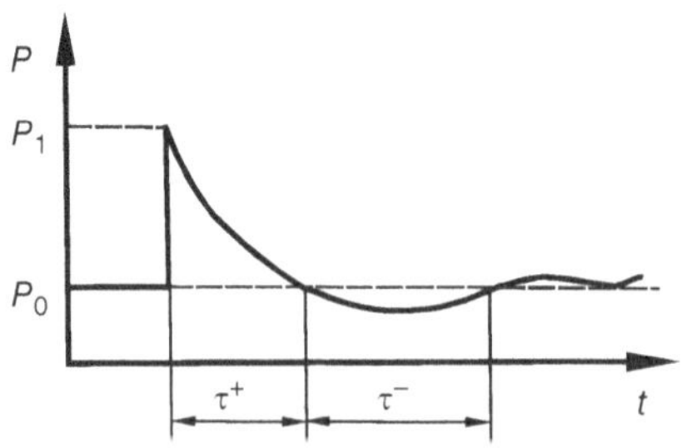

Figure 3.12 Pressure dependence on time at a fixed point of the area; τ^+ – the time of compression phase action in SW. *Source:* From Miropolsky et al. [3].

Specific impulse I_1 and its dependence on the main parameters can be found using an approximation of dependence $\Delta p(t)$, for example, in the following form

$$\Delta p(t) = \Delta p_1\left(1 - \frac{t}{\tau^+}\right)e^{-\frac{t}{\tau^+}}. \tag{3.27}$$

By integrating this dependency, we get

$$I_1 \approx 0.37\Delta p_1 \tau^+ . \tag{3.28}$$

However, in practical engineering calculations, the specific impulse is calculated using formulas based on similarity theory conclusions.

It is obvious that the value I_1 depends on the mass of charge m, on the distance from the point of explosion r, and the specific heat of explosion Q_S. Besides, in explosions at close distances, dependence I_1 on the density of EP $\rho = \rho_{EP}$ can appear.

Bearing in mind that $\dim I_1 = \frac{M}{LT}$, $\dim \rho = \frac{M}{L^3}$ and applying the π-theorem, we get the following dependence:

$$\frac{I_1}{\frac{m\sqrt{Q_s}}{r^2}} = \phi\left(r\sqrt[3]{\frac{\rho}{m}}\right). \tag{3.29}$$

Experimental investigations show that at $r > r_{cr}$

$$\phi(x) = ax. \tag{3.30}$$

For the ground explosion of TNT charges ($\rho_0 = 1.225$ kg/m^3) obtained the following approximation:

$$I_1 = \frac{200m^{2/3}}{r}\ (\text{Pa·s}). \tag{3.31}$$

Substituting the half value of mass, we have an approximation for the values of the specific impulse of SW air explosion at $r > r_{cr}$:

$$I_1 = \frac{126m^{2/3}}{r}\ (\text{Pa·s}). \tag{3.32}$$

The mass of the charge is expressed in kilograms and distance in meters.

Applicability range of formulas for overpressure at SW front, specific pulse, and compression phase duration

$$1 \le \frac{r}{\sqrt[3]{m}} \le 10.$$

The law of energetic similarity allows to calculate coefficients in these formulas in case of the explosion of any explosives in relation to the specific heat of the explosion.

For typical condensed HE charges, the critical distance r_{cr} at explosion corresponds to approximately 8–30 specified radii r_0.

By the specified radius r_0, we mean the radius of the spherical charge, in which mass is equal to the mass of the initial charge. For spherical charges, r_{cr} usually does not exceed $16r_0$.

As air density decreases, the distance r_{cr} increases and the specific SW impulse decreases. Therefore, with increasing altitude, the efficiency of the blast action decreases.

3.2.4 Influence of Conditions of the Explosion of Explosive Charge on Blast Action

3.2.4.1 The Charge Shape

Main parameters at the front of the air SW ($\Delta p_1, I_1, \tau^+$) depend on the shape of the charge. For example, the SW generated by an explosion of a charge of a cubic shape has higher parameters in the direction of the faces of the cube and smaller parameters in the direction of the edges. Analysis of parameters of the explosion of charges of different shapes at sufficiently large distances from the center of the explosion shows that:

1) parameters of the explosion of compact charges of any form at distances of their several characteristic sizes are practically equal to parameters of the explosion of a spherical charge of equivalent mass and can be calculated using the above formulas;
2) parameters of the explosion of charges, the length of which is much longer than two other characteristic sizes, are equivalent to parameters of the explosion of a cylindrical charge of the same mass in the near zone (at distances of the order of average value of smaller sizes of charge) and spherical charge of the same mass in the far zone (at distances of several lengths of charge). It should be taken into account that for cylindrical charges, the parameter decrease with increasing distance is slower than in the spherical case. Parameters of the field of the explosion of elongated charges in the direction of the ends are equivalent to parameters of the explosion of a spherical charge of a smaller mass.

3.2.4.2 Own HE Charge Velocity

If the HE charge velocity is high enough, you can expect an increase of the basic parameters of SW in the direction of its movement. The formula for estimation of the overpressure increase in the direction of charge movement Δp_1^V in relation to the explosion of static charge Δp_1^0 is offered:

$$\frac{\Delta p_1^V}{\Delta p_1^0} = \sqrt{1 + \frac{V^2}{2Q_s}}. \tag{3.33}$$

This formula is valid for quite long distances ($r > 45r_0$).

3.2.4.3 Properties of the Soil

At the explosion of charges on a surface of various materials, a part of the energy of the explosion is used for the destruction of an underlying surface and the formation of a crater. The fraction of this energy can reach 50% of the total energy of the explosion. If it is necessary to take energy losses into account, formulas are used to determine the parameters of SW in the air, substituting instead of mass m the equivalent charge mass of $2\eta_0 m$, where $0.5 \leq \eta_0 \leq 1$ is a pliability factor. The recommended values of this coefficient are presented in Table 3.6.

With the height of the explosion increase, energy losses for crater formation decrease, and at some height, it can be assumed that all the energy of the explosion is used for the formation of SW.

3.2.4.4 The Khariton Layer

When calculating parameters of the explosion of small charges (with mass less than 5–10 kg) of industrial explosives without shells, it is necessary instead of mass m to put the so-called active mass of charge m_a into calculation formulas, i.e. to take into account only that part of explosives which energy is used for the formation of SW. The greatest part of the lost energy is chemical losses in the surface layer of explosives that do not have time to react (the so-called Khariton layer named after Yu.B. Khariton). M.A. Sadovsky proposed the formula for calculating m_a

Table 3.6 Pliability factor values η_0 for some soil types.

Soil type	η_0	Soil type	η_0
Absolutely rigid surface	1.0	Medium ground	0.75
Rocky ground	0.90	Soft ground	0.65
Dense loams, clay	0.80–0.85	Air	0.50

Source: From Miropolsky et al. [3].

$$\frac{m_a}{m} = \left(1 - \frac{h_{cr}}{r_0}\right)^3. \tag{3.34}$$

where h_{cr} is the thickness of the critical layer of HE (in the first approximation we can take $h_{cr} \approx d_{cr}/2$, where d_{cr} is the critical diameter of the HE charge); r_0 is the specified radius of charge.

3.2.4.5 The Shell of the HE Charge

The charge shell has an ambiguous effect on explosion parameters. On the one hand, it noticeably reduces the initial parameters of the shock wave and takes energy for deformation, destruction, and throwing of fragments; on the other hand, it reduces chemical and entropy losses during the explosion and returns some energy to the wave during braking in the environment. Experiments show that with a filling factor $\alpha > 0.4$, the influence of the shell on the parameters of air shock waves can be ignored.

3.3 Reflection of Shock Waves from Barriers and Flow Around Barriers

3.3.1 Reflection of a Shock Wave from a Barrier

To evaluate the destructive effect of shock waves, we need to know the pressure acting on a barrier. The value of this pressure in some cases significantly exceeds the pressure Δp_1 determined by the Sadovsky formulas. This can be explained by the fact that meeting a barrier causes a sharp deceleration of compressed air masses moving behind the SW. As a result, in addition to the pressure, which at the moment of meeting with the barrier the air itself had compressed at the front of the SW (a static component of pressure), the barrier will be subject to additional pressure due to the velocity head pressure. The resulting pressure spike in the form of a so-called reflected shock wave will start to propagate independently in the air compressed by the falling wave. In this case, the reflected shock wave will move in the opposite direction to the falling SW.

3.3.1.1 Normal Reflection

Let us first consider the case of a normal reflection, when an incoming shock wave meets an absolutely rigid barrier of infinite size, the plane of which is perpendicular to the direction of movement of the wave front.

The process of reflected SW formation for this case is shown in Figure 3.13, where sequential pictures of pressure distribution in the falling and reflected SW front at different time moments are presented. It is obvious that the maximum

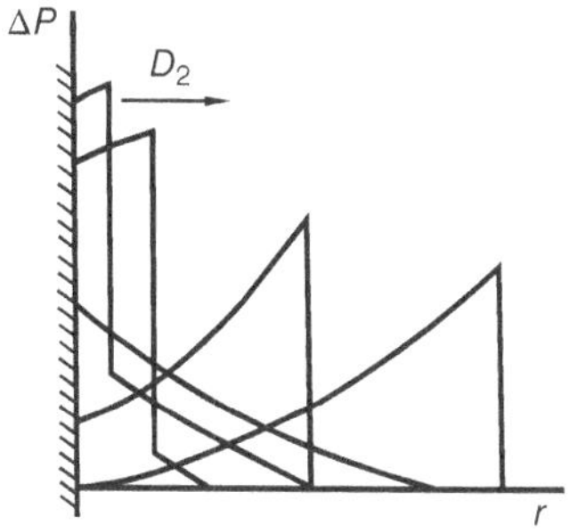

Figure 3.13 Distribution of pressures in the air at different time moments of SW reflection from a barrier. *Source*: From Miropolsky et al. [3].

pressure value of the pressure acting on the barrier (maximum pressure at the reflected SW front Δp_2) will correspond to the moment of complete braking at the falling SW front. The mass velocity of air behind the front of the incident SW u_1 according to (3.5) and (3.8) is equal:

$$u_1 = \frac{(p_1 - p_0)v_0}{D} = (p_1 - p_0)\sqrt{\frac{2v_0}{2kp_0 + (k+1)(p - p_0)}} \tag{3.35}$$

At the moment of formation of reflected SW, mass velocity behind its front $u_2 = 0$, because, as noted earlier, there is complete braking of the incoming air flow on the barrier. In the coordinate system associated with air particles behind the front of the incident SW (i.e. moving at a velocity of u_1 to the barrier), the reflected SW moves in a stationary gas with a pressure of p_1, and, therefore, for the velocity of the particles behind its front can be written expression similar to (3.35):

$$u_2' = (p_2 - p_1)\sqrt{\frac{2v_1}{2kp_1 + (k+1)(p_2 - p_1)}}. \tag{3.36}$$

Going back to the stationary coordinate system, we get $u_2 = 0 = u_2' - u_1$, and therefore, $u_2' = u_1$.

Thus, we have an equation for determining the maximum pressure p_2:

$$(p_2 - p_1)\sqrt{\frac{2v_1}{2kp_1 + (k+1)(p_2 - p_1)}} = (p_1 - p_0)\sqrt{\frac{2v_0}{2kp_0 + (k+1)(p_1 - p_0)}}. \tag{3.37}$$

Using the Hugoniot adiabat equation (3.2), one can obtain

$$\left(\frac{p_2 - p_1}{p_1 - p_0}\right)^2 = \frac{2kp_1 + (k+1)(p_2 - p_1)}{2kp_0 + (k-1)(p_1 - p_0)}. \tag{3.38}$$

By using designations $\Delta p_2 = p_2 - p_0$ and $\Delta p_1 = p_1 - p_0$, one can express Δp_2 from Eq. (3.38):

$$\Delta p_2 = 2\Delta p_1 + \frac{\frac{k+1}{k-1}\Delta p_1^2}{\frac{2kp_0}{k-1} + \Delta p_1}. \tag{3.39}$$

Assuming $p_0 = 1$ atm and $k = 1.4$, we get the following formula:

$$\Delta p_2 = 2\Delta p_1 + \frac{6\Delta p_1^2}{7 + \Delta p_1}. \tag{3.40}$$

According to (3.40), for weak SW ($\Delta p_1 \ll 1$ atm), the overpressure of reflected SW is twice as high as Δp_1. For strong SW ($\Delta p_1 \gg 1$ atm), the maximum value of Δp_2 is eight times more than the overpressure at the front of the falling SW. However, for strong SW due to ionization and dissociation phenomena, the value of k decreases to 1.1, and, consequently, Δp_2 may exceed Δp_1 23 times.

3.3.1.2 Oblique Reflection of SW

Now let us consider the case when the falling SW meets a barrier at some angle α_1 (Figure 3.14). In this case, the dynamic component of pressure (velocity head acting on the barrier) will be determined by the component of mass velocity of air directed to the barrier $u_{1\perp} = u_1 \cos\alpha_1$. Consequently, the total value of the maximum overpressure acting on the barrier is equal to

$$\Delta p_{2\alpha} = \Delta p_1 + \chi\frac{\rho_1 u_1^2}{2}\cos^2\alpha_1, \tag{3.41}$$

where χ – some coefficient that takes into account the "impact" nature of the flow braking on the barrier. At $\alpha_1 = 0$, the pressure, acting on the barrier, can be determined by the formula (3.40):

$$\Delta p_{2\alpha}(\alpha_1 = 0) = \Delta p_1 + \chi\frac{\rho_1 u_1^2}{2} = 2\Delta p_1 + \frac{6\Delta p_1^2}{7 + \Delta p_1}. \tag{3.42}$$

Hence, we get

$$\chi\frac{\rho_1 u_1^2}{2} = \Delta p_1 + \frac{6\Delta p_1^2}{7 + \Delta p_1}. \tag{3.43}$$

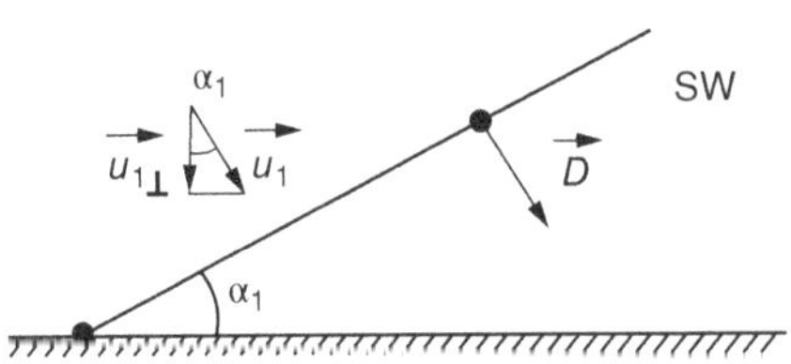

Figure 3.14 Oblique reflection of SW from a barrier. *Source*: From Miropolsky et al. [3].

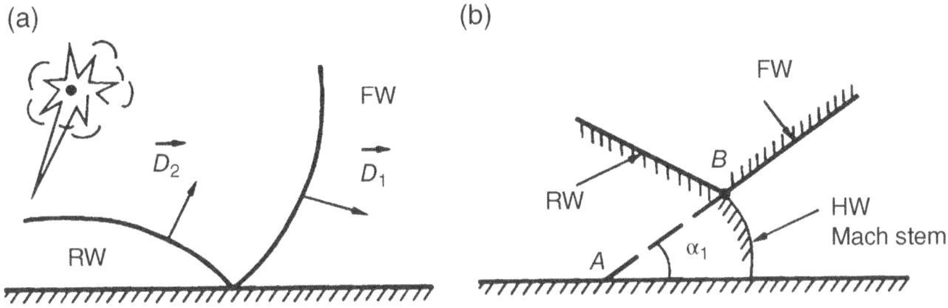

Figure 3.15 Wave pictures of the processes of regular (a) and irregular (b) shock wave reflection. *Source*: From Miropolsky et al. [3].

Using this ratio, the formula for the pressure acting on the barrier $\Delta p_{2\alpha}$ can be represented as follows:

$$\frac{\Delta p_{2\alpha}}{\Delta p_1} = 1 + \left(1 + \frac{6\Delta p_1}{7 + \Delta p_1}\right) \cos^2\alpha_1. \tag{3.44}$$

However, the formula (3.44) can only be used at sufficiently small angles α_1. In this case, the wave picture of the process consists of two waves: a falling wave (FW) and a reflected wave (RW). This reflection is called regular (Figure 3.15a).

When the angle of falling α_1 reaches the value of α^* – the critical angle of falling, the reflection process becomes irregular. In this case, a so-called head shock wave (HW) or the Mach wave (or Mach stem) is formed, and a three-wave configuration consisting of a falling, reflected, and head SW (Figure 3.15b). The critical falling angle α^* is a function of the overpressure in the falling SW (Figure 3.16).

Let's consider the process of head SW formation in details. With regular reflection, the pressure in the reflected SW will be higher than in the falling SW, because

Figure 3.16 Dependence of critical angle of falling α^* on overpressure in the falling SW Δp_1. *Source*: From Miropolsky et al. [3].

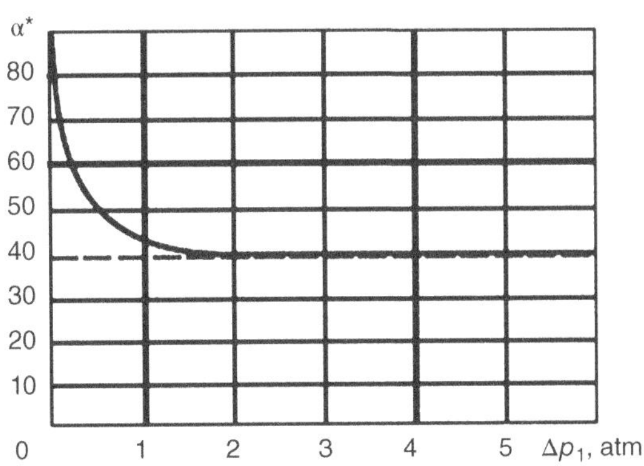

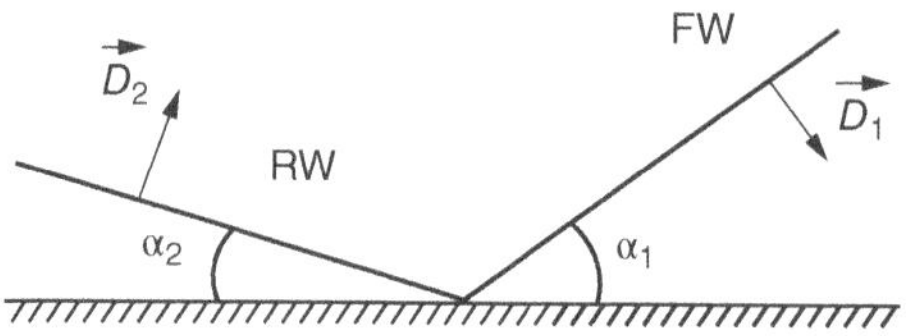

Figure 3.17 Falling angle α_1 and reflection angle α_2. *Source*: From Miropolsky et al. [3].

the reflected SW propagates in the pre-compressed and heated air and, therefore, its velocity relative to air will be greater than the velocity of the falling wave.

However, because of a large component of air velocity directed toward the reflected wave, its velocity relative to the barrier is lower than the velocity of the falling wave, and therefore, the reflection angle α_2 will be smaller than α_1 in this case (Figure 3.17).

Angle α_2 will be the smaller, the higher overpressure Δp_1 in the falling wave is, as well as the component of air velocity directed toward the reflected wave. With the growth of angle α_1, this component decreases, and, consequently, the value of angle α_2 is closer to the value of angle α_1.

Dependency graphs of reflection angle α_2 on a falling angle α_1 for different values of pressure Δp_1 are shown in Figure 3.18. It is interesting to note that regardless of the value of pressure Δp_1, the reflection angle α_2 becomes larger than the angle of falling α_1 at $\alpha_1 > \alpha_0 = 39\,°\,14'$. With further increase in angle α_1, there is a strong increase in angle α_2 and at $\alpha_1 = \alpha^*$, the air strongly compressed at

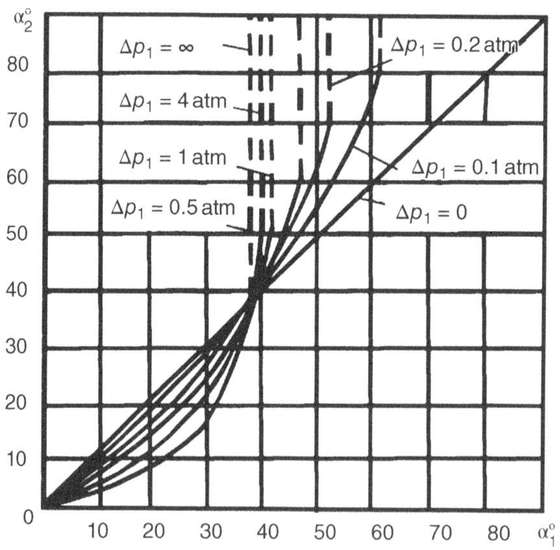

Figure 3.18 Dependence of reflection angle α_2 on the angle of falling α_1, at different values of overpressure in FW Δp_1. *Source*: From Miropolsky et al. [3].

reflection, creating a kind of an air "bag," pushes back the falling SW and, overtaking it, moves forward in the form of the head SW (Mach stem).

Thus, the reflected SW overtakes the falling SW and, merging with it, forms a new SW. It should be noted that with increasing overpressure Δp_1, the value of angle α^* decreases and becomes very close to α_0. At the same time, for reflections of weak SW close to sound waves, the phenomenon of Mach reflection is not observed.

The coupling point of the falling, reflected, and head SW in the Mach configuration is called the triple point (TP). When an HE charge is detonated at some distance from a barrier, the TP has an upward trajectory ending at the zenith above the point of explosion. This is due to the fact that over time, the TP catches up with the falling SW throughout its length and merges with it into a single SW, which is close in its parameters to the SW of the ground explosion (Figure 3.19).

For an approximation of TP trajectory based on experimental data processing, the following dependence has been suggested:

$$\frac{H - z}{r} = \mathrm{ctg}\left(\alpha^* + 1.2\ln\frac{\sqrt{r^2 + (H - z)^2}^2}{1.3\,H}\right), \qquad (3.45)$$

where H is the height of the explosion; z is the TP height; r is the distance from the explosion epicenter to the TP projection to the ground surface; α^* – the critical angle of falling, here it is assumed $\alpha^* = 2/9\pi$.

Formula (3.45) is recommended to determine the height of the TP at $z < H$.

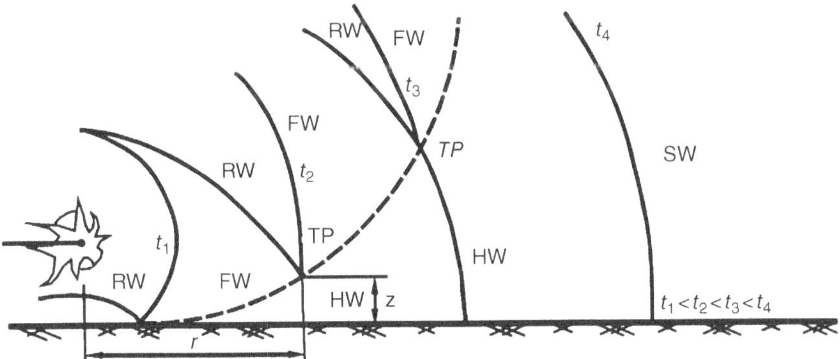

Figure 3.19 Wave pictures of an explosion near a reflective surface at different points in time: RW – reflected wave; FW – falling wave; SW – shock wave; HW – head wave; TP – triple point. *Source*: From Miropolsky et al. [3].

Let us now consider the question of how the pressure in the reflected SW will change, when the angle α_1 is increased. According to (3.44) as α_1 increases, the pressure $\Delta p_{2\alpha}$ will decrease. With the close of α_1 to α^*, the ratio $\Delta p_{2\alpha}/\Delta p_1$ starts to increase rapidly (reflection becomes irregular, and the formula (3.44) is no longer fair). At $\alpha_1 = \alpha^*$, there is a head SW. At a further increase of α_1, the growth $\Delta p_{2\alpha}/\Delta p_1$soon stops. The value of this ratio for the head SW begins to decrease and approaches a certain constant value. The curves of dependence $\Delta p_{2\alpha}/\Delta p_1$on the angle α_1 for different Δp_1 values in the regular reflection zone are shown in Figure 3.20. In the zone of irregular reflection, the pressure at the front of head SW Δp_2 can be determined by the formula

$$\Delta p_2 = 1.06\frac{\sqrt[3]{m}}{r_{TP}} + 4.3\frac{\sqrt[3]{m^2}}{r_{TP}^2} + 14.0\frac{m}{r_{TP}^3}, \quad (3.46)$$

where $r_{TP} = \sqrt{r^2 + (H-z)^2}$ is the distance from the explosion point to the TP (the TP height can be found from (3.45).

At $z > H$, we can consider that the parameters of the head SW are close to those SW of the ground explosion. Therefore, the pressure Δp_2 can be determined by the formula (3.46), where instead of r_{TP} the distance from the explosion epicenter to the calculated point is substituted.

At present, there are no accurate solutions to determine the value of specific impulse in reflected SW and head SW, occurring when air SW reflection from a

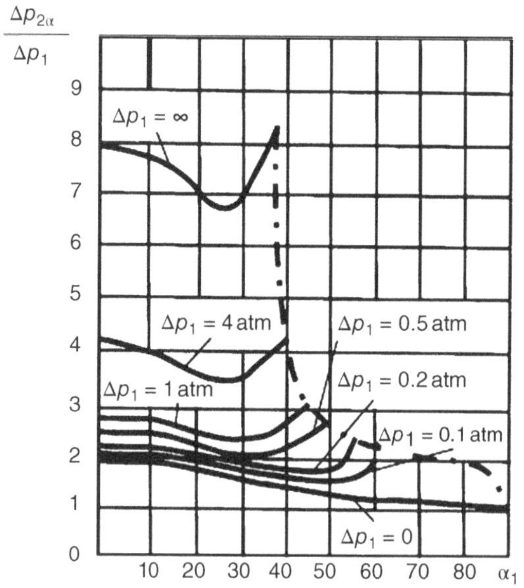

Figure 3.20 The ratio of the pressures in the reflected SW and falling SW versus the angle of incidence in the zone of regular reflection at various values of the overpressure in the falling SW. *Source:* From Miropolsky et al. [3].

rigid surface. Therefore, various approximation methods are used. In particular, in the zone of regular reflection, the impulse in reflected SW by analogy with the formula (3.40) can be calculated using the following formula:

$$I_{2\alpha} = I_1 + I_{1D} \cos^2\alpha_1, \tag{3.47}$$

where $I_{1D} = \int_0^{\tau^+} \rho(t)u^2(t)dt$ is the "dynamic" component of the full impulse of the reflected SW resulting from braking the air flow following the SW on the barrier. In particular, at $\alpha_1 = 0$:

$$I_{20} = I_1 + I_{1D}. \tag{3.48}$$

Experimental research carried out by M.A. Sadovsky allowed him to propose an approximation formula for I_{20} at $r < r_{cr}$:

$$I_{20} = 240\frac{m}{r^2}\,(\text{Pa·s}). \tag{3.49}$$

Here, like earlier, the mass of the charge is substituted in kilograms, the distance in meters. This formula is valid for an air explosion of TNT charges ($Q_s = 4.1$ MJ/kg).

Besides, based on the processing of experimental data, M.A. Sadovsky has obtained the formula for the value of I_{20} at the ground explosion of TNT charges at $r > r_{cr}$:

$$I_{20}^{(g)} = 580\frac{m^{2/3}}{r}. \tag{3.50}$$

The approximation for $I_{20}^{(g)}$ at $r < r_{cr}$ is easily obtained from (3.49) by substituting a double mass of $2m$ instead of mass m:

$$I_{20}^{(g)} = 480\frac{m}{r^2}. \tag{3.51}$$

Using the dependence (3.50) for the half value of the mass of charge, as well as the formula (3.32), it is possible to obtain formulas for I_{20} and I_{1D} for air explosion of TNT at $r > r_{cr}$:

$$I_{20} = 365\frac{m^{2/3}}{r}. \tag{3.52}$$

$$I_{1D} = I_{20} - I_1 = 239\frac{m^{2/3}}{r}. \tag{3.53}$$

Thus, in the zone of regular reflection of SW of an air explosion of a TNT charge at $\sqrt{H^2 + r^2} > r_{cr}$ (Figure 3.19), specific impulse can be calculated using (3.47), (3.32), and (3.53):

$$I_{2\alpha} = \left(126 + 239\cos^2\alpha\right)\frac{m^{2/3}}{\sqrt{H^2 + r^2}}. \tag{3.54}$$

Investigations show that the described methodology is in good agreement with the results of experiments for sufficiently strong SWs. At the same time, for weak SWs, it may give reduced results.

In the zone of irregular reflection, the head SW-specific impulse value is calculated by analogy with the method of Δp_2 determination by the following formula:

$$I_2 = 200\frac{m^{2/3}}{r_{TP}}. \tag{3.55}$$

Formula (3.55) is valid at $r_{TP} > r_{cr}$. Finally, in the remote zone of the explosion, at $z > H$, SW is close to SW of the ground explosion, therefore

$$I_2 = 200\frac{m^{2/3}}{r}. \tag{3.56}$$

In the combat use of ammunition, there are cases of its explosion in a closed volume: the explosion of a projectile inside one of the aircraft sections after piercing the shell, the explosion of an air bomb inside a building or a ship; detonation of a cloud of the fuel–air mixture flowing inside a structure, etc. In this case, the explosion shock wave is repeatedly reflected from the walls of the volume, resulting in the impact of the explosion on the elements of the structure much greater than in an explosion of the same charge outside of the closed volume at the same distance. Each element within a closed volume is affected by a series of SW, the total impulse of which is 5–6 times greater than the impulse of SW explosion in an unlimited space. As a result, the destructive effect of an explosion in a closed volume in some cases significantly exceeds that of an explosion in an open environment.

3.3.2 Flow Around Barriers

If a barrier has limited dimensions (comparable with the thickness of the air layer compressed with SW), the picture of the interaction of SW with the barrier and the behavior of changes in the load acting on the barrier will be somewhat different. In this case, the SW flows around the barrier. Let us consider the process of flow when the SW meets the barrier on the normal. The stages of this process are shown in Figure 3.21. When the SW meets such a barrier on the whole area of the barrier, reflected SW appears, the pressure in the front of which is determined by the

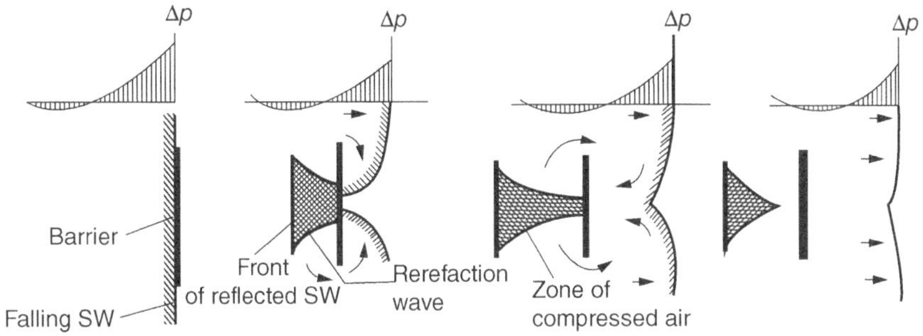

Figure 3.21 Sequential stages of the shock wave flow around the barrier. *Source*: From Miropolsky et al. [3].

formula (3.40). Falling SW, continuing its movement, will pass by the barrier. In this case, the SW front near the edge of the barrier is warped, air compressed by SW moves toward low pressure (behind the barrier the pressure is equal to atmospheric pressure). The velocity of curved SW at the whole length of its front remains the same.

After some time, shock waves coming from the edges will meet. Since air moves behind these shock waves, when they meet, the pressure behind the barrier increases significantly. It is obvious that the pressure acting on the rear surface of the barrier is distributed unevenly over the area of the barrier (as it approaches the edges, the pressure drops). Later, after the rarefaction zone of the passing shock wave approaches the edges of the barrier, the pressure behind the barrier will begin to decrease and will soon become equal to atmospheric.

The front of the passing wave after reaches the distance from the barrier equal to two or three minimum sizes of the barrier (length, width) is aligned and takes its original form.

Let us now consider the processes occurring on the face of the barrier (Figure 3.21). Since the reflected SW generated when contacting the barrier has a higher pressure than the passing SW, from the edges of the barrier on the mass of compressed air rarefaction waves (RW) will begin to propagate. As a result, the air near the edges of the barrier is involved in the movement toward the edges of the barrier and, flowing around the barrier, increases the pressure acting on its rear surface. RW movement occurs at a variable velocity equal to the local speed of sound in the air compressed by reflected SW. The pressure behind the RW is decreasing. The so-called unloading takes place. As the reflected SW is going away from the barrier, the volume of compressed air gradually decreases due to the action of the rarefaction waves. If we measure the pressure acting on the barrier at any point located at a distance l from its edge, then until the arrival of RW, pressure at this point will be the same as in the case of the action of SW on the barrier of infinite dimensions. After the arrival of the RW, the pressure in this point decreases. It is obvious that the further from the edge, the point under consideration is the later the RW will reach it, and the longer the increased pressure arising from SW reflection from the barrier will remain in the point.

When determining the effectiveness of the damaging effect of the blast ammunition, the explosive load acting on the structure should be reduced. With the same mechanical strength of structures and the same area accepting the explosive load, the structure with the relative elongation of this area will be more explosion-resistant. In other words, all other things being equal, due to the phenomenon of flows around barriers, the pressure acting on tall and relatively narrow structures (factory pipes, cylindrical towers, and other structures with a well-streamlined form) will act less than the average pressure on the buildings, the height of which is approximately equal to their width.

3.4 Determination of Parameters of an Air Shock Wave During Detonation of Fuel–Air Mixtures

3.4.1 General Information About Fuel–Air Mixtures

Fuel–air mixtures (FAM) are mixtures of flammable gases or vapors of fuel with air, as well as suspensions of small liquid droplets or solid particles of flammable materials in air. In the former case, fuel assemblies are gas (homogeneous), in the latter, heterogeneous.

It is known that gaseous FAM can ignite and burn steadily only under certain conditions. In the theory of combustion, there are concepts of the so-called lower and upper concentration limits of ignition (respectively, LCLI and UCLI). When the concentration of combustible gas or vapors in a mixture with air below the LCLI or above the UCLI, there is no ignition with subsequent development of combustion. Thus, one of the conditions for a chemical reaction of the fuel is its mixing with air in a certain ratio. The values of the LCLI and UCLI for several FAMs are given in Table 3.7.

Another condition for burning of FAMs is the presence of an ignition source in the volume of the reactive substance. It should be noted that the spark energy arising on the contacts of the electric bell, light switch, refrigerator relay, and other electrical appliances at the moment of their switching on or off is usually sufficient to ignite the prepared FAM. This explains a large number of accidents related to household gas explosions.

The ignition of FAM in the local zone leads to the development of the combustion process; its initial velocity does not exceed 0.3–0.4 m/s, and the temperature in the combustion front can take values close to 2000 °C. Combustion of a relatively

Table 3.7 Concentration (volumetric %) limits of detonation and ignition of FAMs in an open atmosphere and confined volumes.

Fuel	Detonation				Ignition	
	Open atmosphere		Confined volume			
	Lower	Upper	Lower	Upper	Lower	Upper
Acetone C_3H_6O	—	—	—	—	2.2	13.0
Acetylene C_2H_2	—	—	4.2	50.0	2.0	81.0
Gasoline $C_{7.17}H_{14.0}$	—	—	~5.6	~9.4	0.79	5.16
Butane C_4H_{10}	2.5	5.2	1.98	6.18	1.9	9.1
Hydrogen H_2	—	—	18.3	58.9	4.0	75.0
Propane C_3H_8	3.0	7.0	2.57	7.37	2.1	9.5
Propylene C_3H_6	3.5	8.5	3.55	10.40	2.4	11.0
Ethane C_2H_6	4.0	9.2	2.87	12.20	2.9	15.0
Ethylene C_2H_4	—	—	3.32	14.70	2.7	36.0
Benzene C_6H_6	—	—	1.6	5.55	1.3	7.9
Xylene C_6H_{10}	—	—	—	—	1.1	6.4
Cyclohexane C_6H_{12}	—	—	—	—	0.57	7.8
Methane CH_4	—	—	—	—	5.0	15.0
Ammonia NH_3	—	—	—	—	15.5	27.0
Carbon monoxide CO	—	—	—	—	12.5	74.2
Hydrogen sulfide H_2S	—	—	—	—	4.3	45.5

Source: From Nettlenton [5].

small volume of FAM (up to several cubic meters) in free air space can have only some thermal effect on objects located in proximity, and the compression waves arising, in this case, are comparable to wind gusts.

The picture changes during the ignition of FAM inside a closed volume. This is due to changes in physical and chemical processes occurring in the FAM under these conditions. As a result of a chemical transformation during combustion, the volume of heated reaction products is several times greater than the volume of the original gas mixture, which causes the formation of compression waves propagating at the speed of sound. The compression waves are reflected from the surfaces limiting the volume of FAM, which, in turn, causes an increase in pressure of the entire gas medium. The kinetics of chemical reactions is such that the increase in pressure leads to an intensive acceleration of the combustion front, the speed of which for hydrocarbon mixtures can reach tens or even hundreds of meters per second. At the same time, all reactive mixture is operated with the effect of explosion regardless of the volume of the fuel assembly and another form of chemical transformation is realized: explosive combustion with the formation of strong compression waves (deflagration) or, in some cases, shock waves with overpressure at the front to several atmospheres.

Explosion burning may occur when FAM is ignited in an open space; however, the linear dimensions of the fuel–air cloud should be at least several tens or even hundreds of meters, and the volume should reach tens or hundreds of thousands of cubic meters.

Another form of explosive transformation of the FAM is detonation. Detonation is the main form of explosive transformation in the case of a volumetric explosion ammunition operation. The main distinctive condition of this process is the presence of a source capable of generating a shock wave with certain parameters in the prepared mixture. Such a source can be a condensed explosive charge explosion or a powerful spark discharge. In this case, the minimum energy of the source of detonation initiation is many times greater than the minimum energy required to cause combustion. If the minimum ignition energy is usually about $0.3 \cdot 10^{-3}$ J, the value of the minimum mass of TNT charge, the explosion of which can cause a detonation in the fuel assembly is in the range from hundreds of grams to several kilograms.

The Institute of Hydrodynamics of the Siberian Branch of the Russian Academy of Sciences has proposed an original method of detonation initiation in fuel–air mixtures using a shaped charge operating in the "reverse" cumulation mode [6]. This provides a throwing velocity of about 4 km/s, and the diameter and mass of the thrown body are comparable with the initial diameter and mass of the shaped charge liner.

Parameter values at the detonation wave front in FAMs are significantly higher than those at the combustion front. Thus, the detonation velocity in hydrocarbon-based mixtures is 1500–1800 m/s, the pressure is 15–20 atm, and the mass velocity of the gas flow directed in the direction of the wave motion is 600–800 m/s. Detonation wave with the specified parameters results in the subsequent propagation of the air shock wave outside the FAM cloud. This is the reason for the significant blast effect of such explosions not only within the volume of the mixture but also at significant distances outside it. These features were the reason why many countries have developed and adopted high-efficiency ammunitions based on fuel–air detonating mixtures.

It should be noted that, as in the case of ignition, there are also lower and upper concentration limits within which the mixture is capable of detonating stably. The values of these limits for several FAMs are given in Table 3.7. The concentration limits of detonation of FAMs, as a rule, are narrower than the concentration limits of ignition.

Regardless of the form of transformation, the greatest energy of the chemical reaction is released at such a ratio of fuel and oxidizer, when all carbon and hydrogen atoms are able to fully oxidize to CO_2 and H_2O, while there is no oxygen in the mixture. Such fuel assemblies are called stoichiometric.

The formal formula for air represented as a mixture of oxygen and nitrogen in the volume ratio 21:79 with the molar mass in $\mu_a = 29$, is $O_{0.43}N_{1.58}$. Then the chemical reaction of the transformation of a stoichiometric gaseous mixture of hydrocarbons with air can be written as

$$C_aH_b + dO_{0.43}N_{1.58} = aCO_2 + 0.5bH_2O + 0.79dN_2, \tag{3.57}$$

where a and b are, respectively, the number of carbon and hydrogen atoms in a fuel molecule or in the formal chemical formula of a hydrocarbon mixture; $d = (2a + 0.5b)/0.43$ is a coefficient [7].

According to (3.57) relative mass concentration of hydrocarbon fuel in the stoichiometric gas mixture with air,

$$\delta_{st} = \frac{\mu_f}{\left(\mu_f + d\mu_a\right)}, \tag{3.58}$$

where $\mu_f = 12a + b$ is the molar mass of the hydrocarbon fuel or mixture of hydrocarbons.

The volumetric concentration of the fuel in the fuel mixture Δ_{st} is related to the mass ratio

$$\Delta_{st} = \delta_{st}/\left(\delta_{st} + (1 - \delta_{st})\mu_f/\mu_a\right). \tag{3.59}$$

Example 3.1.

For a mixture of propane (C_3H_8) with air according to formulas (3.57)–(3.59) we will have:

$$d = \frac{(2{\cdot}3 + 0.5{\cdot}8)}{0.43} = 23.26;$$

$$\mu_f = 12{\cdot}3 + 8 = 44;$$

$$\delta_{st} = \frac{44}{(44 + 23.26{\cdot}29)} = 0.0612;$$

$$\Delta_{st} = \frac{0.0612}{(0.0612 + (1 - 0.0612){\cdot}44/29)} = 0.0412.$$

Thus, the stoichiometric concentrations of propane are 6.12% by mass and 4.12% by volume.

Calculated values for some flammable gases and vapors are presented in Table 3.8. Formulas (3.57)–(3.59) can be used for calculations of any hydrocarbon fuels or their mixtures.

Table 3.8 Properties of some combustible substances and their mixtures.

Fuel	Molar mass μ_f, g/mol	Mass stoichiometric concentration δ_{st}, %	Specific heat of combustion $Q_s \cdot 10^{-3}$, kcal/kg
Methane CH_4	16	5.60	11.94
Ethane C_2H_6	30	5.97	11.36
Propane C_3H_8	44	6.12	11.09
Butane C_4H_{10}	58	6.20	10.94
Acetylene C_2H_2	26	7.16	11.53
Gasoline $C_{7.17}H_{14.0}$	100	6.50	10.37
Acetone C_3H_6O	58	9.71	7.37

Source: From Zarubin et al. [7].

If the fuel is a mixture of several gases, it is necessary first of all to determine its nominal chemical formula, i.e. to calculate the coefficients a and b included in the expression (3.57) according to the relations

$$a = \sum_{i=1}^{N} \alpha_i a_i; \quad b = \sum_{i=1}^{N} \alpha_i b_i, \tag{3.60}$$

where N is the number of components in the mixture; a_i, b_i are, respectively, the number of carbon and hydrogen atoms in the molecule of the i-th component; α_i is the molar concentration of the i-th component in the combustible mixture.

As a rule, the mass concentrations of individual components in the initial combustible mixture are known. The molar concentrations required to calculate the coefficients a and b can be determined by the expression

$$\alpha_i = \frac{\delta_i}{\mu_{fi} \sum_{i=1}^{N} \delta_i / \mu_{fi}}, \tag{3.61}$$

where μ_{fi} and δ_i are, respectively, the molar mass and mass concentration of the i-th combustible component.

It should be noted that the sum of the molar (as well as mass) concentrations of the mixture components is always equal to 1.

Example 3.2.
Let's give calculations by formulas (3.60) and (3.61) for a combustible mixture consisting of 30% methane (CH_4) and 70% butane (C_4H_{10}) by mass, and $N = 2$.

$$\text{For methane}: \quad i = 1; \quad \delta_1 = 0.3; \quad a_1 = 1; \quad b_1 = 4; \quad \mu_{f1} = 12 + 4 = 16;$$

$$\alpha_1 = \delta_1/\mu_{f1}/\left(\delta_1/\mu_{f1} + \delta_2/\mu_{f2}\right) = 0.3/16/(0.3/16 + 0.7/58) = 0.608.$$

$$\text{For butane}: i = 2; \quad \delta_2 = 0.7; \quad a_2 = 4; \quad b_2 = 10; \quad \mu_{f2} = 12{\cdot}4 + 10 = 58;$$

$$\alpha_2 = \delta_2/\mu_{f2}/\left(\delta_1/\mu_{f1} + \delta_2/\mu_{f2}\right) = 0.7/58/(0.3/16 + 0.7/58) = 0.392.$$

$$\text{For the mixture}: \quad a = \alpha_1 a_1 + \alpha_2 a_2 = 0.608 + 0.392{\cdot}4 = 2.18;$$

$$b = \alpha_1 b_1 + \alpha_2 b_2 = 0.608{\cdot}4 + 0.392{\cdot}10 = 6.35.$$

Thus, the conditional chemical formula of this mixture of hydrocarbons is $C_{2.18}H_{6.35}$. Its average molar mass is $\mu_f = 12 \cdot 2.18 + 1 \cdot 6.35 = 32.51$.

Now by formulas (3.57)–(3.59), one can calculate the values of mass and volume stoichiometric ratios of this fuel mixture with air, similarly as it was done for propane.

3.4.2 Parameters of an Explosion of Fuel–Air Mixtures in the Detonation Mode

As noted earlier, fuel–air mixtures can explode in deflagration (explosive combustion) or detonation modes [7]. Here, we will consider calculations of fuel–air mixture explosion parameters only in the detonation mode since this mode is the most characteristic of volumetric explosion ammunition. Readers interested in calculating the parameters of fuel mixture deflagration mode are referred to the sources given at the end of this chapter.

3.4.2.1 Parameters of FAM Detonation Inside the Cloud

Parameters at the detonation wave front inside the FAM cloud are fully determined by the specific heat of explosion of the mixture Q_m, J/kg, the adiabatic index of explosion products k_{EP}, the initial pressure p_0, Pa, and the mixture density ρ_m, kg/m^3

$$D = \sqrt{2\left(k_{EP}^2 - 1\right)Q_m};$$

$$\Delta p_D = p_D - p_0 = \frac{\rho_m D^2}{(k_{EP}+1)} = 2(k_{EP}-1)\rho_m Q_m; \tag{3.62}$$

$$u_D = \frac{D}{(k_{EP}+1)};\ c_D = \frac{k_{EP} D}{(k_{EP}+1)},$$

where D is the velocity of propagation of the detonation wave front (detonation velocity), (m/s); p_D, u_D, c_D are pressure, (Pa), mass velocity, (m/s), and speed of sound, (m/s), in the detonation products behind the detonation wave front.

The formulas (3.62) obtained for a strong detonation wave are valid with an error of no more than 5% for pressure at the front $p_D > 10p_0$.

The density of the mixture is related to the density of the surrounding air by the relation

$$\rho_m = \rho_a \frac{1}{1-\delta\left(1-\mu_a/\mu_f\right)}, \tag{3.63}$$

where $\delta = M_f/(M_a + M_f)$ is the relative mass concentration of the fuel; M_a and M_f are the masses of air and fuel in the mixture, respectively.

The specific heat of explosion Q_m depends on the mixture composition and can be calculated by the specific heat of explosion per unit mass of fuel Q_s, if the assumption is correct that in poor FAMs the whole mass of fuel reacts, and in rich ones only a part corresponding to the presence of oxygen in the mixture does. In this case, it is not difficult to obtain

$$Q_m = \begin{cases} \delta Q_s \text{ if } \delta \le \delta_{st}, \\ \delta Q_s(1/\delta - 1)/(1/\delta_{st} - 1) \text{ if } \delta > \delta_{st}, \end{cases} \tag{3.64}$$

where δ_{st} is the mass stoichiometric concentration.

The specific heat of explosion per unit mass of fuel is determined by thermodynamic calculations. In addition, as Q_s it is possible to take the standard heat of combustion of fuel (see Table 3.8).

With the known explosion heat and detonation velocity of the FAM, the adiabatic index of the explosion products is calculated by the formula that is a consequence of the first relation from (3.62):

$$k_{EP} = \sqrt{\frac{1+0.5D^2}{Q_m}}. \tag{3.65}$$

Analysis of the results of thermodynamic calculations shows that the adiabatic index of explosion products k_{EP} takes a minimum value during detonation of fuel mixtures with stoichiometric composition and, in first approximation, in the area of the concentration limits of the explosiveness of the mixture can be described by a linear dependence

$$k_{EP} = \begin{cases} k_{st} - (k_a - k_{st})(\delta - \delta_{st})/\delta_{st} \text{ if } \delta \leq \delta_{st}, \\ k_{st} + (k_f - k_{st})(\delta - \delta_{st})/(1 - \delta_{st}) \text{ if } \delta > \delta_{st}, \end{cases} \tag{3.66}$$

where $k_a = 1.4$ – adiabatic index of the air; k_f – calculation constant.

For hydrocarbon gaseous and liquid fuels, the values of k_{st} and k_f change insignificantly ($k_{st} = 1.23–1.25$; $k_f = 1.5–1.7$) and can be taken equal to the average values of $k_{st} = 1.24$; $k_f = 1.6$.

With a point source of initiation (explosion of a condensed explosive charge, powerful spark discharge, etc.), the detonation wave formed has spherical symmetry (in fuel mixtures with constant fuel concentration by volume) and the parameters in the explosion products behind its front are described by a self-similar solution. In particular, for the mass and sound velocities in the products of the explosion, the relations are valid:

$$u = \begin{cases} 0 \text{ when } \lambda < \lambda_q; \\ u_D \left\{ 1 - \left[1 - \left(\frac{\lambda - \lambda_q}{1 - \lambda_q} \right)^{1.1} \right]^{0.5} \right\} \quad \text{when } \lambda_q \leq \lambda \leq 1; \end{cases} \tag{3.67}$$

$$c = \begin{cases} c_q \text{ when } \lambda < \lambda_q; \\ c_q + (c_D - c_q) \left\{ 1 - \left[1 - \left(\frac{\lambda - \lambda_q}{1 - \lambda_q} \right)^{1.1} \right]^{0.5} \right\} \quad \text{when } \lambda_q \leq \lambda \leq 1, \end{cases} \tag{3.68}$$

where $\lambda = r/Dt$ is the relative coordinate behind the detonation wave front (r is the distance from the initiation point; t is the time since the initiation began), changing from zero at the center of symmetry to one at the detonation wave front; λ_q, c_q are the coordinates of the central boundary of the quiescent area and the sound speed in it.

The last two values are determined by the dependence

$$\lambda_q = \frac{c_q}{D} = \frac{1}{D}\left(c_D - 1.174 \frac{k_{EP} - 1}{2} u_D \right), \tag{3.69}$$

which, taking into account (3.62), can be rewritten as

$$\lambda_q = \frac{c_q}{D} = \frac{k_{EP}}{k_{EP} + 1} - 0.587 \frac{k_{EP} - 1}{k_{EP} + 1}.$$

For mixtures of hydrocarbon fuels with air ($k_{EP} \cong 1.24$) from the last ratio follows $\lambda_q \cong 0.49 \approx 0.5$; $c_q \approx 0.5D$.

The pressure in the explosion products behind the detonation wave front is calculated through the speed of sound (3.68) according to the isentropic relation

$$p = p_D \left(\frac{c}{c_D} \right)^{2k_{EP}/(k_{EP} - 1)}. \tag{3.70}$$

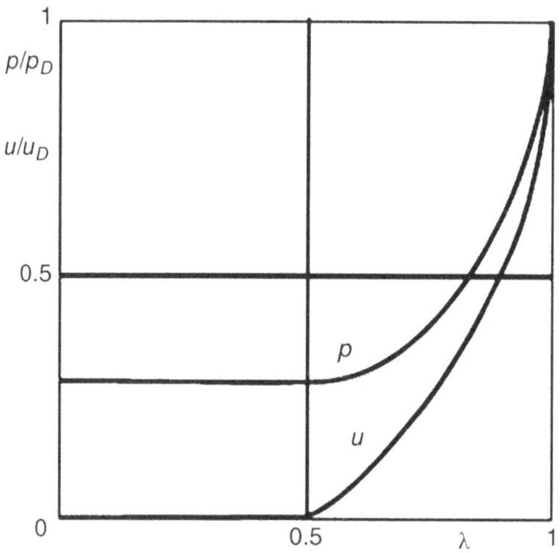

Figure 3.22 Distribution of pressure (p) and mass velocity (u) in detonation products behind the detonation wave front in a stoichiometric hydrocarbon FAM. *Source*: From Zarubin et al. [7].

The distributions of pressure p and mass velocity u in the explosion products behind the detonation wave front in a stoichiometric hydrocarbon fuel–air mixture, plotted using formulas (3.67)–(3.70), are shown in Figure 3.22. Assuming that $r = \text{const}$, $t = \text{var}$, it is possible to plot the time dependences of p and u at the location of the object inside the region of fuel–air mixture, where it is not influenced by the rarefaction wave coming from the cloud boundary.

3.4.2.2 Parameters of a Detonation Explosion at the Boundary of the FAM Cloud

The results of mathematical modeling of the spherical explosion of gas and heterogeneous FAMs show that there are two characteristic areas: inside and outside the cloud. The parameters in them depend on their respective characteristics and change in different ways. In this case, the parameters at the boundary of the cloud, which relate to the explosion field of both regions, are of particular importance.

The initial overpressure of the air shock wave at the cloud boundary Δp_i can be found from the solution of the decay of an arbitrary discontinuity problem. In consideration of the relatively low intensity of the detonation wave in the FAMs, it is not difficult to obtain an approximate analytical solution:

$$\Delta p_i \cong 1.025 \Delta p_D \sqrt[3]{\frac{\rho_a}{(k_{EP} + 1)\rho_m}}, \tag{3.71}$$

where ρ_a, ρ_m is the density of air and mixture, respectively, kg/m^3.

Since ρ_m differs little from ρ_a, for stoichiometric hydrocarbon FAMs, ($k_{EP} \cong 1.24$) ratio (3.71) gives $\Delta p_i \cong 0.78 \Delta p_D$.

For the specific overpressure impulse at the boundary of the FAMs cloud

$$I_0 = \int_0^{\tau^+} \Delta p dt \approx \tau^+ p_i,$$

where τ^+ is the duration of the compression phase in the wave, which for spherical symmetry is approximately equal to $3.3r_0/D$ (r_0 is the cloud radius).

According to the results of the analysis of numerical results, we obtained

$$I_0 = 0.1472\tau^+ \Delta p_i = \frac{0.486 r_0 \Delta p_i}{D}. \tag{3.72}$$

For stoichiometric hydrocarbon fuel mixtures, (3.72) gives

$$I_0 \cong \frac{0.38 r_0 \Delta p_D}{D}.$$

The maximum overpressure Δp_1 inside the FAM cloud coincides with the overpressure at the detonation wave front Δp_D and is calculated by the corresponding relation of (3.62).

For the specific impulse of overpressure around the point of initiation, taking into account the self-similar solution for the spherical detonation wave, we can obtain a functional dependence

$$I = \frac{r_0 \Delta p_D}{D}\left(a - b\frac{r}{r_0}\right),$$

where a and b are constants, which according to the results of numerical calculations are 0.96 and 0.47, respectively. Consequently,

$$I = \frac{r_0 \Delta p_D}{D}\left(0.96 - 0.47\frac{r}{r_0}\right). \tag{3.73}$$

In the vicinity of the cloud boundary, due to the rapid arrival of the rarefaction wave, the value of the specific impulse of the compression phase can deviate appreciably from the dependence (3.73). Generalizing this formula for the entire region inside the FAM, taking into account the numerical results, we can write

$$I = I(r) + [I_0 - I(r_0)]\left(\frac{r}{r_0}\right)^4,$$

where $I(r)$ and $I(r_0)$ are the impulse values calculated by (3.73) for the considered point and the cloud boundary r_0.

With the help of (3.72) and (3.73), the last relation is converted to the form

$$I = \frac{r_0 \Delta p_D}{D}\left(0.96 - 0.47\frac{r}{r_0}\right) + \left(0.486\frac{r_0 \Delta p_i}{D} - 0.49\frac{r_0 \Delta p_D}{D}\right)\left(\frac{r}{r_0}\right)^4. \quad (3.74)$$

In the case of stoichiometric hydrocarbon FAMs, formula (3.74) is simplified and takes the form

$$I = \frac{r_0 \Delta p_D}{D}\left[0.96 - 0.47\frac{r}{r_0} - 0.11\left(\frac{r}{r_0}\right)^4\right].$$

3.4.2.3 Parameters of the Air Shock Wave During FAM Detonation

The main parameters determining the propagation of explosion waves are usually taken as the energy of the explosion E, as well as the pressure p_0 and density ρ_a of the surrounding air. However, it is physically more reasonable to choose not the energy of the explosion, but the energy going into the shock wave E_{SW}.

Under the explosion of condensed explosives, the energy of the shock wave is of the order of $0.9E$ and changes weakly when changing from one explosive to another.

During FAM detonation, the work of the expanding explosion products over the surrounding atmosphere is about half of the released energy E and can vary appreciably depending on the mixture composition and the parameters of the surrounding air. Therefore, the shock wave energy E_{SW} was chosen as the main defining parameter of shock waves during FAM detonation, which can be calculated by the formula

$$E_{SW} = \eta\, \xi\, E_0, \quad (3.75)$$

where $E_0 = M_f Q_s$ is the total energy that can be released during the reaction of the entire mass of fuel; ξ is the coefficient of completeness of the reaction of fuel; and η is the explosion efficiency, i.e. the ratio of energy escaping into the shock wave to that released during the explosion.

In the case of an ideal detonation, i.e. the release of all the energy in the wave front, according to (3.64), the coefficient of completeness of the fuel reaction for FAMs with a constant concentration by volume

$$\xi = \begin{cases} 1 \text{ when } \delta \le \delta_{st}, \\ \dfrac{1/\delta - 1}{1/\delta_{st} - 1} \text{ when } \delta > \delta_{st}. \end{cases} \quad (3.76)$$

For typical hydrocarbon FAMs at the upper concentration limit of explosiveness, the combustible reaction completeness factor ξ is about 0.25; in poorly mixed mixtures with regions of combustible concentration beyond the explosive limit – 0.02–0.10.

The efficiency of a detonation explosion of FAMs can be calculated quite accurately by the formula

$$\eta = 1 - \left(\frac{p_0}{p_D}\right)^{(k_{EP}-1)/k_{EP}}, \tag{3.77}$$

which follows from the model of equilibrium expansion of explosion products from the detonation pressure to the pressure of the surrounding atmosphere.

For stoichiometric hydrocarbon fuel mixtures with pressure at the detonation wave front $p_D = (15\text{–}20)p_0$, the explosion efficiency varies in the range $\eta = 0.41\text{–}0.44$.

The maximum overpressure at a point at distance r, (m), from the center of the detonation explosion of a spherical cloud of FAMs corresponds to the pressure at the shock wave front in air and can be calculated by the approximation dependence

$$\Delta p_1 = p_0\left[0.249\left(\frac{\sqrt[3]{E_{SW}/p_0}}{r}\right) + 0.393\left(\frac{\sqrt[3]{E_{SW}/p_0}}{r}\right)^2 + A\exp\left(-\frac{r-r_0}{0.15r_0}\right)\right] \text{ (Pa)}. \tag{3.78}$$

The third term in the square bracket (3.78) describes a very sharp pressure drop at the air shock front near the FAM cloud and noticeably affects the calculation results only at $r < 1.5r_0$. At larger distances, it can be neglected. The constant A is determined from (3.78) by the known initial pressure of the shock wave (3.71), provided that $r = r_0$.

The specific overpressure impulse I of a shock wave at distance r, (m) from the center of the explosion is calculated by the formula

$$I = 0.058\sqrt{p_0\rho_a}\sqrt[3]{\frac{E_{SW}}{p_0}}\frac{\sqrt[3]{E_{SW}/p_0}}{r}\left[B\left(\frac{r_0}{r}\right)^3 - \left(\frac{r_0}{r}\right)^2 + 1\right] \text{ (Pa·s)}. \tag{3.79}$$

The expression in the square bracket (3.79) describes the nonmonotonic change of I in the vicinity of the FAM cloud and becomes practically equal to one at $r > 5r_0$. The constant B is calculated from the known value of the specific impulse at the boundary of the FAM cloud (3.72) from (3.79) at $r = r_0$.

For hydrocarbon FAMs at any distance ($r > r_0$), the value of the expression in square brackets differs from one by no more than 10%. Therefore, with the mentioned error, to estimate the specific impulse of the air shock wave, it is possible to use the dependence

$$I \approx 0.058\sqrt{p_0\rho_a}\sqrt[3]{\frac{E_{SW}}{p_0}\frac{\sqrt[3]{E_{SW}/p_0}}{r}}\ (\text{Pa·s}). \tag{3.80}$$

Writing expressions for determining the maximum overpressure and the specific impulse (3.78)–(3.80) using p_0 makes it possible to consider the influence of the surrounding atmosphere pressure (in explosions at different heights above sea level), and the introduction of ρ_a into the formulas for calculating the specific impulse makes it possible to also take into account the temperature and humidity of the air.

In the case of the formation of a FAM cloud directly near the ground surface, at a first approximation, it can be represented as a hemispherical volume. In this case, in dependencies (3.78)–(3.80), the doubled energy of the shock wave calculated by relation (3.75) should be substituted, and the cloud radius r_0 should be calculated as for a hemisphere.

3.5 Evaluation of the Damaging Effect of Shock Waves on Various Objects

3.5.1 Criteria of the Damaging Effect of Shock Waves

The results of the action of shock waves on various objects are determined mainly by the maximum overpressure Δp_m and the specific impulse of the compression phase I in the wave [7]. For a shock wave where length of the compression zone is much greater than the characteristic dimensions of the target object, the loading has a "quasi-static" behavior. For a very short wave, the response of objects to loading is determined by the specific impulse of the compression phase ("impulse" loading).

The choice of the kind of loading when evaluating the damaging effect of shock waves is related to the ratio of the duration of the compression phase in the wave τ^+ and the natural oscillation period of the object T. If the duration of the compression phase $\tau^+ \leq 0.25T$, then the load can be considered impulsive and the condition of the destruction of the object will be written in the form of the criterion for the specific impulse:

$$I \geq I_d. \tag{3.81}$$

At $\tau^+ \geq 10T$, the loading becomes quasi-static and the destruction criterion is written through overpressure

$$\Delta p_m \geq \Delta p_d. \tag{3.82}$$

The relations (3.81) and (3.82) are called partial criteria for damage caused by shock wave action, and the values I_d, Δp_d are the critical values of the specific impulse and maximum overpressure in the wave.

In the range $0.25T < \tau^+ < 10T$, both the maximum overpressure and the specific impulse in the wave affect the behavior of the object. In this case, as shown by experiments and theoretical solutions of model problems, the joint effect of overpressure and the specific impulse in the wave can be accounted for by the generalized criterion of damage, which is written in the form

$$(\Delta p_m - \Delta p_d)(I - I_d) = K, \tag{3.83}$$

where K is a certain constant; I_d, Δp_d are critical wave parameters.

The function (3.83) in the plane Δp_m I has the form of a hyperbola with two asymptotes and corresponds to the limiting values of maximum overpressure and the specific impulse of the compression phase in the wave, ensuring the infliction of a certain kind of damage on the object (Figure 3.23). If the shock wave parameters Δp_m, and I correspond to the point lying above the curve of the generalized criterion (3.83), then the object will be inflicted a given type of damage.

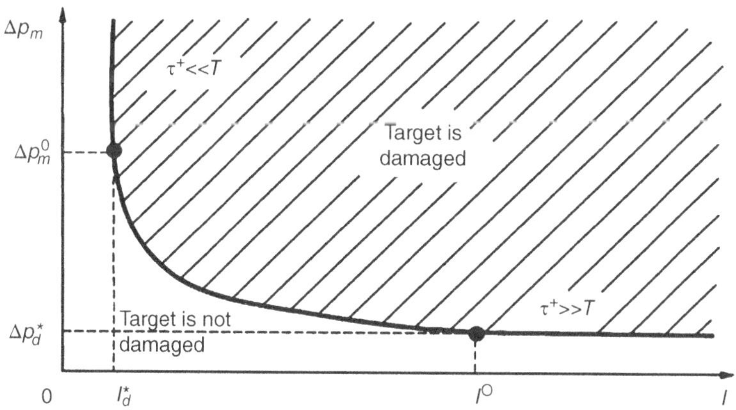

Figure 3.23 Geometric representation of the criterion for the destruction of a target by the blast effect in the phase plane $\Delta p_m - I$. *Source*: From Miropolsky et al. [3].

From the generalized criterion, it is not difficult to obtain partial criteria for the shock wave action.

Thus, when $\Delta p_d = K = 0$ from (3.83) follows the criterion for the specific impulse (3.81), and when $I_d = K = 0$ – for the maximum overpressure (3.82). The critical parameters Δp_d, I_d, and K constant (3.83) are usually determined experimentally for each object and its type of damage.

3.5.2 Characteristics of Target Vulnerability to Blast Effects

3.5.2.1 Parameters of the Destruction of Buildings and Other Objects

In many buildings, windows collapse at the weakest explosive loads. Glass is a brittle material that crashes as soon as the stress in it reaches its elastic limit. Windows usually have small horizontal dimensions between supports and are therefore the first to respond to blast loads, and therefore, most often collapse under quasi-static loading when subjected to blast waves. For such cases, the application of a partial criterion for maximum overpressure is justified.

The critical overpressure at which windows of a building begin to break down is related to the surface area of the window pane S, (m^2), its thickness h, (mm), and the ratio of the maximum and minimum dimensions L/B as follows [7]:

$$\Delta p_d = \frac{10^{(0.175h\,+\,0.0634L/B-0.1939)}}{S}\ (\text{kPa}). \tag{3.84}$$

When a long shock wave slides along the glass surface, the value calculated by (3.84) is compared with $\Delta p_m = \Delta p_1$, and when the wave falls normally, with the overpressure of normal reflection $\Delta p_m = \Delta p_2$.

For buildings and other objects, there are four degrees of damage: complete, strong, medium, and weak. In the case of complete damage, most of the walls, columns, and ceilings are destroyed. Strong is characterized by partial damage to walls, columns, and ceilings; light elements (doors, partitions, roofs) collapse completely or partially. Medium is characterized by deformation (sagging) of the main enclosing and supporting structures, and mainly secondary structures are destroyed. Weak is characterized by damage or serious deformations of some light elements of the enclosure (windows, doors, roofs).

The strength calculations and the evaluation of the level of target damage depend on the most important characteristic, such as the natural oscillation period of target T. The literature contains extensive data on the testing of structures on the effects of stimulated or free oscillations with small amplitudes, the purpose of which is to determine the first few forms of natural oscillations and the corresponding damping coefficients. Periods of natural oscillations of buildings with

Table 3.9 Periods of natural oscillations of buildings and structures.

Building type	Number of floors	Transverse *T*, s	Longitudinal *T*, s
Residential with load-bearing stone walls	3	0.15	0.16
	5	0.26	0.22
	6	0.37–0.39	—
	8	0.43	0.43
School with load-bearing brick walls	3	0.22	0.21
	4	0.30	0.27
	5	0.32	—
Residential large-panel	4	0.16	0.15
	5	0.30	0.22
	6	0.36	—
	9	0.40	0.32
Residential prefabricated frame-panel	14	0.86	0.76
	16	1.20	0.76
Residential with lower frame and upper large-panel floors	4	0.28	0.30
	10	0.64	0.44
	18	1.14	1.05
Administrative building with brick filling:			
Reinforced concrete frame	12	0.69–0.96	0.62–0.89
Steel frame	12	1.17	1.12
Administrative frame building	22	1.10	1.16

Source: From Kochetkov et al. [8].

rigid structural scheme T, (s), as a rule, are in the range of 0.1–1.0 seconds and can be estimated by empirical formulas [7]:

$$T = \alpha n; \quad T - 0.017H; \quad T = 0.0905\frac{H}{\sqrt{B}}; \quad T - 0.3H\sqrt{Bg}, \tag{3.85}$$

where n – the number of floors; α – coefficient depending on the design of the building and the type of foundation (for the average density of soil $\alpha = 0.04$–0.09); H – building height, (m); B – width or length of the building, (m); $g = 9.81\ \text{m/s}^2$ – acceleration of gravity.

Experimental values of T for different types of buildings are given in Table 3.9.

At the duration of the shock wave $\tau^+ \geq 10T$, i.e. in the explosion of large-scale clouds of FAM, the degree of damage to objects can be estimated by the critical overpressure, the values of which are given in Table 3.10.

To estimate the combined effects of overpressure and the specific impulse of the blast wave on typical buildings and industrial structures, one can use data on bombing damage in World War II obtained in Great Britain, which for brick buildings are approximated by the equation [7]

Table 3.10 Values of overpressure of SW (Δp_m, кПа), characterizing the degree of destruction during a ground explosion of buildings, structures, machinery, and equipment.

Category of objects	Destruction		
	Strong	Medium	Weak
Industrial buildings			
Industrial buildings with heavy metal or reinforced concrete frame	40–50	30–40	20–30
Industrial buildings with light metal frame and buildings without a frame	35–45	25–35	10–20
Multi-floor reinforced concrete buildings with large glass areas	40–90	20–40	8–20
Buildings of nuclear and hydroelectric power plants	200–300	150–200	25–35
Warehouse brick buildings	30–40	20–30	10–20
Multi-floor brick buildings	20–30	12–20	8–12
Wooden houses	12–20	8–12	6–8
Destruction of glass in industrial and residential buildings	1.5–3.0	1–1.5	0.5–1.0
Industrial buildings and equipment			
Technological pipelines and auxiliary facilities at industrial objects	60–70	40–60	30–40
Reinforced concrete overpasses with a span of up to 20 m	120–150	110–120	100–110
Power transformers, openly located	50–60	30–50	10–30
Oil field derricks	—	60–100	40–60
Ground fuel storage tanks	70–90	50–70	30–50
Semi-underground field fuel and lubricant storages	60–100	40–60	20–40
Underground and bunded storage tanks for fuel and lubricants	150–200	100–150	80–100
Public utility networks			
Underground steel welded pipelines up to 350 mm in diameter	1500–2000	1000–1500	600–1000
Underground cast iron and ceramic pipelines	1000–2000	600–1000	200–600
Steel pipes, buried by 20 cm	—	250–350	—
Manholes and gate valves on utility networks	600–1000	400–600	200–400
Electric networks			
Underground cable lines	1000–1500	700–1000	500–700
High-voltage overhead lines	80–120	50–70	30–50
Low-voltage overhead lines on wooden piers	60–100	40–60	20–40
Power lines of electrified railroads	70–120	50–70	30–50

Table 3.10 (Continued)

Category of objects	Destruction		
	Strong	Medium	Weak
Means of communication			
Car radio stations	45–55	20–30	15–20
Telephone and telegraphic equipment	—	60–90	—
Radio relay lines	—	50–70	—
Permanent overhead lines	—	70–90	—
Bridges, dams, wharfs			
Metal bridges with a span of 30–45 m	200–300	100–200	50–100
The same with a span of 45–100 m	150–200	80–120	40–80
The same with a span of more than 100 m	100–200	60–100	40–60
Reinforced concrete bridges and overpasses with spans of 20 m	130–200	110–130	50–80
Dams	100–150	—	—
Aerodromes, highways, and railroads			
Airport ground facilities	35–45	25–35	15–25
Runways with a concrete surface	1500–2000	400–1500	300–400
The same with metal covering	400–700	150–400	100–150
Asphalt concrete paved highways	1500–2000	400–1500	300–400
Railroad tracks	300–500	200–300	100–200
Means of transport, engineering machines			
Rolling stock	70–90	40–70	20–40
Diesel and electric locomotives	100–150	70–100	50–70
Trucks and tank trucks	50–60	40–50	20–40
Passenger cars and buses	30–50	20–30	15–20
Crawler tractors, tractors, armored vehicles	70–80	50–70	30–40
Transport ships	80–100	60–80	30–60
Jet aircrafts	45–55	30–40	13–15
Piston airplanes, helicopters	10–13	8–10	7–8
Protective facilities			
Basements in single-story wooden houses	70–100	40–70	20–40
Basements in industrial buildings	80–110	60–80	35–60
Underground structures of shallow subway type	—	400	—

Source: From Kochetkov et al. [8].

Table 3.11 Parameters of the generalized criterion for buildings and industrial structures.

K_t	3.8	5.6	9.6	28	56
Δp_d, kPa	77.50	38.10	16.40	4.21	1.96
I_d, Pa·s	743.00	504.00	294.00	101.00	50.40
K, kPa·Pa·s	921.00	346.00	106.00	11.50	3.00

Source: From Zarubin et al. [7].

$$r = \frac{K_t m_T^{0.33}}{\left[1 + (3180/m_T)^2\right]^{0.167}}. \tag{3.86}$$

where r is the distance from the explosion place, (m); m_T – the mass of explosive charge in TNT equivalent, (kg); K_t is the coefficient (3.8 – the complete destruction of buildings; 5.6 – buildings half-destroyed; 9.6 – buildings uninhabitable; 28 – moderate destruction, damage to internal partitions of low strength; 56 – buildings suffered minor damage, more than 10% of glass is broken).

Using (3.86) for each type of damage, it is possible to construct limit diagrams in the plane $\Delta p_m - I$ having a characteristic hyperbolic form with two asymptotes Δp_d and I_d, whose values together with the constant K in the generalized criterion (3.83) are given in Table 3.11.

Table 3.12 shows the characteristics of limit diagrams for the case of damage of military equipment.

3.5.2.2 Parameters of Human Damage

Human damage is associated with trauma to various parts of the body. At the same time, individual organs may react differently to the effects of a shock wave. For example, the eardrums have a very small period of their own oscillations, so at any size of the exploding cloud of FAM, their response to the effects of the shock wave can be estimated by the criterion of maximum overpressure in the wave.

The relationship between the maximum overpressure in the wave Δp_m, kPa, and the probability of damage to human eardrums Ω is described by [7]

$$\Omega = 0.88 \cdot \lg(\Delta p_m) - 1.323 \tag{3.87}$$

By multiplying the probability (3.87) by a hundred, we obtain the percentage of damage to the eardrums in a group of people equally distant from the place of explosion. The boundary of temporary hearing loss corresponds to the pressure $\Delta p_m = 1.5–2.0$ kPa, the lower boundary of eardrum damage is 34–35 kPa, and 50% probability of eardrum rupture is about 100 kPa.

Under explosions of large-scale clouds of FAMs, a person can be considered as a single object and as a criterion of damage use the critical pressure in the SW Δp^*, the values of which are shown in Table 3.13 depending on the probability of a lethal outcome.

Table 3.12 Vulnerability characteristics of elementary ground targets under the action of air SW generated by a munition explosion.

Target	Damage degree	Δp_d, MPa	I_d, Pa·s	K, MPa·(Pa·s)
Launcher with a "Lance" missile	*A*	0.060	280	35
	B	0.045	200	35
	C	0.030	120	35
Launcher with air defense missiles "Patriot" and "Hawk"	*A*	0.070	500	42
	B	0.055	370	42
	C	0.040	240	42
AN/MPQ-33 and AN/MPQ-39 radars	*A*	0.050	450	23
	B	0.040	350	23
	C	0.030	250	23
AN/MPQ-3 radar	*A*	0.040	340	21
	B	0.030	250	21
	C	0.020	180	21
A-10A aircraft	*A*	0.060	350	40
	B	0.045	250	40
	C	0.030	150	40
AH-1G helicopter	*A*	0.020	250	20
	B	0.015	180	20
	C	0.010	120	20
Launcher with "Lars" missiles	*A*	0.055	400	30
	B	0.040	300	30
	C	0.025	200	30
Car "Unimog"	*A*	0.055	400	30
	B	0.040	300	30
	C	0.025	200	30
Special truck of van type	*A*	0.040	340	21
	B	0.030	250	21
	C	0.020	180	21
Antiaircraft missile system "Roland-2" (damage of antenna systems)	*A*	0.050	450	23
	B	0.040	350	23
	C	0.030	250	23
Armored personnel carrier M113A1	*A*	0.170	1200	220
	B	0.110	1000	220
	C	0.060	700	220

Source: From Balagansky et al. [1].

Table 3.13 Critical pressure of shockwave damage of a person.

Probability of lethal outcome	0.99	0.75	0.50	0.25	0.10
Δp^*, kPa	500	370	320	280	250

Source: From Zarubin et al. [7].

The most sensitive to the damaging action of the shock wave are the human organs that have a large difference in the density of neighboring tissues, primarily the lungs. At the same time, the orientation of the person relative to the falling wave and the surrounding objects has a significant influence. The greatest damage at minimum values of pressure and impulse of incoming SW is sustained by a person standing or lying near a flat reflecting surface along the normal to which the blast wave is traveling. However, accounting for the shape and type of the reflecting surface, the distance of the person from it, and the angle of approach of the blast wave presents considerable difficulties.

In the USA, $\Delta p_m - I$ diagrams were obtained for different degrees of human respiratory injury in the standing position with the body positioned along the front of the incoming shock wave. The studies have shown that the shockwave action results depend on the initial ambient pressure p_0, and in the case of impulse loading also on the human mass M. In this case, critical parameters of the shock wave, which are included in generalized criterion (3.83), can be found from relations:

$$\Delta p^* = p_0 \Delta \overline{p}^*; \quad I^* = p_0^{0.5} M^{0.33} \overline{I}^*. \tag{3.88}$$

The values of the listed critical parameters $\Delta \overline{p}^*$, $\overline{I}^*$ for different degrees of damage are shown in Table 3.14.

The constant K in the generalized criterion (3.83) is calculated by the relation

$$K = 1.2 p_0^{1.5} M^{0.33} \Delta \overline{p}^* \overline{I}^* \cdots \left[\text{Pa}^{1.5}\text{kg}^{0.33}\right]. \tag{3.89}$$

When a standing person is subjected to a blast wave, the body is transported as a whole, followed by a decelerating impact upon falling. This results in traumas, the degree of seriousness of which depends on several factors such as the velocity of impact, deceleration time, type of surface, and impact area.

In a decelerating impact, the head is the most vulnerable to mechanical damage. The degrees of human damage when hitting the head against a hard surface, depending on the velocity, are shown in Table 3.15.

The impact with other parts of the body is safer for a person, but it can also lead to various traumas (Table 3.16).

It is interesting that the levels of practically safe impact velocity of both the head and other parts of the body coincide. Using these data, Baker (the USA, 1975) calculated the overpressure Δp_m and the specific impulse I of the incoming shock

Table 3.14 Critical parameters of the generalized criterion of respiratory damage.

Percentage of fatal outcome	1	10	50	90	99
$\Delta \overline{p}^*$, dimensionless	1.78	2.01	2.42	2.83	3.40
$\overline{I}^*$, $\text{Pa}^{0.5} \cdot \text{s/kg}^{0.33}$	0.251	0.306	0.373	0.472	0.575

Source: From Zarubin et al. [7].

Table 3.15 Human damage from a head impact.

Category of human trauma	Relative impact velocity, m/s
Practically safe	3.05
Lower boundary, or threshold of acceptable damage	3.96
50% chance of death	5.49
Nearly 100% chance of death	7.01

Source: Adapted from Wite [9].

Table 3.16 Human damage when struck by various parts of the body.

Category of human trauma	Relative impact velocity, m/s
Practically safe	3.05
Lower boundary, or threshold of acceptable damage	6.40
50% chance of death	16.46
Nearly 100% chance of death	42.06

Source: Adapted from Wite [9].

wave, which ensures the transfer of the human body at critical velocities. The obtained results can be represented as approximate dependences:

- when hitting by the head

$$\frac{\Delta p_m I}{M^{0.33}} = A; \tag{3.90}$$

- when hitting by other parts of the body

$$\frac{\Delta p_m I}{M^{0.33}} = B, \tag{3.91}$$

where M is the human body mass, (kg); the values of parameters A and B depending on the acquired velocity are given in Table 3.17.

The values of parameters A (3.90) and B (3.91) are calculated by the value of the maximum overpressure Δp_m and the specific impulse I in the shock wave at the location of the standing person, as well as by his mass M, according to

Table 3.17 Parameters of human impact on a rigid surface.

Impact velocity, m/s	3.05	3.96	5.49	6.40	7.01	16.46	42.06
$A \cdot 10^{-5}$, $\text{Pa}^2 \cdot \text{s/kg}^{0.33}$	400	546	748	—	964	—	—
$B \cdot 10^{-5}$, $\text{Pa}^2 \cdot \text{s/kg}^{0.33}$	392	—	—	801	—	2043	5383

Source: From Zarubin et al. [7].

which the throwing velocity and the degree of damage of the person at impact are determined from Tables 3.15 to 3.17.

3.5.2.3 Determination of the Degree of Damage of Enemy Personnel

A slightly different approach is taken when estimating the degree to which enemy troops are damaged by high-explosive munitions. In this case, the data is used to estimate the enemy's *A*, *B*, or *C* damage type.

The main parameters of the explosion, which determine the possible seriousness of damage to enemy troops, are the air velocity head, overpressure, and the duration of the compression phase. The characteristics and seriousness of the damage also depend on the position of the personnel relatively the explosion, the kind of explosion, and other conditions. Considering the relationship between the parameters of the explosion and the different position of the personnel, for approximate calculations, we can assume that the seriousness of the damage is characterized by critical values of overpressure in EP and air SW Δp^*, the values of which correspond to a certain probability of this degree of damage are given in Table 3.18. The data are for the duration of the compression phase of the order of 10^{-2} s.

Increasing the duration of the compression phase to about 10^{-1} second leads to a significant decrease in the overpressure required to damage the personnel. Reducing the duration of the compression phase to 10^{-3} second requires increasing the overpressure to achieve the same degree of damage to the personnel, and when the duration of the compression phase SW is less than $\cdot 2 \cdot 10^{-3}$ second, the damage to the personnel is determined by the value of the specific impulse of the SW. The fatal outcome, in this case, occurs at values of $I^* \geq 900$ Pa · s.

Figure 3.24 presents a generalization of the results on the damaging effect of the shock wave by four categories of damage:

1) Lethal outcome.
2) Extremely serious damages 2*a* and heavy damages 2*b*, often accompanied by complications and lethal end; 50–60% of the damaged need medical aid as

Table 3.18 Characteristics of the vulnerability of personnel to the action of air SW.

Parameters	Degree of damage								
	A			*B*			*C*		
Probability of damage	1	0.5	0.1	1	0.5	0.1	1	0.5	0.1
Δp^*, MPa	0.50	0.32	0.25	0.36	0.21	0.16	0.29	0.17	0.13

Source: From Balagansky et al. [1].

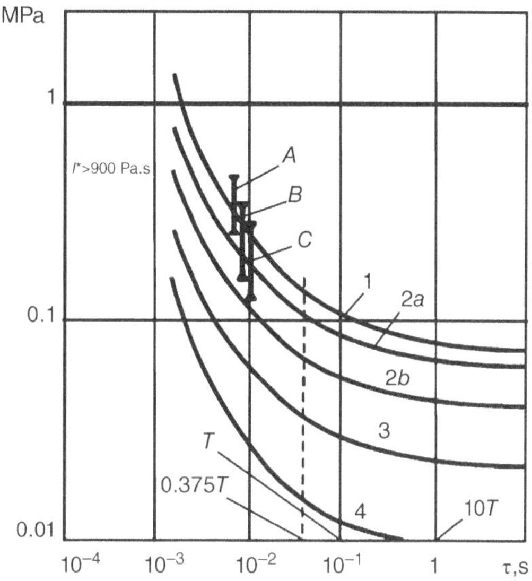

Figure 3.24 Generalization of results on the vulnerability of personnel when exposed to a blast field of damage. *Source*: From Balagansky et al. [1].

soon as possible and are transported recumbently; about 50% of people affected become non-transportable at the first stage of medical evacuation; a treatment period of patients from 2 to 12 months with the long loss of working capacity (the lower limit of degree *A* damage).

3) Damages of medium serious degree, which in most cases are not life-threatening, and 10–12% of those damaged need urgent medical care with transportation to medical facilities in a prone position; period of hospital treatment is 2–3 months with the short loss of ability to work after treatment (degree *B*).
4) Light damages not associated with life-threatening or disability-threatening consequences do not require emergency medical care; those damaged are transported to medical facilities in a sitting position, and about 50% of them can walk and return to normal activities within 1–60 days (type *B* to type *C*).

Obviously, when the duration of the compression phase exceeds 10^{-1} second, the damage will be determined by the overpressure. As the duration of the compression phase decreases, the level of critical overpressure increases. Consequently, the longer the duration of the compression phase, the lower the overpressure causes the same degree of damage (we are talking about shock loads).

In the functioning of conventional high-explosive munitions, filled with condensed explosives, in the near blast zone, the duration of air SW action is counted in milliseconds and the SW action has an impulsive behavior. Such SW acts on the personnel in the case of an explosion of charges of small mass when $\tau^+ \leq 20$ ms. With increasing the mass of the charge, the time of action of SW increases up to

20–200 ms. In this case, the main influence on the type of damage has pressure at the front of the shock wave. The human body is very resistant to the action of the SW and does not have serious damage at the pressure at the front of the SW up to 0.01 MPa.

If the duration of the air SW exceeds 200–250 ms, then both the pressure and the velocity of the air flow moving behind the wave front are dangerous for a person, which can be approximated by the formula:

$$u = 2000\ \Delta p_1, \tag{3.92}$$

where u is the air velocity, m/s; Δp_1 is the overpressure at the SW front, MPa. For example, at an overpressure of 0.02 MPa, the air flow velocity reaches 40 m/s. At such flow velocity, a person cannot remain standing on their feet and gets subjected to traumas when falling. Together with the air flow moving behind the front of the SW, various elements of the destroyed objects and other objects move, posing a serious hazard to people and structures. In this case, the greater the flow velocity and time of action of the shock wave, more objects are involved in the movement.

3.6 Explosion in Water

3.6.1 The Physical Picture of an Explosion in the Water

The physical picture of an explosion of a HE charge in water is qualitatively similar to an explosion in the air. However, since water has low compressibility and is a denser medium than air, the numerical values of the explosion parameters differ from the same parameters in air.

Consider an explosion in a boundless liquid medium of a spherical HE charge with central initiation of detonation. At the moment when the detonation wave approaches the boundaries of the charge, a strong shock of the explosion products onto water occurs, and, as in any other compressible medium, a shock wave develops in it. In this case, the initial pressure at the shock wave front is of the same order as the pressure at the detonation wave front. The initial velocity of the shock wave in water is less than the detonation velocity of the explosive D and is about $0.85D$. The density of water at the shockwave front increases at most by a factor of 1.5. The initial mass velocity behind the SW front is 2.5–3.0 times lesser than the SW velocity. Therefore, the velocity of the boundary of the media is significantly less than the initial SW velocity, the front of which immediately breaks away from the explosion products.

As the SW propagates in water, all its parameters rapidly decrease. Thus, at a distance equal to twice the charge radius r_0, the velocity of the shock wave decreases more than twice, and at a distance $r = (8\text{–}12)r_0$, it becomes almost equal to the velocity of sound in water under normal conditions $c_0 = 1500$ m/s.

Irreversible losses during an explosion in water are small, and the energy of the explosion transferred to the SW is spent mainly on mechanical work associated with the involvements in the motion of the mass of liquid, through which the SW passes. Therefore, its temperature increases insignificantly (at the explosion of a TNT charge, for example at a distance of $r = 2r_0$ the temperature increases by 38 °C). This reason leads to the fact that the damping of the SW in water is slower, i.e. at the same relative distances, the maximum pressure in the SW in water will be much higher than in air. At the same time, a steeper pressure drop is observed behind the shock wave front in water than in air, and the thickness of the layer of water compressed by the shock wave is less than the thickness of the layer of air in the air shock wave.

After separation of the SW from the EPs, the latter continue to expand, increasing the volume of the gas cavity (see Figure 3.25). At the same time, the expansion velocity of the gas bubble (GB), which initially was of the order of 2.2–2.5 km/s, will gradually decrease. The pressure inside GB also decreases and at some point, becomes equal to the hydrostatic pressure of water at this depth. Water displaced by the expanding bubble by inertia will continue to move away from the center of the explosion, increasing the size of the GB. By the end of the expansion, the EP pressure inside the GB becomes less than the hydrostatic pressure, and it begins to reduce at an ever-increasing rate. This changes the direction of the fluid flow, and it moves toward the center of the explosion, forming a rarefaction wave (RW) in water, which propagates after the SW.

The GB is compressed until the pressure inside the bubble becomes balanced by the dynamic pressure of the converging water flow. As a result, by the end of compression, the pressure in the GB becomes higher than the hydrostatic pressure and the process of its expansion begins again. The second phase of expansion is also accompanied by the compression of water and the emergence of an increased

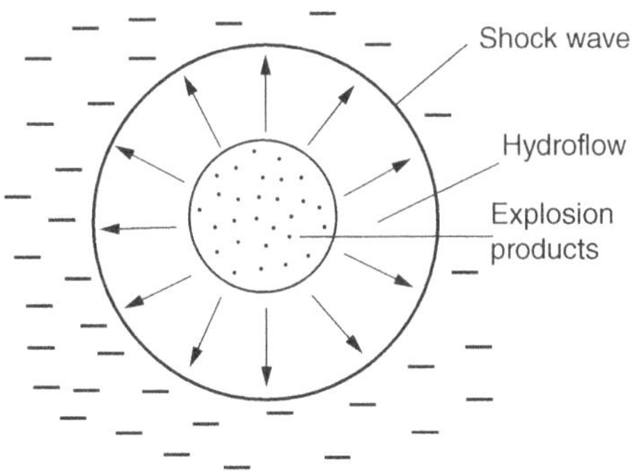

Figure 3.25 Movement of SW and expansion of EP during an explosion in the water. *Source*: From Miropolsky et al. [3].

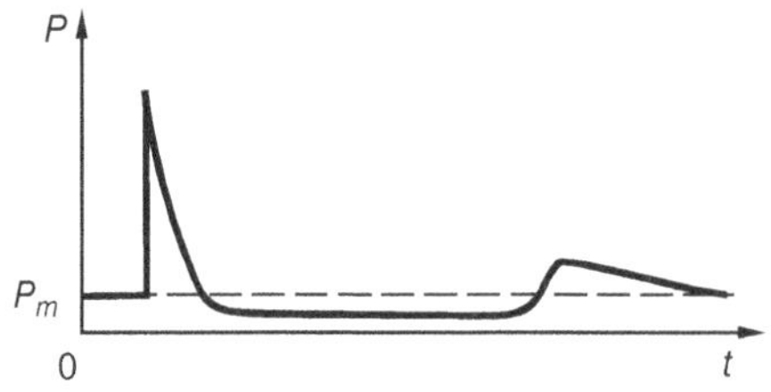

Figure 3.26 Pressure dependence versus time at a fixed point in an explosion of a charge in water. *Source*: From Miropolsky et al. [3].

pressure zone in it. The so-called secondary SW is formed, the maximum pressure in which is about 10% of the pressure in the primary SW. However, since the duration of the secondary expansion phase is much longer than the action time of the primary SW, the specific impulses of both waves are of the same order. The behavior of pressure changes in time at a fixed point when the primary and secondary waves pass through it is shown in Figure 3.26.

The periodic expansion and compression of the EP occurs repeatedly (about ten such oscillations can be identified) and is called gas bubble pulsation. Figure 3.27 shows how the radius of a pulsating bubble changes over time. During the process of GB pulsation, the Archimedes force acts on it. Since the EP density is less than the water density, the GB floats to the water surface. Since the volume and density of the GB change during the pulsations, the surfacing occurs at a variable velocity. In addition, during the surfacing, the resistance force from the water acts on the GB. The maximum surfacing velocity is observed when the GB reaches the minimum sizes. The bubble surfacing velocity is much less than the velocity of the processes occurring during the explosion and the formation of the pressure field, as a result of which these processes are over by the time of the surfacing of the GB.

When the surface of the bubble touches the free boundary of the water, the EP bursts into the air environment. The cavity formed in the water is intensively slammed, and a cumulative splash of water appears above the surface in the form of a sultan.

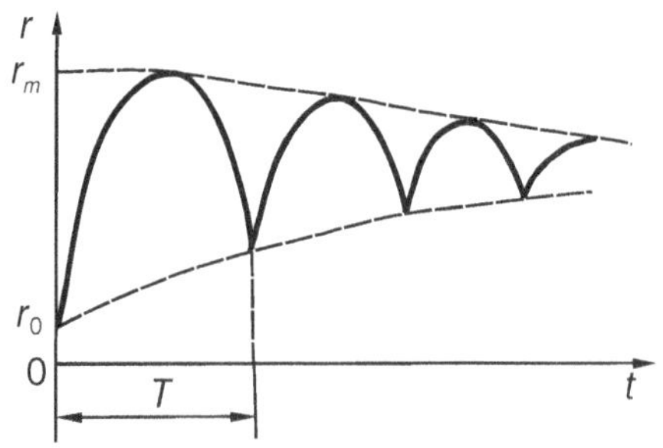

Figure 3.27 Dependence of gas bubble radius on time: T – period of the first pulsation; r_m – maximum radius of the gas bubble. *Source*: From Miropolsky et al. [3].

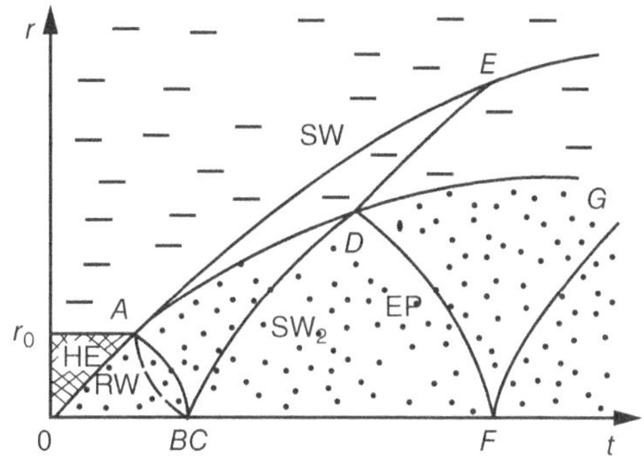

Figure 3.28 *r–t* diagram of the initial stage of a spherical charge explosion in the water. *Source*: From Miropolsky et al. [3].

Numerical calculations of the explosion process in water made it possible to detect the secondary SW going into the EP. After reflection from the center of the explosion, the secondary SW moves to the boundary of the GB and, reaching it, disintegrates into two SWs: in water and in the EP. Such a process is repeated with decaying amplitude. In Figure 3.28, the *r–t* diagram of the described process is shown. Here, *OA* is the DW front, *AE* is the SW in water, *AD* is the contact discontinuity (CD) between water and EP, *AB* is the rarefaction wave in EP, *ACD* is the secondary SW in EP, and *DE* and *DF* are the SW formed during refraction of the secondary SW at the CD.

This is in general terms the physical picture of the explosion process in water.

3.6.2 Basic Parameters of an Underwater Explosion

Parameters describing the process of GB pulsation: the pulsation period T and the maximum bubble radius r_m can be found using the π-theorem. In this case, the desired values are functions of four parameters: the mass of explosive charge m, specific explosive heat Q_s, water density ρ, and hydrostatic water pressure p_H at a given depth H. Application of the π-theorem gives the following types of functional dependencies:

$$T = \sqrt[3]{\frac{m}{\rho}}\,\frac{1}{\sqrt{Q_s}}\,f_t\left(\frac{\rho Q_s}{p_H}\right), \tag{3.93}$$

$$r_m = \sqrt[3]{\frac{m}{\rho}}\,f_r\left(\frac{\rho Q_s}{p_H}\right). \tag{3.94}$$

The results of processing the experimental data allow us to represent the functions f_t and f_r as

$$f_t(x) = a_t x^{5/6}, \tag{3.95}$$

$$f_r(x) = a_r x^{1/3}, \tag{3.96}$$

where a_t and a_r are constant coefficients determined by experiments.

After substitution of dimensionless parameters, we obtain the following formulas:

$$T = a_t \sqrt{\rho} \sqrt[3]{m Q_s} p_H^{-5/6}, \tag{3.97}$$

$$r_m = a_r \sqrt[3]{\frac{m Q_s}{p_H}}. \tag{3.98}$$

If in these formulas p_H is expressed in atmospheres and dimensions of other quantities correspond to the system of SI units, then the coefficients are equal: $4.4 \cdot 10^{-5}$, $a_r = 9.6 \cdot 10^{-3}$.

The hydrostatic pressure of water increases with depth according to the law

$$p_H = p_a(1 + 0.1H), \tag{3.99}$$

where H – depth, m; p_a – water surface pressure (atmospheric pressure).

To determine the basic parameters of SW in water ($\Delta p_1, I_1$), we can use formulas (3.17) and (3.29) obtained for air SW (here $\Delta p_1 = p_1 - p_H$). In this case, pressure p_0 is a sum of molecular pressure of water compression $p_m = 10^4$atm and hydrostatic pressure at a given depth p_H.

Keeping in mind that for real explosion depths $p_m \gg p_H$, we can substitute the value $p_0 \approx p_m = 10^4$atm into the formula for Δp_1. Thus, because of the enormous molecular pressure of water and its low compressibility, hydrostatic pressure practically does not affect the value of pressure at the front of the SW. Experimental studies conducted by the American researcher Robert H. Cole [10] made it possible to obtain calculation formulas for TNT charges ($Q_s = 4.1$ MJ/kg):

$$\Delta p_1 = 533 \left(\frac{\sqrt[3]{m}}{r} \right)^{1.13} \text{ (atm)}, \tag{3.100}$$

$$I_1 = 5900 \frac{m^{0.63}}{r^{0.89}} \text{ (Pa·s)}. \tag{3.101}$$

Here the mass of the explosive is given in kilograms and the distance from the place of the explosion in meters.

The formula (3.100) for Δp_1 is valid in the range of 30–900 atm, and the formula (3.101) for I_1 in the range of distances corresponding to overpressures of 30–500 atm. For other kinds of explosives, the following formulas can be used:

$$\Delta p_1 = 533 \left(\frac{\sqrt[3]{k_T m}}{r} \right)^{1.13}, \tag{3.102}$$

$$I_1 = 5900\sqrt{k_T}\frac{m^{0.63}}{r^{0.89}}, \tag{3.103}$$

where k_T is the TNT equivalent of the explosive charge.

The duration of the compression phase of SW in water in seconds can be approximated by the following formula:

$$\tau^+ \approx \frac{\sqrt[3]{k_T m}}{c_0}. \tag{3.104}$$

3.6.3 The Damaging Effect of an Underwater Explosion

The main factors determining the destructive effect of an explosion of a charge in water are SW (primary and secondary), hydro flow (created by the movement of the mass of water displaced by the GB and moving behind the SW front), an EP of the GB (in the case when the target is at a distance $r < r_m$). At present, there are no theoretical methods to estimate the destructive effect of the explosion of the explosive charge in water. Since up to 50–60% of the explosion energy is concentrated in the main SW, the calculation of the effect of an underwater explosion is commonly limited to the evaluation of the SW action. It is assumed that the more intense the main SW is, the stronger will be the effect of other associated damaging factors (secondary wave, water flow, explosion products), all other things being equal.

An underwater explosion is used to act on such targets, for which, as a rule, the ratio $\tau^+ \ll T$ is valid, where T is the period of natural oscillations of the target object. Under these conditions, the specific impulse should be taken as a characteristic of the destructive action. However, since the duration of the compression phase in the SW is practically independent of distance, we can assume that

$$I_1 \approx k\Delta p_1, \tag{3.105}$$

where k is a constant value. Therefore, the overpressure in the SW Δp_1 can also be taken as a characteristic of the damaging action.

Let us briefly consider the peculiarities of an underwater explosion in real reservoirs, when the influence of the bottom and the side walls of the reservoir, as well as the free surface of the water, is observed.

The effect of side walls and the bottom depends to a large extent on their strength. If the depth of the reservoir (distance from the charge to the walls) is small and its bottom (wall) is sufficiently strong, the resulting reflection SW increases the explosive load on the object. The pressure in the reflected wave increases approximately twice. In addition, an underwater explosion may also have an irregular reflection. The nature of the explosive load acting on the object depends on the relative location of the charge, the object, and the bottom (walls) of

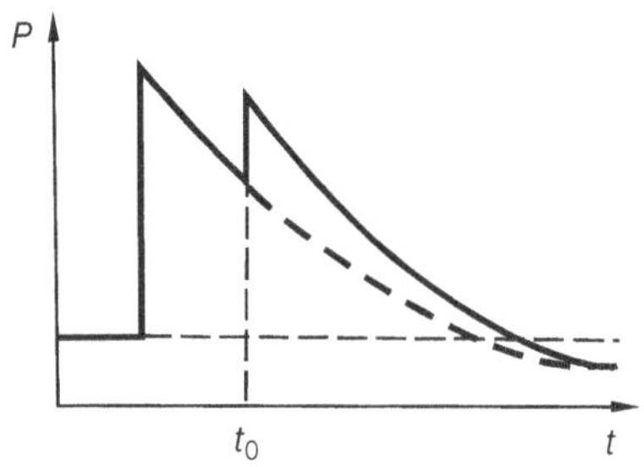

Figure 3.29 The behavior of the explosive load acting on an object near a reflecting surface. *Source*: From Miropolsky et al. [3].

the reservoir and, for example, with a regular reflection is determined by the moment t_0 of arrival to the object of the reflected SW (Figure 3.29).

Also, it should be taken into account that during the pulsation process, the GB is as if "attracted" to a rigid surface. As a result, the GB can come into contact with the object, even if they were originally at a distance greater than the maximum bubble size r_m. In this case, the pulsating GB will remain in contact with the surface of the object being destroyed all the time, and thus the object will be subjected to an additional pressure of strongly compressed EPs, which will lead to an increase in its degree of destruction.

The effect of the free surface is determined by the fact that the reflection of the SW from it generates a rarefaction wave, which propagates deep into the water and can cut off part of the impulse acting on the object (Figure 3.30). This is possible in case of a small depth of explosion. If the depth of the explosion is such that the time of arrival of the rarefaction wave will be longer than the time of action of the shock wave, then the presence of a free surface has almost no effect on the magnitude of the explosive load (Figure 3.31).

The air cavity that is formed behind a penetrating aerial bomb reduces the destructive effect of the explosion because part of the energy is directed along the line of least resistance for compression and heating of the air in the cavity. The use of air cavities to protect underwater structures from the action of the

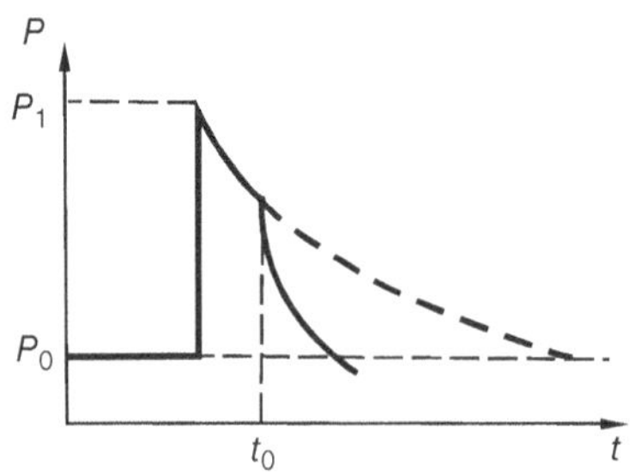

Figure 3.30 The behavior of explosive load acting on the object near the free surface. *Source*: From Miropolsky et al. [3].

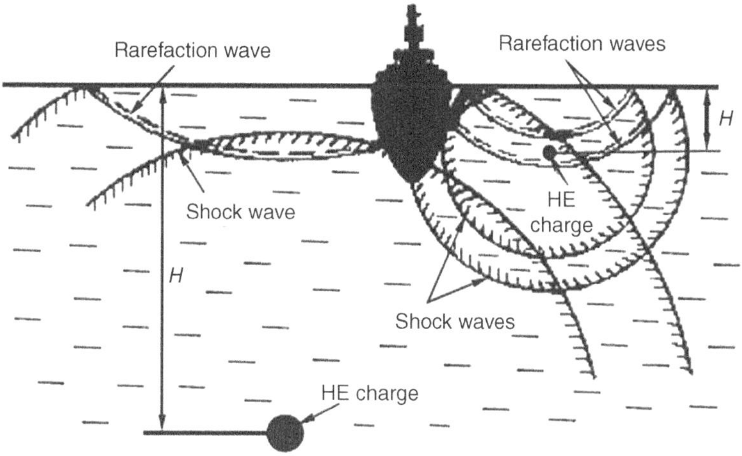

Figure 3.31 Influence of the free surface on the destructive effect of an underwater explosion. *Source*: From Miropolsky et al. [3].

explosion leads to the need to increase the explosive charge required for destruction by several times.

3.7 Underground Explosion

3.7.1 The Physical Picture of an Underground Explosion

The explosion of a HE charge under ground, or an underground explosion, is one of the main explosion actions in the military use of high-explosive munitions filled with condensed explosives. Some important targets are located at considerable depths to protect them from powerful explosions. Also, an explosion in the ground can create conditions for damage or destruction of above-ground or underground structures.

Unlike air and water, the physical properties (density, strength, elasticity, water saturation, porosity, plasticity, etc.) of soils are extremely diverse. For example, the velocity of sound varies from 100 m/s for peat to 3000 m/s for limestone. The main distinguishing feature of soils is their porosity (i.e. the presence of inclusions in the form of water and trapped air). The volume of such inclusions can vary from 0.1% in granite to 70% in clay. When compressing the porous soil, at first it is compressed due to the elimination of porosity. At the same time, the density of the soil increases significantly. A further increase in pressure leads to compression of the solid particles of the soil.

In addition, the soil, like any solid medium, has residual deformation. After the load is removed, the density of the ground remains almost the same as at the moment of its compression, and the ground particles do not return to their original position. Finally, a considerable part of the explosion energy is spent on irreversible losses associated with the creation of plastic deformations and destruction (crushing, loosening) of soil particles.

The theoretical investigation of the phenomenon of the explosion of a charge in the ground encounters difficulties due to the lack of a general equation of the state for it. The issues determining the destruction of the ground under explosive loads have been insufficiently investigated. Besides, when naturally occurring, soil and rocks are usually layered media with variable layer thicknesses. The properties of each layer are essentially individual, which makes studying the propagation of wave processes in such media a difficult and uncertain task. The diversity of soils and their physical properties and characteristics makes it necessary to use various simplified models that are valid only for certain types of soils. When solving practical problems, experimental data are widely used.

Consider the explosion of a spherical explosive charge in unlimited homogeneous soil (Figure 3.32). After completion of the detonation process, strongly compressed explosion products make a sharp impact on the adjacent layers of soil, compressing them and creating a SW in the ground. In addition, the expanding EP displaces the surrounding soil, driving it in radial directions. The initial pressure at the front of the generated SW depends on the properties of the soil. For rocky materials, it can be higher, and for low-strength soils, it is significantly lower than the pressure at the front of the detonation wave.

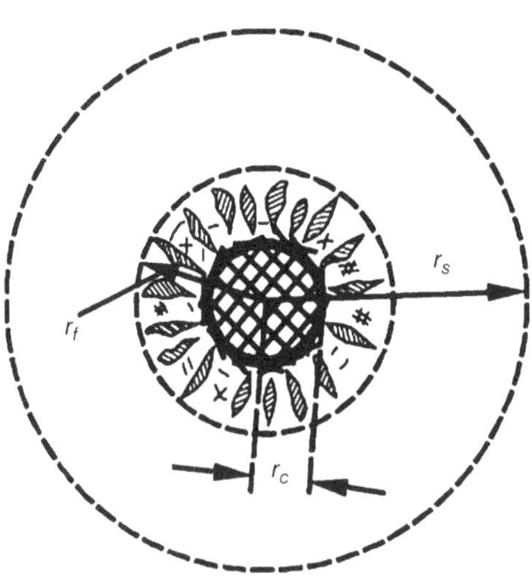

Figure 3.32 Characteristic zones formed by a charge explosion in boundless soil. *Source*: From Miropolsky et al. [3].

The less compressible the medium (i.e. the higher the sound velocity in the medium) is, the earlier the SW detachment from the EP occurs. For example, for rocky and water-saturated soils, the SW detaches from the EP almost immediately; for porous soils, the more porous they are, the later the detachment occurs. The initial velocity of propagation of the SW only slightly exceeds the velocity of sound in the ground and already at distances equal to 2–3 specified radii of the charge becomes equal to the velocity of sound. The behavior of the time dependence of the pressure at the shock wave front is largely determined by the plastic properties of soils.

For media with low plasticity, a sharp jump in pressure at the wave front, which is common for SWs, is typical. In plastic soils, the pressure increases less sharply, so that at some distance from the point of the explosion, the shock wave transforms to a compression wave, and the maximum pressure corresponds approximately to the middle of the compression phase. In low-plastic media, such transition takes place at greater distances from the center of the explosion at pressures below 10 atm. The products of the explosion, displacing, in the process of expansion, the ground already compacted by the SW, additionally compress it in the radial direction and stretch it in the tangential one. As a result, a system of radial cracks is formed in a certain volume of soil, called the fracture zone and characterized by the radius r_f, while in the area adjacent to the EP, the soil is fragmented, crushed, and pressed into a monolithic mass.

The volume occupied by gaseous EP at the termination of the expansion process is called the expansion zone and is characterized by the compression radius r_c. The zone of soil destruction is followed by the zone of its plastic and then elastic deformations called the shaking zone and characterized by the radius r_s. The presence of this zone is a consequence of radial displacements of soil particles under the action of passing SW and expanding EP. In this case, in the zone of plastic deformations, ground particles do not return to their original position, while in the zone of elastic deformations the oscillations occur within the limits of ground elasticity. Theoretically, the size of the shaking zone in an unlimited solid medium is very large. The objects located in this zone are subjected to the so-called seismic action of the explosion. Therefore, in practice, the value of r_s is conditioned by the critical amplitude, velocity, or acceleration of displacement in the analysis of the destruction of a particular object. The dimensions of the zones arising from an explosion in the ground are determined by the mass of the explosive charge, its explosive heat, and the properties of the ground.

3.7.2 Parameters Characterizing the Explosion Process in the Ground

Modern theoretical methods of calculation do not allow to estimate the parameters of the underground explosion field with the accuracy required for practice. These data are obtained by experimental investigations, which can be carried out both in laboratory conditions and in natural soils.

For example, it was found by experience that the radius of the fracture zone in meters can be estimated by the formula

$$r_f = k_f\sqrt[3]{m_T}, \tag{3.106}$$

where m_T is the TNT equivalent of the explosive charge, (kg); k_f is the coefficient that depends on the properties of the soil.

The values of k_f for some media are presented in Table 3.19.

The radius of the compression zone is proportional to the radius of the fracture zone:

$$r_c = 0.36r_f. \tag{3.107}$$

When determining the parameters of the shaking zone r_s, it should be taken into account that the sizes of this zone are determined on the basis of the assigned critical amplitude of ground particles oscillations.

Analysis of experimental data shows that the maximum pressure p_m at the front of spherical SW and compression waves in the ground can be represented based on similarity theory and dimensions in the form of

$$p_m = f\left(\frac{\sqrt[3]{m}}{r}\right). \tag{3.108}$$

Table 3.19 Values of the k_f – the coefficient for some media.

Medium	k_f
Freshly sprinkled soil	1.40
Regular soil	1.07
Dense sand	1.04
Stony soil	0.96
Clay	0.94
Limestone	0.92
Granite	0.77
Concrete	0.77
Reinforced concrete	0.65

Source: From Miropolsky et al. [3]

Table 3.20 Constants determining the pressure dependence on the distance in the soil.

Soil	r_0, g/cm^3	Δ	*W*, %	A	n	Δp_m, atm
Water-saturated						
Water	1.00	—	—	533	1.13	70–1800
The sand of natural formation	1.98	0	—	600	1.05	10–200
Same	1.98	$5{\cdot}10^{-4}$	—	450	1.5	20–220
Same	1.97	0.008–0.012	—	250	2.0	10–150
Dense clay of natural formation	2.05	0.010–0.020	—	120	2.5	1–180
Non-water-saturated						
The sand of natural formation	1.58	—	8–10	7.5	3.0	1–40
Loam of natural formation	1.62	—	10–12	8.0	3.0	1–125
Bulk sand	1.50	—	3–6	2.8	3.3	1–30
Same	1.50	—	5–7	6.0	3.2	1–40
Loess of natural formation	1.36	—	12–15	4.5	2.8	1–46
Granite			—	320 150	2.0 1.2	50–250 <50

Source: From Lyakhov [11].

So, for example, the pressure in freshly sprinkled soil can be calculated by the formula

$$p_m = 6\left(\frac{\sqrt[3]{m}}{r}\right)^4 + 1.32\frac{\sqrt[3]{m}}{r}. \tag{3.109}$$

For most soils, the basic formula is

$$p_m = A\left(\frac{\sqrt[3]{m}}{r}\right)^n. \tag{3.110}$$

Here, the mass of the explosive m is given in kilograms, the distance from the place of detonation r – in meters, and the pressure p_m is given in atmospheres. The values of coefficients A and n determined experimentally are given in Table 3.20.

Here are also the values of soil density ρ_0, the fraction of soil volume occupied by air Δ, mass humidity (ratio of the mass of water contained in the soil to the mass of solid particles) W, as well as the pressure range Δp_m, in which the wave parameters in the soil were measured.

The constants presented in Table 3.20 are true for camouflet explosions when the charges are placed at such a depth h that there is no visible deformation on the surface of the ground. The measurements were taken at the depth of the explosion. In this case, the value of $h > 0.6\sqrt[3]{m}$. With decreasing the charge depth, the wave parameters decrease due to rarefaction waves propagating from the ground surface. The influence of the free surface can be taken into account using the following formula:

$$p_m = k\varphi\left(\frac{h}{\sqrt[3]{m}}\right)\frac{m}{r^3}, \tag{3.111}$$

$$\varphi\left(\frac{h}{\sqrt[3]{m}}\right) = \begin{cases} 0.2 + 1.33\dfrac{h}{\sqrt[3]{m}}, & \text{for } h < 0.6\sqrt[3]{m}, \\ 1, & \text{for } h \geq 0.6\sqrt[3]{m}, \end{cases} \tag{3.112}$$

where h – depth of explosion, (m); k – coefficient depending on soil properties (for loess $k = 3.5$, for loam $k = 8.75$, for silt clay $k = 22.2$, for dry clay $k = 65.5$, for wet clay $k = 440$).

To account for the effect of the free surface of the ground in an approximate manner, the method of fictitious charges is often used. In this case, it is assumed that the effect of the rarefaction wave from the free surface is equivalent to the explosion of a fictitious charge located symmetrically to the real charge relative to the surface of the ground (Figure 3.33). The fictitious charge creates the same pressure field as the real charge, but with the opposite sign. In this case, the maximum pressure can be determined by the formula

$$p_m = f\left(\frac{\sqrt[3]{m}}{r_1}\right) - f\left(\frac{\sqrt[3]{m}}{r_2}\right), \tag{3.113}$$

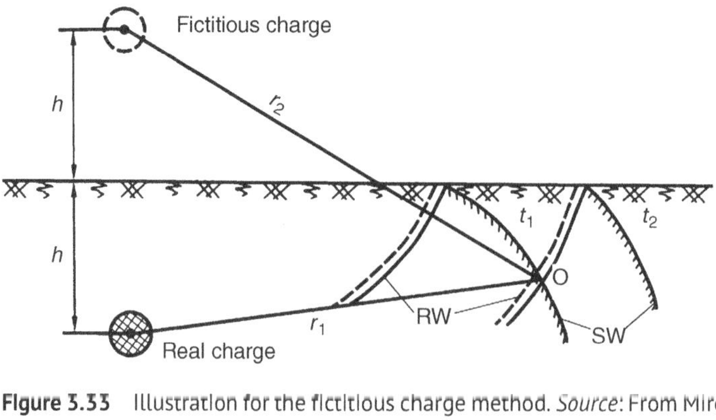

Figure 3.33 Illustration for the fictitious charge method. *Source:* From Miropolsky et al. [3].

where r_1 and r_2 are, respectively, distances in meters from the object O (Figure 3.33) to the real and fictitious charges.

The specific impulse acting on underground structures, based on the theory of similarity and dimensions, can be determined by the formula for TNT charges with a mass of 0.2–1000 kg:

$$I_1 = B\sqrt[3]{m}\left(\frac{\sqrt[3]{m}}{r}\right)^{\mu} \text{ (Pa·s)}, \tag{3.114}$$

where B and μ – coefficients that depend on the properties of the soil and presented in Table 3.21.

To account for the influence of the free surface, we can use the formula

$$I_1 = B\varphi\left(\frac{h}{\sqrt[3]{m}}\right)\sqrt[3]{m}\left(\frac{\sqrt[3]{m}}{r}\right)^{\mu}, \tag{3.115}$$

where the function $\varphi\left(\frac{h}{\sqrt[3]{m}}\right)$ is defined from (3.112).

For the time of action of the SW, dependence has been proposed:

$$\tau^+ = \sqrt[3]{m}\left(a + b\frac{r}{\sqrt[3]{m}}\right) \text{ (s)}, \tag{3.116}$$

where a and b are the empirical coefficients given in Table 3.22.

A ground explosion can be of two types:

- A crater-forming explosion, sometimes called a ground ejection explosion;
- An explosion at relatively great depth, where no visible changes on the surface of the ground are formed; such an explosion is called a camouflet explosion.

Table 3.21 Constants determining the dependence of impulse on the distance in the soil.

Soil	ρ_0, g/cm³	Δ	W, %	B	μ
Water	1.0	0		5888	0.89
Water-saturated sand	1.89–1.98	0 $5\cdot10^{-4}$ 10^{-2} $4\cdot10^{-2}$		8000 7500 4500 4000	1.05 1.10 1.25 1.40
Non-water-saturated sand	1.45–1.55 1.50–1.55	— —	3–6 10–12	3200 4000	1.50 1.50
Loamy non-water-saturated soil	1.60–1.65	—	—	4000	1.45
Clay soil	1.70–1.75	—	15	4500	1.40
Loess-like non-water-saturated soil	1.34–1.38	—	—	5000	1.50

Source: From Lyakhov [11].

Table 3.22 Constants determining the action time of the compression wave in non-water-saturated soil.

Non-water-saturated soil	ρ_0, g/cm^3	W, %	$a \cdot 10^3$, s/kg$^{0.33}$	$b \cdot 10^3$, s/m
Sandy	1.52–1.60 1.45–1.50	8–10 3–6	4 16	16 24
Loamy	—	—	10	10
Loess-like	—	—	14	14
Granite	—	—	0.23	0.13

Source: From Lyakhov [11].

Let us consider a ground ejection explosion. When the charge is exploded near the surface, the nature of the process does not differ from an explosion in unlimited soil until the shock wave reaches the surface. Then, after the SW reaches the surface, it is reflected from it in the form of a rarefaction wave. The rarefaction wave moves deep into the ground, and the accompanying tensile stresses erode the ground mass adjacent to the surface. Therefore, the displacement and fracture zones are deformed, stretching toward the surface and contracting in the opposite direction.

If the explosion depth is small and the charge mass is large, part of the soil will be ejected upward. At the initial moment, a crater is formed, the shape of which is a combination of a cone and a deformed compression sphere (Figure 3.34). A part of the ejected soil falling down comes back to the crater filling the volume of the compression sphere and forming a ridge at the edges of the crater. The final shape of the crater is close to a truncated cone, the radius of the smaller base of which is half the radius of the crater.

In this case, the depth of the crater h is equal to the explosion depth of the charge. The ratio of the crater radius R to the explosion depth h is called the ejection ratio n. Obviously, the volume of the crater W_0 can be expressed in terms of its depth h and the ejection index n:

$$W_0 = \frac{7}{12}\pi h R^2 \approx 1.83 n^2 h^3. \tag{3.117}$$

A charge with $n > 1$ is called a charge of increased ejection; if $n = 1$, such a charge is called a charge of normal ejection; a charge with $n < 1$ is called a charge of reduced ejection.

The general formulas linking the mass of the TNT equivalent charge to the sizes of the crater can be determined using the theory of similarity and dimensions, taking into account the strength of the ground and the gravity force

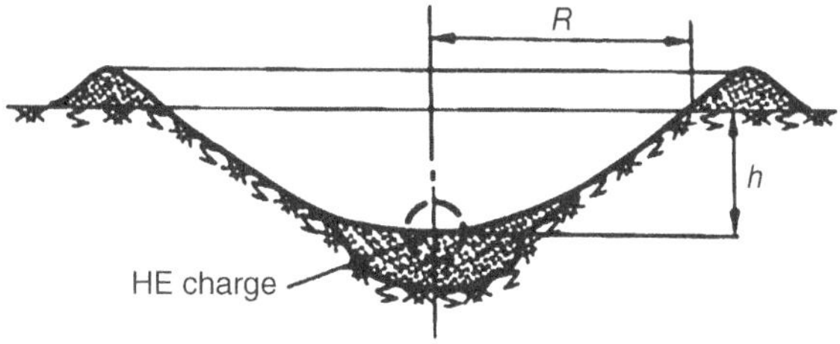

Figure 3.34 Crater scheme. *Source*: From Miropolsky et al. [3].

$$m_T = Kh^{\alpha}f(n), \tag{3.118}$$

where $f(n)$ – ejection function; $\alpha = 3$–4 is a coefficient depending on the kind and depth of the explosion. The form of the ejection function

$$f(n) = A + Bn^3$$

was proposed based on empirical data; the empirical coefficients: $A = 0.4$, $B = 0.6$. K is a coefficient that depends on soil properties; its values are given in Table 3.23.

If the depth of charge placement is $h < 25$ m, then factor $\alpha = 3$. Such a ratio is used to calculate high-explosive charges of the increased ejection (i.e. charges that produce craters with $n > 1$. For charges of the reduced ejection, the formula (3.118) with $\alpha = 3$ gives overestimates of the charge masses.

Table 3.23 The values of the coefficient K.

Kind of soil	K	Kind of soil	K
Basalt	1.75–2.30	Chalk	0.80–0.95
Granite	1.50–2.25	Loess	0.90–1.30
Limestone	1.30–2.10	Clay	1.00–1.30
Limestone–shellstone	1.50–1.75	Loam	1.00–1.15
Gypsum	1.00–1.30	Sand dense and wet	1.20–1.30
Regular soil	0.95	Sand	1.50–1.70
Freshly poured soil	0.43	Concrete	3.15–3.60

Source: From Miropolsky et al. [3].

3.7.3 The Damaging Effect of an Explosion in the Ground

Analyzing the physical picture of the explosion in the ground, we can distinguish three kinds of destructive action:

1) explosion for the ejection of soil;
2) destruction of underground structures by the explosion;
3) seismic action of the explosion.

Let us consider ways to assess the various kinds of destructive actions of an explosion in the ground.

3.7.3.1 Explosion for Ejection

This kind of explosion is used for both civilian and military purposes (to damage objects such as airfield runways, highways, railroad tracks). In this case, it is necessary to achieve the formation of the craters of the increased ejection, having a large volume. As research shows, the most effective is the ejection ratio $n = 1.5\text{–}3.0$. Sizes of craters are calculated by formulas (3.117) and (3.118). Table 3.24 gives an example of experimental data on crater sizes produced during bombing from altitudes of 1200–3500 m by bombs of various calibers with a fuze deceleration of 0.2 second.

3.7.4 Destruction of Underground Structures

In this case, as a rule, a greater depth of penetration is required. Therefore, the use of charges of reduced ejection ($n = 0.75\text{–}1.00$) is advantageous. The destruction of underground objects occurs when the object is within the fracture zone $r_f = k_f\sqrt[3]{m_T}$. The method of equivalent layers is used to calculate the damaging effect. The material layer of the object of destruction with thickness b is replaced by an equivalent ground layer in terms of resistivity be b_e:

Table 3.24 Data on crater sizes.

Index of an aerial bomb	Soil kind	Crater dimensions, m	
		Diameter	Depth
FAB-100	Limestone	1.5–2.0	0.7–1.0
FAB-250	Limestone	4.5–5.5	1.2–1.5
FAB-500	Limestone	6.0–8.0	1.5–2.0
FAB-500	Loam	11.0–12.0	4.1–4.5

Source: From Miropolsky et al. [3].

$$\frac{b_e}{k_f} = \frac{b}{k_{f1}}, \tag{3.119}$$

where k_f, k_{f1} are coefficients determining the size of the fracture zones in the ground and the object material, respectively. Hence, the radius of destruction of the object can be calculated by the formula

$$r_f = k_f \sqrt[3]{m_T} - b_e = k_f \left(\sqrt[3]{m_T} - \frac{b}{k_{f1}} \right). \tag{3.120}$$

The mass of the charge required to fracture an object at a distance of r_f is calculated as follows:

$$m_T = \left(\frac{r_f}{k_f} + \frac{b}{k_{f1}} \right)^3. \tag{3.121}$$

3.7.4.1 Seismic Action of the Explosion

After the shock wave reaches the surface of the ground, surface waves begin to propagate from the epicenter of the explosion. The reason for the appearance of surface waves is the mass velocity of the soil particles behind the shock wave front. After the SW passage, the soil particles, moving by inertia, pass their equilibrium position and return to their original position.

As a result, there are longitudinal oscillations of soil particles along the direction of the SW movement. At the same time, in the directions perpendicular to the SW movement, transverse oscillations of soil particles arise, connected with the resistance of solid bodies to change their shape.

The joint action of longitudinal and transverse vibrations of soil particles causes a surface seismic wave (Rayleigh wave) to be propagated along the free surface of the ground from the explosion epicenter. The point is that soil particles located on the ground surface (Figure 3.35a) experience less resistance from the surroundings during their movement than particles located at depth (Figure 3.35b) because in the first case the contact area of particles with the ground is almost twice as small.

The seismic wave causes oscillations of soil particles outside the fracture radius, i.e. in the elastic deformation zone. Recording the horizontal displacements of soil particles under the action of seismic waves makes it possible to obtain the so-called explosion seismogram (Figure 3.36). In the beginning, the recorder will record the forced oscillations of the soil under the action of the SW. The period of these oscillations corresponds to the elastic deformations of the soil at the shock front ($\tau = 0.005$–0.050 seconds). This phase of oscillations is called onset. After 1–10 seconds after the beginning of the entry phase, the main phase of free soil oscillations, which is a seismic wave, begins.

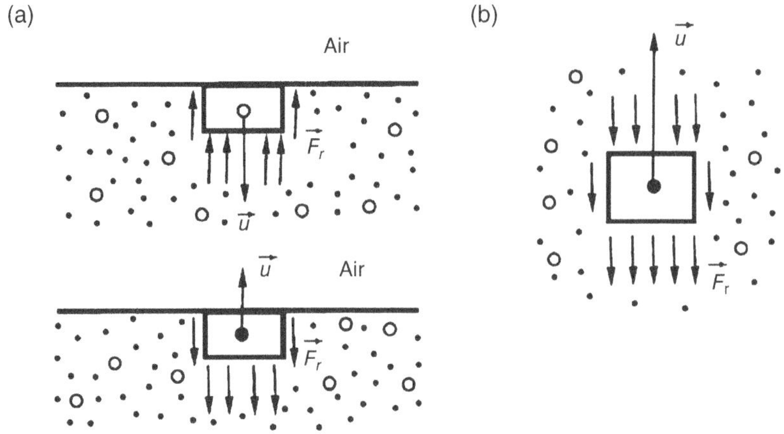

Figure 3.35 Resistance forces $\vec{F}_r$ to the movement of soil particles located: (a) at the surface; (b) at depth; $\vec{u}$ – velocity of particles. *Source*: From Miropolsky et al. [3].

The phenomenon of seismic action of the explosion was investigated by M.A. Sadovsky. The total period of oscillations of ground particles in the main phase can be estimated by the formula

$$T_s = \tau_s \lg r \text{ (s)}, \tag{3.122}$$

where r is the distance from the explosion point, (m); τ_s is the coefficient depending on soil properties (Table 3.25).

The value of the maximum vibration velocity of soil particles u_m for an arbitrary ejection index n can be determined by a formula obtained using the theory of similarity:

$$u_m = \frac{2}{\sqrt[3]{f(n)}}\left(\frac{\sqrt[3]{m}}{r}\right)^{1.5}, \tag{3.123}$$

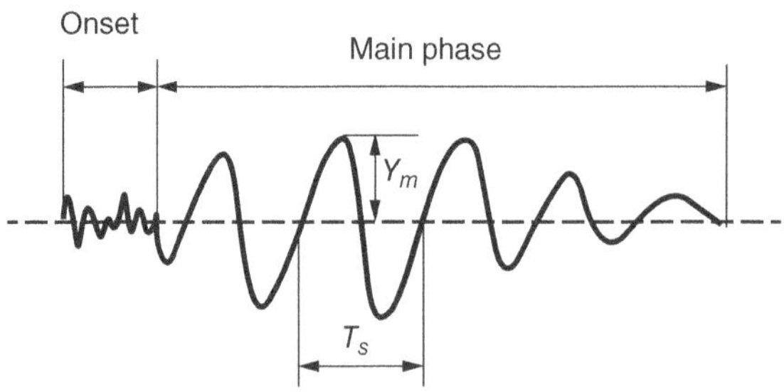

Figure 3.36 Seismogram of an explosion. *Source*: From Miropolsky et al. [3].

Table 3.25 The values of τ_s.

Soil kind	τ_s, **s**
Water-saturated soil (peat bogs, quicksand)	0.11–0.13
Sedimentary soil of medium strength	0.06–0.09
Rocky soil	0.01–0.03

Source: From Miropolsky et al. [3].

where *f(n)* is the ejection function. Assuming that the oscillations of the soil particles are sinusoidal

$$y = y_m \sin \frac{2\pi}{T_s} t, \tag{3.124}$$

it is possible, knowing u_m and T_s, to determine the amplitude y_m and acceleration a_m of oscillations. The amplitude y_m, in this case, is determined by the formula

$$y_m = \frac{u_m T_s}{2\pi}, \tag{3.125}$$

and the maximum acceleration a_m

$$a_m = \frac{4\pi^2}{T_s^2} y_m. \tag{3.126}$$

Substituting the values of u_m and T_s from formulae (3.122) and (3.123) into (3.125) and (3.126), we obtain

$$y_m = \frac{0.32\tau_s \lg r}{\sqrt[3]{f(n)}} \left(\frac{\sqrt[3]{m}}{r} \right)^{1.5}. \tag{3.127}$$

$$n_m = \frac{1.28}{\tau_s \lg r \sqrt[3]{f(n)}} \left(\frac{\sqrt[3]{m}}{r} \right)^{1.5}, \tag{3.128}$$

were $n_m = \frac{a_m}{g}$.

The seismic effect of an explosion is evaluated by comparing the values of u_m, y_m, or n_m with their critical values determined by the dynamic properties of the object in question. For example, some researchers recommend the use of a destruction scale for earthquakes. Catastrophic destructions correspond to the value of $n_m = 0.3$–0.4, strong destructions – $n_m = 0.05$–0.10, weak destructions – $n_m = 0.02$–0.05. It should be taken into account that the degree of damage increases significantly if the periods of free vibrations of the object and the soil are close to each other.

3.8 Self-assessment Questions

1) What is meant by the blast (high-explosive) effect?
2) What targets are high-explosive ammunition designed to damage?
3) Describe the distinguishing features of the volumetric explosion ammunition.
4) List the main damaging factors of high-explosive ammunition in air, water, and ground explosions.
5) Why is the pressure in the shock wave higher in an explosion on the Earth's surface than in an explosion of a charge at an altitude?
6) Why does the reflection of a shock wave from an object increase the pressure at its front?
7) What are regular and irregular shockwave reflections?
8) Why can the irregular reflection of shock waves occur?
9) Which of the two barriers, other conditions being equal, will be subjected to a higher pressure when the shock wave meets it if both barriers have the same area, equally oriented with relation to the shock wave, but the cross-section of one barrier is close to the square, while the second barrier has an elongated shape?
10) What are the upper and lower concentration limits of ignition and detonation for fuel–air mixtures?
11) What are the differences between deflagration and detonation of a fuel–air mixture?
12) Can a fuel–air mixture detonate in free air space from a flame source?
13) What conditions are necessary to ensure that the cloud of a fuel–air mixture can explode in the detonation mode?
14) In which cases should the effects of blast waves on targets be evaluated by overpressure, in which cases by specific impulse, and in which cases by the combined effects of pressure and impulse?
15) Write down and explain the expression for the generalized criterion for high explosive action.
16) In what cases, when assessing the damage of targets by a high-explosive action, can the values of the parameters be used directly at the front of a shock wave in the air?
17) Why would the shockwave's damaging effect increases dramatically when a person is near a building or fence?
18) Describe the typical characteristics of a condensed explosive charge explosion in the water.
19) Why does a gas bubble formed by an explosion in water pulsate?
20) Why is the specific impulse usually used as a criterion for evaluating the damaging effect of an underwater explosion?

21) How do the walls and the bottom of a reservoir affect the shock wave parameters of an underwater explosion?
22) For what kinds of targets the damaging effect of the high-explosive charges in the ground is used?
23) What is the fictitious charge method and what is it used for?
24) What is a camouflet explosion?
25) What should be understood by the charge of increased, normal, and reduced ejection?
26) List and characterize the main kinds of destructive action of the explosive charge when it explodes in the ground.

References

1 Balagansky, I.A. and Merzhievsky, L.A. (2004). *Effects of Means of Destruction and Ammunition*. Novosibirsk: Novosibirsk State Technical University [in Russian].

2 Babkin, A.V., Veldanov, V.A., Gryaznov, E.F. et al. (2008). High explosive ammunition. In: *Means of Damage and Ammunition: Textbook* (ed. V.V. Selivanov). Moscow: Bauman Moscow State Technical University [in Russian].

3 Miropolsky, F.P., Sarkisian, R.S., Vishniakov, O.L. et al. (1996). *Aircraft Warheads and Their Investigation*. Moscow: Zhukovsky Academy [in Russian].

4 Sadovsky, M.A. (1952). Mechanical action of air shock waves according to experimental research. *Physics of Explosion* **1** [in Russian].

5 Nettlenton, M.A. (1987). *Gaseous Detonation: Their Nature, Effects, and Control*. Chapman and Hall.

6 Vasiliev, A.A., Kulakov, B.I., Mitrofanov, V.V. et al. (1994). Initiation of explosive gas mixtures by a fast-flying body. *Dokladi Russian Academy of Science* **338** (2): 188–190. [in Russian].

7 Zarubin, V.N., Eliseev, K.V., Martinov, V.V. et al. (2002). *Methodology for Determining the Parameters of an Explosive Device by Destruction of the Environment, Typical Building Structures, and Damage to Biological Objects at the Location of the Accident* (ed. V.A. Khimichev). Moscow: Special Technics and Communications [in Russian].

8 Kochetkov, K.E., Kotlyarevsky, V.A., and Zabelin, A.V. (ed.) (1995–1998). *Accidents and disasters. Prevention and Elimination of Consequences: Textbook. In Four Parts*. Moscow: Publishing house ASV [in Russian].

9 Wite, C.S. (1971). *The Nature of the Problems Involved in Estimating the Immediate Casualties from Nuclear Explosions, CEX-71.1, Civil Effects Study*. U.S. Atomic Energy Commission DR-1886.

10 Cole, R.H. (1948). *Underwater Explosions*. Princeton: Princeton University Press.
11 Lyakhov, G.M. (1964). *Fundamentals of the Dynamics of an Explosion in Soils and Liquid Media*. Moscow: Nedra [in Russian].

4

Penetrating Ammunitions

4.1 Basic Concepts and Definitions. General Information

Penetrating weapons and ammunition represent a broad group, the common feature of which is their ability to pierce the target due to their kinetic energy (kinetic energy projectiles). Depending on the kind of target and its protection, such weapons may have different designs and use different physical mechanisms to damage the target.

To make this clear, here are some examples of targets against which penetrating weapons and ammunition are used.

Like shaped-charge ammunition, armor-piercing munitions (kinetic energy projectiles) are used to damage armored and lightly armored targets such as tanks, self-propelled artillery, infantry fighting vehicles, and armored personnel carriers, among others. The mission of such munitions is to penetrate armor protection and then hit vulnerable units and crew located inside the target.

Concrete-piercing ammunition is used to destroy material and manpower in fortifications.

Special penetrating ammunition is developed also to destroy aircraft in reinforced concrete shelters, military equipment, and manpower located in especially strong shelters, to disable airfield runways, and to destroy command and communications points and other facilities.

Small arms ammunition may also be considered as penetrating ammunition with good reason.

4.1.1 Armor-piercing Artillery Projectiles

Armor-piercing projectiles are designed to shoot at armored targets (tanks, armored vehicles, aircraft, ships) from artillery systems of various calibers. Armor-piercing projectiles in naval artillery appeared almost simultaneously with

Damaging Effects of Weapons and Ammunition, First Edition. Igor A. Balagansky.

rifled guns. Armor-piercing projectiles evolved with the growth of target protection – velocities increased, their design became more complicated and improved.

According to the purpose of armor-piercing projectiles, they have the following requirements:

1) A powerful impact action on armor, which is characterized by the greatest thickness of the armor pierced and, other things being equal, depends on the kinetic energy of the projectile at the moment of meeting the barrier, the shape of the head part, the angle of hitting the armor, and the strength of the projectile body upon impact with the armor.
2) High destructive effect behind the armor, manifested in the form of impact, blast, fragmentation, and incendiary effects.
3) High firing accuracy, as it is necessary to ensure a direct hit on the target to destroy it. Note in passing that all armor-piercing projectiles necessarily have a tracer that provides visibility of their trajectory and the ability to visually correct the fire.

Armor-piercing artillery projectiles are subdivided into caliber and sub-caliber projectiles.

4.1.2 Armor-piercing Caliber Projectiles

Caliber armor-piercing projectiles got their name as opposed to sub-caliber ones and accordingly have a caliber that coincides with the caliber of the gun. They may or may not contain HE. The weight of the explosive that such ammunition may be filled with is 0.4–1.4% of the weight of the round.

According to the design of the warhead, caliber armor-piercing projectiles are subdivided into projectiles with a sharp head; projectiles with a blunted head and a ballistic tip; projectiles with an armor-piercing and a ballistic tip (Figure 4.1).

Sharp-headed projectiles (Figure 4.1a) are the earliest design form of armor-piercing projectiles and are intended for firing at homogeneous armor of small thickness. On the outer surface of this group of projectiles, there are one to two circular grooves called undercuts-localizers. When the projectile strikes the armor, the undercuts localize the propagation of cracks, which reduces the probability of the explosive-filled cavity being destroyed when the projectile penetrates the armor.

Projectiles with blunted head part and ballistic tip (Figure 4.1b) appeared before World War II when heterogeneous armor with external cemented layer and homogeneous armor of high hardness were used for tank armor. Their main advantage is that projectiles with a blunted head part ricochet much less when they meet the armor at large angles from the normal to it.

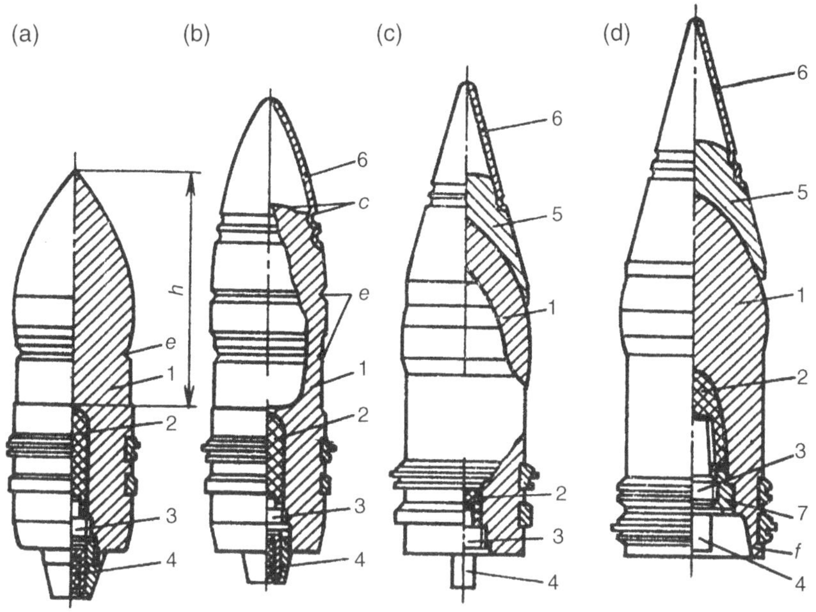

Figure 4.1 Armor-piercing caliber projectiles: (a) sharp-headed projectile; (b) a projectile with a blunted head and a ballistic tip; (c) a projectile with armor-piercing and ballistic tips; (d) a projectile with armor-piercing and ballistic tips and a deepened bottom; 1 – body; 2 – explosive charge; 3 – fuse; 4 – tracer; 5 – armor-piercing tip; 6 – ballistic tip; 7 – screw-in bottom; *e* – localizers; *f* – skirt; *c* – cutting edges; *h* – head height. *Source:* From Balagansky et al. [1].

Projectiles with armor-piercing and ballistic tips (Figure 4.1c,d) were adopted in the late 1940s. These projectiles damage both cemented and homogeneous armor well along the normal and at large angles from the normal to its surface.

The body of these projectiles is made of alloyed steel, which, after complex heat treatment, provides high hardness and resistance to impact on armor. The armor-piercing tip is mounted on the head part of the body and, when the projectile collides with the armor, it does the following:

- partially destroys the outer layer of armor over an area similar to the projectile's midsection;
- reduces ricochets and promotes normalization of the projectile;
- reduces stresses in the head part of the body at the initial stage of penetration;
- protects the head part of the body from destruction.

The hardness of the armor-piercing tip does not exceed the hardness of the head part of the projectile body.

To improve the ballistic shape of the projectiles, a ballistic tip is rolled onto the armor-piercing tip. It is made of thin-sheet low-carbon steel.

Ammunition filled with explosives has bottom fuzes with deceleration time to ensure that the projectile explodes behind the armor to increase damage inside the target.

4.1.3 Sub-caliber Armor-piercing Projectiles

Modern armies have subcaliber armor-piercing projectiles for rifled and smoothbore artillery systems that are stabilized in flight by rotation or fins. The point of development of subcaliber projectiles is that they have lower weight in comparison with caliber ones and acquire a higher initial velocity when fired from the same artillery system.

Sub-caliber rotating projectiles were first developed in the USSR and adopted for the 20-mm aircraft gun in 1938. For ground artillery, spool-shaped sub-caliber armor-piercing projectiles were created in 1942.

Three types of rotating projectiles were developed: spool-shaped (Figure 4.2a), streamlined (Figure 4.2b,c), and with a discardable sabot (Figure 4.3).

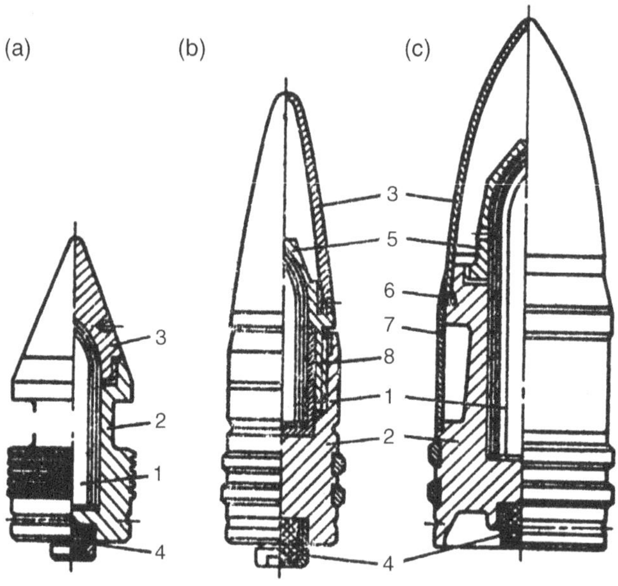

Figure 4.2 Armor-piercing sub-caliber projectiles: (a) 57-mm spool-shaped; (b) 57-mm streamlined; (c) 85-mm streamlined; 1 – armor-piercing core; 2 – body with driving band; 3 – ballistic tip; 4 – tracer; 5 – head; 6 – crimping ring; 7 – fairing; 8 – cup. *Source*: From Balagansky et al. [1].

The armor-piercing core is the main part of the projectile. It is made of tungsten carbide or tungsten powder with the addition of nickel, cobalt, and iron by powder metallurgy methods and, after sintering, has a density of about 14 g/cm^3, high hardness, and strength.

To increase the armor-piercing effect of a projectile, the core must have the greatest mass at the smallest cross-section and provide strength during armor penetration. In spooled and streamlined sub-caliber projectiles, the core mass usually does not exceed 30% of the projectile mass, and the relative length of the core is 3–4 of its diameter. Heavier and longer cores in identical calibers have streamlined projectiles, which gives them, other things being equal, a greater armor-penetrating effect.

The body of a sub-caliber projectile serves to guide the projectile through the barrel of the gun, to protect the core from damage during service and firing, and to increase the strength of the core before it penetrates the armor. In addition, when the projectile impacts the armor, the body transfers some of its kinetic energy to the core.

The ballistic tip serves to give the head of the projectile a streamlined shape. In addition, in spool-shaped projectiles, the ballistic tip keeps the core in the body during service and firing. The head is used in streamlined projectiles to keep the core in the sabot in service and firing conditions. The cup is used to center the core in a 57-mm streamlined projectile. The fairing improves the ballistic shape of the sub-caliber projectile and is a thin-walled steel ring covering an annular slot on the body. The fairing is held in place by a crimp ring. Streamlined projectiles have less air resistance in flight and therefore retain their velocity better when fired.

Projectiles with discardable sabot (Figure 4.3) have a carbide-tungsten core, which is mounted in a case with a welded head. Such assembly is called an active projectile. A tracer device is mounted in the lower part of the body. The active projectile is mounted in a steel sabot to which a driving band is welded. The sabot serves to guide the projectile through the gun's barrel channel. In the lower part of the projectile, the case and the sabot form a chamber for the accumulation of gases with high pressure during the shot. After the projectile leaves the barrel, the gases expand and separate the active projectile from the sabot. The sabot removal is also facilitated by the pressure of the incoming air stream.

Having a considerable mass (0.6–0.7 of the total projectile mass) and a small midsection (0.25–0.30 of the projectile midsection), as well as a good ballistic shape, the active projectile retains a high flight velocity and can damage targets with powerful armor protection at a considerable range.

Sub-caliber fin-stabilized projectiles (Figure 4.4) were first developed in the USSR and adopted by the Soviet Army in early 1960. They were designed to shoot at armored targets with heavy armor protection.

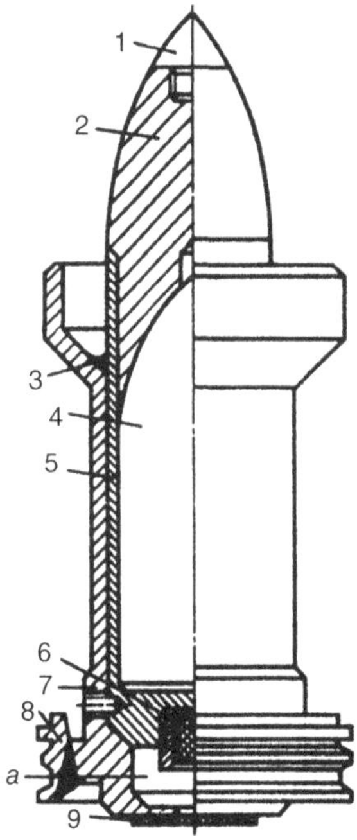

Figure 4.3 Sub-caliber projectile with discardable sabot: 1 – ballistic tip; 2 – head; 3 – cannon lubricant; 4 – armor-piercing core; 5 – body; 6 – tracer; 7 – sabot; 8 – driving band; 9 – celluloid circle; *a* – chamber. *Source:* From Balagansky et al. [1].

The fin-stabilized sub-caliber projectiles can be of two types: without armor-piercing cores and with cores.

Armor-piercing sub-caliber projectiles with cores (Figure 4.4b,c) have, also, an armor-piercing core 9 placed in the head part of the projectile and a damping-localizer 10 instead of an armor-piercing tip.

The body, the armor-piercing and ballistic tips, the core, the stabilizer, and the tracer form the active part of the projectile, which serves to directly damage the target.

The body is the main damaging part of the projectile and is made of high alloyed steel. The body has an ogival or threaded part at the head end to attach the armor-piercing tip, a thread at the tail end to attach the stabilizer, and annular grooves at the center of the body to attach the drive ring sectors. After specific heat treatment, the body has a high hardness that decreases from the head to the tail of the projectile.

The armor-piercing tip has the same purpose as in caliber armor-piercing projectiles. In addition, some projectiles with tips have an armor-piercing core placed in and attached to the armor-piercing tip. The damper-localizer performs the same tasks as the armor-piercing tip and is made of quality steel or a tungsten alloy. The armor-piercing core serves to increase the lethality of the active projectile and is made of tungsten carbide. The ballistic tip is designed to improve the ballistic

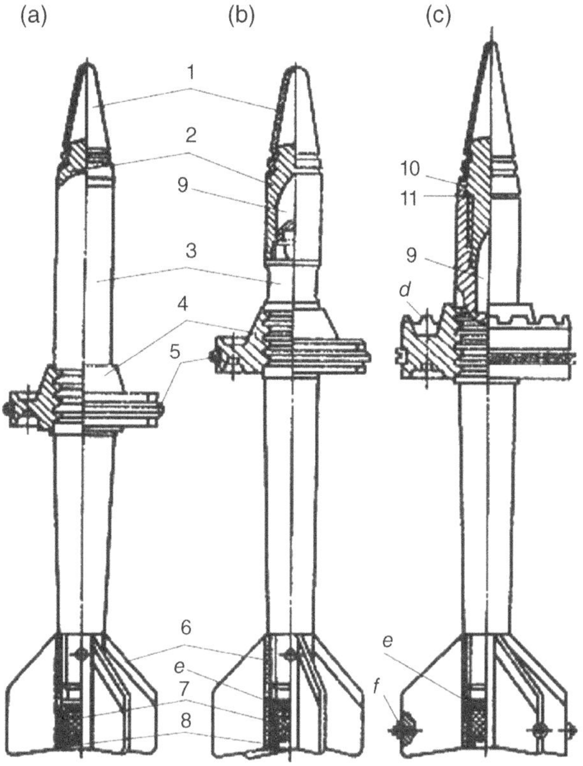

Figure 4.4 Armor-piercing sub-caliber projectiles without cores (Figure 4.4a) consist of a body 3, a driving ring 4 with a copper obturating band 5, an armor-piercing tip 2, a ballistic tip 1, a caliber stabilizer 6, a tracer 7, and a tracer nut 8.Figure 4.4 Sub-caliber fin-stabilized projectiles for smoothbore guns: (a) projectile BM2 without an armor-piercing core; (b) projectile BM1 with an armor-piercing core; (c) projectile BM22 with an armor-piercing core; 1 – ballistic tip; 2 – armor-piercing tip; 3 – body; 4 – driving ring; 5 – obturating band; 6 – stabilizer; 7 – tracer; 8 – tracer nut; 9 – armor-piercing core; 10 – damper-localizer: 11 – ring; *d* – protrusions on the sectors of the driving ring; *e* – hole for gas passage to and from the tracer area; *f* – centering pin on the chord of the blade. *Source:* From Balagansky et al. [1].

shape of the projectile, is made of thin sheet soft steel, and is attached to the armor-piercing tip by rolling.

The stabilizer serves to center the projectile in the cartridge case and during its movement through the barrel channel, and its main purpose is to ensure the stability of the projectile flight on the trajectory. The stabilizer blades have one-sided bevels that serve to maintain the rotation of the projectile on the trajectory due to the impact of the lateral component of the air resistance force. Such rotation at a relatively low angular velocity leads to improved accuracy.

The driving ring accepts the pressure of the powder gases when firing and serves to center the projectile in the barrel channel. The ring consists of three sectors with an angle of 120 each. The groove on the side surface of the sectors is fitted with an obturator band. In addition to the obturation of powder gases during a firing, the obturator band also binds the sectors together into a ring. The internal surfaces of the sectors are stamped with a comb-like projection to connect them to the case. Each sector has two holes, each 3–4 mm in diameter. The axes of the holes are inclined in a tangential plane and when the projectile moves in the gun barrel, the powder gases, escaping through these holes, create a force that makes the projectile rotate. The design characteristics of the projectiles are given in Table 4.1.

When the projectile is ejected from the barrel channel, under the action of the centrifugal force and the opposing air stream, the sectors destroy the obturating band and at a distance of about 35 calibers completely separate from the projectile body and fly away at an angle of no more than 4° from the firing direction to a range of up to 700 m.

The presence of discardable sectors made it possible to create projectiles with the mass of the active part being up to 80% of the mass of the projectile. The low total mass made it possible to give the projectiles a high initial velocity when fired (over 1500 m/s), and the good ballistic shape of the projectile and a small midsection with a large cross-loading made it possible to maintain the resulting velocity on the trajectory. All this led to an increased direct fire range (a range at which the height of the trajectory does not exceed the height of the target) with low flight time and high specific energy of the projectile hitting the armor, which led to a highly damaging effect of such projectiles. The streamlined shape of the projectile

Table 4.1 Constructive characteristics of sub-caliber fin-stabilized projectiles.

Projectile Index	Caliber of projectile, mm	Caliber of active projectile, mm	Length of projectile		Mass of the projectile, kg	Mass of the active part, kg	Presence of core
			mm	clb			
BM1	100	42	543	12.9	4.30	3.38	Yes
BM2	100	38	525	13.8	4.30	3.38	No
BM3	115	42	543	12.9	5.55	4.05	Yes
BM4	115	40	571	14.3	5.55	4.05	No
BM5	115	42	554	13.2	5.34	3.90	No
BM6	115	42	523	12.5	5.34	3.90	No
BM9	125	42	519	12.4	5.67	3.62	No
BM12	125	42	545	13.0	5.67	3.62	No
BM15	125	42	—	12.5	5.86	3.84	Yes
BM22	125	44	560	12.8	6.55	4.47	Yes

Source: From Balagansky et al. [1].

and its extremely high velocity mean that if it misses the target or ricochets off the ground, it can fly another 30–50 km, which creates certain difficulties for training firings.

It is important to note that the armor-piercing tungsten-carbide cores provide a high damage effect of such projectiles in the armored space because after penetration of the armor, these cores are destroyed with the formation of a large number of fragments flying with high velocity.

In recent decades, many countries have developed armor-piercing projectiles with cores made of depleted uranium U 238. Uranium is a silvery shiny metal, unlike tungsten carbide, it is much less hard, easy to process, and has a density of 18–20 g/cm^3. Uranium heated to high temperature in the space of the armor burns intensively when combined with air.

4.1.4 Concrete Piercing Artillery Projectiles

Concrete piercing projectiles appeared after the World War I when the development of fortification technology made it possible to use concrete structures on a large scale, and the experience of using large-caliber high-explosive projectiles to destroy structures showed their insufficient efficiency. Because of the insufficient strength of the body, high-explosive projectiles when hitting a reinforced concrete barrier were destroyed without penetrating its thickness.

Concrete piercing projectiles were widely used in the 1939–1940 war between USSR and Finland, as well as in the Great Patriotic War of 1941–1945 when storming fortified bases of the enemy. Thus, concrete piercing projectiles are intended for firing at reinforced concrete fortifications, strong stone and brick buildings, adapted for defense, as well as fortified bunkers of the enemy. If necessary, concrete-piercing projectiles can be used for direct fire against tanks and other armored targets.

The concrete-piercing action of the projectiles is ensured by the appropriate design of their body, the use of metal with high mechanical characteristics and heat treatment, as well as the use of powerful explosives for equipping. The device of concrete-piercing projectiles is shown in Figure 4.5.

The body of a concrete-piercing projectile has a long-range streamlined shape and an apex with a slight bluntness. In the head part, the body has a comparatively thick vault (0.7–1.5 clb), which provides the necessary strength of the head part of the projectile when penetrating reinforced concrete barriers. The driving band is attached to the projectile body and the chamber is closed by a screw bottom with a fuse.

Concrete-piercing projectile bodies are made of alloyed steel with good hardenability and toughness. After heat treatment, the high hardness of the body is

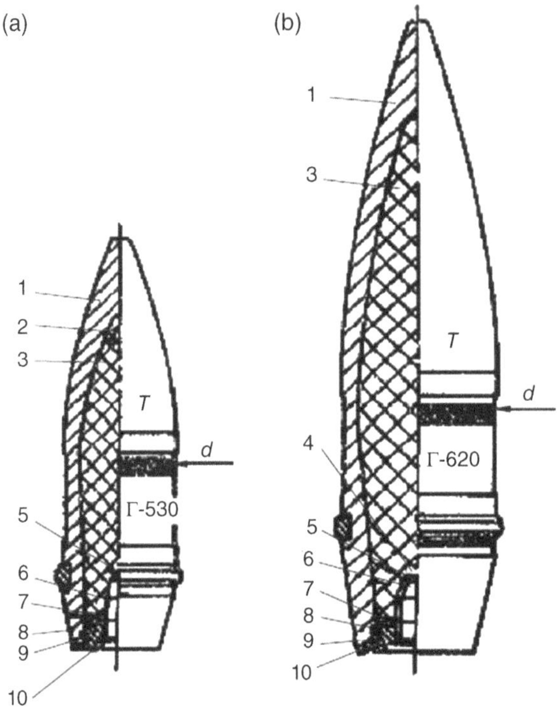

Figure 4.5 Concrete-piercing projectiles: (a) 152-mm projectile; (b) 203-mm projectile; 1 – body; 2 – inert insert; 3 – explosive charge; 4 – brass case; 5 – cloth gaskets; 6 – bottom fuse; 7 – gasket set; 8 – bottom; 9 – lead gasket under bottom: 10 – lead gasket under fuse; (d) distinctive blue stripe. *Source:* From Balagansky et al. [1].

achieved. A decrease in the hardness of the body from the head to the bottom ensures that the necessary strength of the body is obtained when hitting and penetrating a target.

To reduce stresses in the explosive material when the projectile hits a reinforced concrete barrier, inert inserts are installed at the top of the chamber of some projectiles. Gaps between the surface of the explosive and the bottom are eliminated using a set of gaskets. The same gaskets are used to hold the explosive charge by the bottom. To isolate the explosive charge from the bottom threads, the fuzing hole is sealed with a brass case. The breech block with the explosive charge is sealed at the threaded joints of the case to the bottom and the bottom to the fuzes by crimping the lead shims.

Concrete-piercing projectiles have bottom fuzes with settings for delayed and inertial action.

4.1.5 Weapons and Ammunition for Damaging Extremely Resistant Targets

As noted earlier, there are special weapons and penetrating munitions for destroying aircraft in reinforced concrete shelters, objects of military equipment and manpower located in especially strong shelters, disabling airfield runways, as well as for destroying command and communication points and other objects.

By now many armies around the world have such munitions in service. As a rule, they are aviation munitions: unguided or guided missiles, air bombs, monoblocks, or cluster munitions. All such munitions have solid propellant engines and are accelerated to high velocities when launched or dropped from the carrier, ensuring the penetration of a barrier utilizing their kinetic energy. Naturally, the experience gained in the development of armor-piercing and concrete-piercing projectiles was taken into account in their design.

Here are some examples of the characteristics of several penetrating munitions for hitting extremely resistant targets.

The S-8BM unguided aviation missile is designed to damage equipment and manpower in fortifications. A specially designed concrete-piercing warhead of the missile pierces a reinforced concrete barrier and explodes, producing a large number of damaging fragments.

Main tactical and technical characteristics of the S-8BM.

Caliber, mm	80
Length, mm	1540
Mass, kg:	
Missile	15.2
Warhead	7.41
HE	0.6
Thickness of the pierced reinforced concrete barrier, m	0.8
Launch range, m	1200–2200
Initial velocity of the missile, m/s	450

S-13, S-13T unguided aircraft missiles

S-13 missile is designed to destroy aircraft in reinforced concrete shelters, as well as military equipment and manpower located in especially hard shelters. S-13T missile is intended to destroy aircraft located in various types of shelters, including reinforced shelters, to disable airfield runways, as well as to destroy command and control posts, communication points, and other objects.

The S-13T is designed to increase the effectiveness of the S-13 missile by using two independent separable warheads, the first of which is a penetrating warhead (analog of the S-13 warhead) and the second is a high-explosive fragmentation warhead. The design solutions used in the development of the S-13T missile system not only increased the amount of explosive carried behind the barrier but also made it possible to separate the blast zones of the warheads behind the barrier, which increased the effectiveness of the warheads.

Main tactical and technical characteristics of missiles.	S-13	S-13T
Caliber (engine/munition), mm	122/90	122/90
Length, mm	2540	3100
Mass, kg:		
Missile	57	75
Warhead	21	21 + 16.3
HE	1.82	1.8 + 2.7
Damaging effect	Penetrates 3 m Soil and 1 m Reinforced concrete	Penetrates 6 m Soil and 1 m Reinforced concrete, Damage area of Runway 20 m^2
Launch range, m	1100–3000	1100–4000
Initial velocity of the missile, m/s	650	500

4.1.6 Ammunition of Small Arms

The primary use of small arms is to damage enemy personnel. The ammunition of small arms is usually in the form of cartridges. A cartridge contains all the elements necessary to produce a shot: bullet, powder charge, and primer. Since all these elements are combined within a cartridge case, it is called unitary ammunition. Since the target is damaged by the bullet fired from the firearm, we will speak here specifically about bullets.

The main type of bullet for military weapons is the so-called ordinary bullet. This bullet affects the target only mechanically, hitting it due to its kinetic energy. The construction of an ordinary bullet is very simple. Usually, such bullets consist of a lead core enclosed in a shell of a harder metal. (There are exceptions – the core may be of a different metal or composite, or it may be absent at all.) The larger the length of the bullet and the greater its cross load, the higher the energy saving on the trajectory is, along with the smoothness of the trajectory, and the kinetic energy.

The first samples of ordinary bullets for smokeless powder cartridges were exclusively cylindrical with a rounded head part. Over time, a somewhat lighter, pointed bullet type was developed. The trajectory of such bullets due to lower weight and hence less cross load and because of this increase in initial velocity turned out to be more gentle at the beginning and more steep at the end. Pointed ogival-shaped bullets, having a high velocity, showed the ability to spread the impact force in a circle to the sides, thereby increasing their lethal effect.

Variations of bullets that affect targets only mechanically are so-called heavy bullets (usually elongated) or especially light bullets, e.g. with aluminum cores. Varying the mass of the bullet makes it possible to achieve better results when hitting different types of targets. Sometimes the center of mass is shifted to the tail of the bullet to increase the lethality of the bullet.

For this purpose, for example in the British Mk-VII bullet (1914), the lead core is not made completely of lead, but with an aluminum or fiber head; in the Japanese bullet, due to the unequal thickness of the shell walls, the main mass of the lead core is concentrated at the tail end of the bullet; finally, a heavy core may not fill the head section of the bullet, leaving it hollow and, therefore, also light. When colliding with an object, especially in that part of the trajectory where it has already deviated noticeably from the throwing line, such bullets exhibit the ability, due to the phenomenon of precession, to change their position abruptly and penetrate an object not by the head part, but by its side.

Special bullets, used most often for machine-gun fire, are designed primarily to hit enemy combat equipment. In addition to hitting action on the target, they may also have special effects on it. Thus, they can be, for example, incendiary (filled with ignition composition), armor-piercing (with a very strong, usually steel core), and high explosive (with explosive material inside). During the flight, all the bullets mentioned above are invisible, and when firing, you can rely only on sighting devices to ensure accuracy.

Tracer bullets, which leave a clearly visible trail of light or smoke behind them in flight, are very helpful in correcting fire and making adjustments to targeting. The aforementioned exploding bullets may also serve as fire correction targets. On hitting an object such bullets explode, forming a cloud of smoke that is visible at a considerable distance. They also have an incendiary effect.

Special bullets can be combined, e.g. armor piercing, armor-piercing-incendiary, armor-piercing-incendiary-tracer. Cartridges with special bullets have distinctive features. Most often they have colored head parts of bullets.

To conclude our discussion of special bullets, we should return to the name "exploding" bullet. Not only bullets filled with an explosive substance but also "Dum-dum" bullets with a lead core and a shell-free head of some shape or another are called like that.

These bullets, first used by the British during the Boer War, caused particularly serious wounds because they deformed strongly ("exploded") when encountering a target. The Hague Convention of 1899 has forbidden their use in military weapons. The bullets used in hunting rifles – with a lead head without a shell – are called expansive bullets.

Figure 4.6 shows samples of bullets for rifle and automatic rifle cartridges.

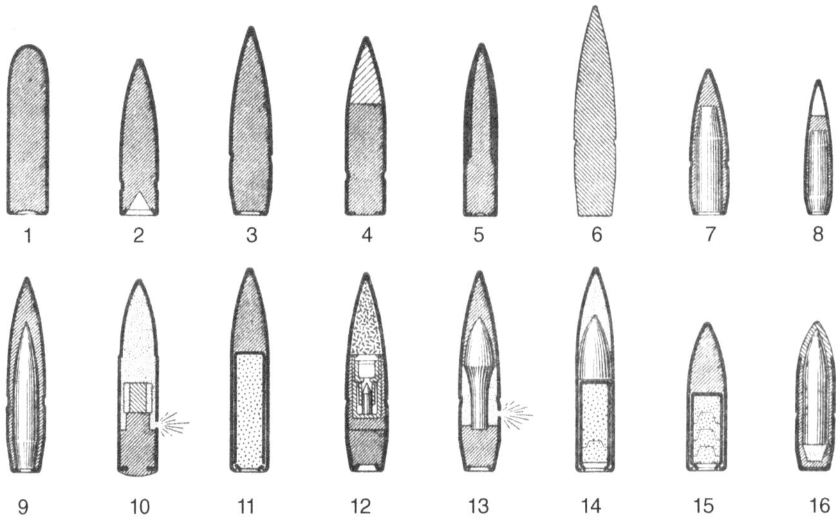

Figure 4.6 Samples of bullets for rifle and automatic rifle cartridges. ***Rifle ordinary bullets:*** 1 – blunt-pointed bullet, 1888, for 8 × 50R cartridge (Austria-Hungary); 2 – pointed bullet, 1910, for 7.62 × 53R cartridge (Russia, USSR); 3 – heavy pointed SS to 7.92 × 57 cartridge (Germany); 4 – Mk VII to 7.71 × 56R cartridge (UK) 5 – to the 6.5 × 50 cartridge (Japan); 6 – "D" to the 8 × 50R cartridge (France). ***Automatic rifle ordinary bullets:*** 7 – for 1943 7.62 × 39 cartridge (USSR); 8 – for 5.45 × 39 cartridge (USSR). ***Rifle special bullets:*** 9 – armor-piercing (USSR); 10 – incendiary (Great Britain); 11 – tracer (USSR); 12 – targeting "B" (Germany); 13 – armor-piercing incendiary "PMK" (Germany); 14 – armor-piercing incendiary tracer "BZT" (USSR). ***Automatic rifle special bullets:*** 15 – tracer to 7.62 × 39 cartridge (USSR); 16 – armor-piercing incendiary to 7.62 × 39 cartridge (USSR). *Source:* From Zhuk [2].

4.2 Interaction of Impactors with Targets

Without dwelling on the design features of the penetrating munitions described above, to understand the basic laws of their interaction with barriers, we will consider two main groups of impactors: compact impactors in which all dimensions are close together (typical examples are a ball, cube, cylinder with a height equal to the base diameter) and elongated impactors in which one of the dimensions significantly exceeds the others; we will call the latter "rods." Such a division makes it possible to analyze the mechanisms of interaction of the main types of impactors with barriers and to estimate their damaging (destructive) effect. Examples of typical schematization of impactors are shown in Figure 4.7.

We will distinguish between **piercing** and **penetration** based on the effect of the impactors on the barrier. Penetration is defined as the entry of a body into a

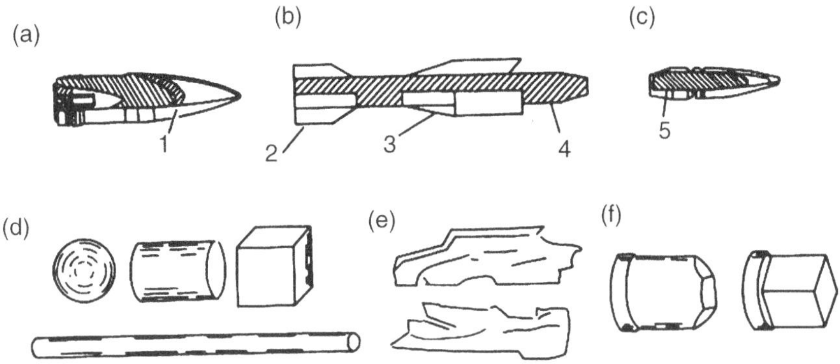

Figure 4.7 Examples of types of impactors: (a) armor-piercing projectile; (b) projectile with stabilizer and discardable sabot; (c) projectile of 12.7 mm caliber; (d) solid bodies of regular shape; (e) fragments of irregular shape; (f) fragment imitators. 1 – armor-piercing tip; 2 – stabilizer; 3 – sabot; 4 – penetrating body; 5 – chromium–molybdenum steel core. *Source*: From Zukas et al. [3]. Reprinted with permission of Wiley.

target without forming a through hole in it. In a general case, the impactor deforms to a greater or lesser degree and gets stuck in the target, in which a cavity is formed. If the impactor bounces off the surface of the target or penetrates it along a curvilinear trajectory and then leaves it at a slower speed on the same side from which the impact was made, this phenomenon is called a **ricochet**.

In **piercing**, a body or its fragments pass completely through the target. It is clear that the difference between penetration and piercing is conditional, and depending on the problem to be solved, some cases of interaction may pass from one category to the other.

Let us consider the basic physical processes occurring in the projectile (impactor) and the target (barrier) during the impact. Regardless of the geometric parameters of the impactor during its impact with the target, shock waves are formed in both bodies and propagate from the impacted place. When they reach free surfaces, rarefaction waves are generated. The rarefaction waves, meeting inside the impactor in the area adjacent to its axis, create a zone of tensile stresses, which may lead to the destruction of the impactor material. This effect is more intense if there are cavities, pores, or other inhomogeneities near the impactor axis. The phenomenon becomes much more complicated if the direction of impact does not coincide with the normal to the surface of the barrier.

A rarefaction wave also forms in the barrier as the shock wave enters the surface and moves into the barrier. When the initial shock wave reaches the rear free surface of the barrier, another rarefaction wave is formed, moving toward the first one. In a certain combination of tensile stresses, their duration, and the strength of the material, the material crashes in the unloading waves (front and rear spall).

Particularly, favorable conditions for spalling occur when rarefaction waves propagate from the front and rear surfaces of a barrier. The phenomenon of spalling can be used to damage a barrier space even in the absence of piercing or penetration.

4.2.1 Classification of Dynamic Penetration Conditions. Main Factors

Let us briefly consider the main factors influencing the characteristics of the penetration process during the interaction of deformable bodies.

4.2.2 Impact Velocity

Table 4.2 gives a classification of the prevailing material properties depending on the impact velocity V_s and deformation velocity $\acute{\varepsilon}$.

4.2.3 Mechanical Properties

Depending on its strength properties and the properties of the barrier, the impactor is usually considered to be absolutely rigid or plastically deformable. The head part of a deformable impactor acquires a shape close to a hemisphere in the process of interaction. Robust impactors are practically undeformable, but they can produce brittle fractures at certain impact velocities and angles.

The mechanical properties of the barrier at medium impact velocities significantly affect the behavior of fracture. At high impact velocities, the main characteristics of the barrier are density and compressibility. At ultrahigh velocities, the thermal–physical properties of the barrier and the impactor become determinative.

Table 4.2 Material properties depending on impact conditions.

V_s	$\acute{\varepsilon}$, s^{-1}	**Material properties (behavior)**
<50 m/s	10^0	Elastic deformations prevail. Zones of localization are possible
50–500 m/s	10^2	Plastic deformations prevail
0.5–1 km/s	10^4	The appearance of viscoplastic and viscous properties. The strength of the material remains essential
1–3 km/s	10^5	Hydrodynamic behavior of solids. The defining parameter – density
3–12 km/s	10^6	Compressibility must be taken into account
>12 km/s	10^8	Evaporation of material

Source: From Balagansky et al. [1].

4.2.4 The Geometry of the Impactor and the Barrier

A distinction is made between thin, medium, and semi-infinite (in relation to the diameter of the impactor) barriers. The thickness of the barrier in combination with its mechanical properties influences the behavior of fracture and the complexity of analytical or numerical modeling of the process. Thin and semi-infinite barriers have been studied to a greater extent.

The shape of the impactor also affects the behavior of the penetration process. There are two basic types of impactor head shapes: pointed and blunted. According to their length, impactors are subdivided into elongated ($l/d > 5$–10), slightly elongated ($l/d \cong 2$–3), and compact ($l/d \cong 1$) (l and d are the length and diameter of the impactor, respectively). At high impact velocities, elongation plays an important role.

4.2.5 The Angle of Impact

When pointed impactors are penetrated at an angle between the normal to the barrier and the impactor axis $\varphi_0 \leq \varphi_0^*$ (where φ_0^* is the ricochet angle), denormalization of the impactor may be observed as it enters the barrier and a return to normalization as it exits the barrier.

At $\varphi_0 > \varphi_0^*$, the ricochet phenomenon is observed. Note the great influence of the initial angle between the velocity vector of the impactor's center of mass and its symmetry axis on the penetration process. When the angle from the velocity vector is counted clockwise, the initial positive angle contributes to ricocheting, and the negative angle contributes to "bite" and penetration.

4.2.6 Other Factors

Technological factors have a large and difficult to estimate the influence on the behavior of deformation and especially on the fracture of metal barriers.

All of the mentioned factors, taken together, form the conditions of collision. Considering their various combinations, it is possible to observe processes completely different in the behavior of deformation and fracture, which differ from each other so much that it is impossible to analyze them based on a single approach. Examples of different types of fractures when hitting on the barrier are shown in Figure 4.8.

Real destruction is a combination of these types, so when solving specific problems, it is important to identify the main mechanism of destruction, neglecting others or partially considering them. The schematization of the process makes it possible to set a mathematically observable problem since taking into account all factors is very difficult and practically impossible.

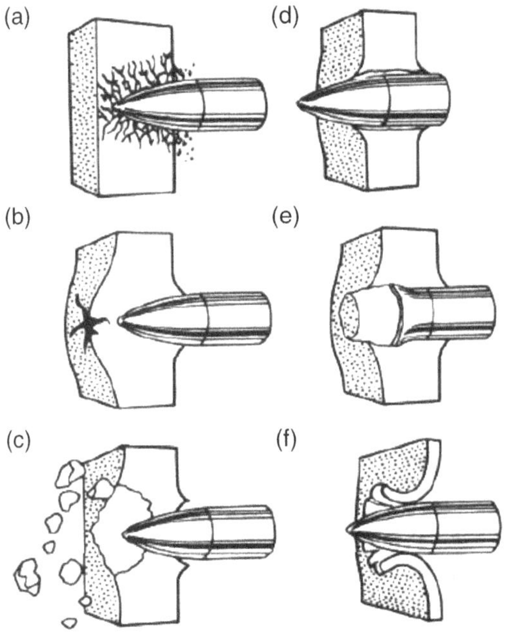

Figure 4.8 Types of target destruction: (a) brittle fracture; (b) fracture with the formation of radial cracks; (c) fragmentation; (d) plastic expansion of the hole; (e) punching out of the plug; (f) formation of a petal hole. *Source*: From Backman [4].

As an attempt at a general approach to the penetration conditions, we note the analysis based on the methods of similarity theory. Comparison with experimental data allows us to conclude that the penetration process is modelable if the corresponding similarity criteria are fulfilled.

It is convenient to present the results of experimental studies of the collision of impactors with barriers graphically in the form of dependences between variables such as the velocity of the impactor, barrier thickness, collision angle, kinetic energy of the impactor. Such graphs are plotted for fixed physical and geometrical characteristics of the impactor and the barrier. Often the dependencies are plotted in a dimensionless form, then the corresponding characteristics are referred to the diameter or length of the impactor, sound velocity in the material, some critical energy values, etc.

Figure 4.9 shows an example of a phase diagram characterizing the behavior of a 6.35-mm diameter impactor with an ogival head during an impact on a 6.35-mm thick aluminum alloy barrier. The result of the impact is determined by the impact velocity and the angle of collision. It is clear that for a particular projectile and target, much more complex phase diagrams can be constructed from the results of experiments and can be used in engineering applications.

Let us dwell in more detail on the two main types of fracture of barriers at medium impact velocities. Experiments show that the increase in resistance of

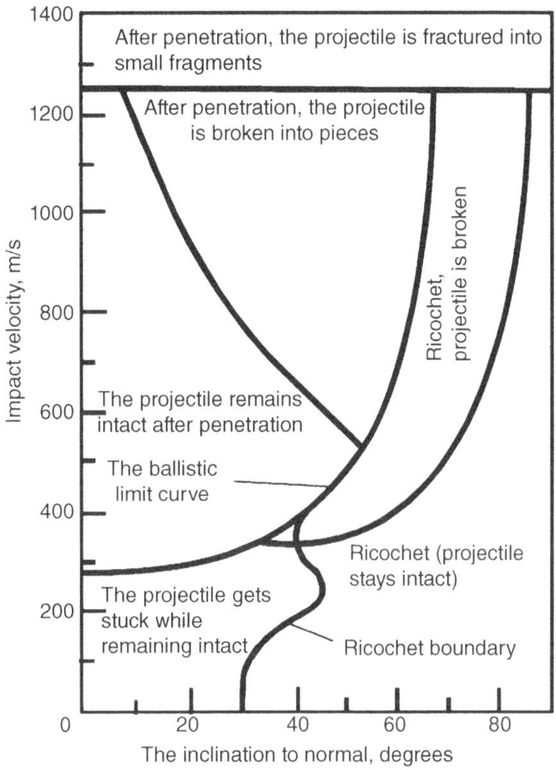

Figure 4.9 Phase diagram characterizing the collision of the impactor with the barrier. *Source*: From Backman and Goldsmith [5]. Reprinted with permission of Elsevier.

metallic barriers with an increase in their hardness occurs only to the limit of viscous formation of the crater and the beginning of plug formation.

A further increase in strength properties is associated, as a rule, with a decrease in plasticity and fracture toughness of the metal, which leads to easier plug formation, brittle fracture, and weakening, ultimately, the resistance of the obstacle to penetration. This is the reason for the practical interest in the two main types of target fracture: piercing and plug formation.

4.2.7 Plug Formation

At the moment of impact of the blunted impactor, a shock wave propagates from the contact surface, the interaction of which with radial rarefaction wave in the impactor leads to its thickening near the contact surface, and in the barrier – to the formation of an area of developed plastic shear deformation. This region in the penetration process is localized near the conical surface, extending to the backside of the barrier.

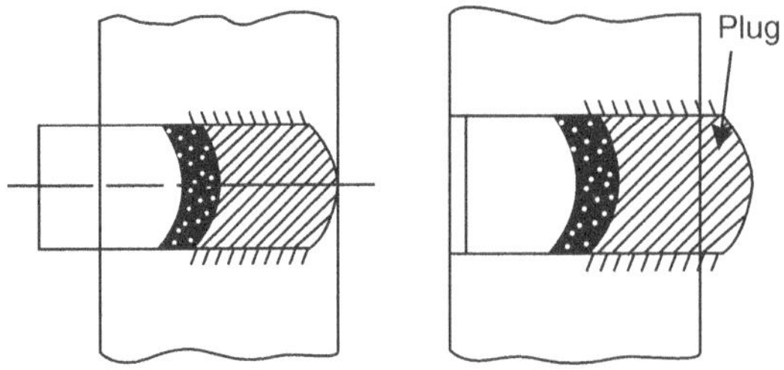

Figure 4.10 Scheme of the mechanism of plug formation. *Source:* From Balagansky et al. [1].

Temperature growth on this surface in conditions of limited heat dissipation leads to softening of the material and greater localization of plastic deformations – the phenomenon of thermo-shear instability. The release of a shear crack to the rear surface of the barrier corresponds to the formation of a plug. Further deformation consists of pushing out the plug and deformation of surrounding layers of the barrier (Figure 4.10).

In practical engineering calculations of penetration for plug formation, the impact is usually assumed to be inelastic and the deformation of the surrounding layers of the barrier is neglected. Based on the energy balance equation, a simple relationship for the final velocity of the impactor after penetration is derived,

$$V_f = \frac{m}{m + m_0}\left(V_0^2 - V_l^2\right)^{0.5},$$

where m and m_0 are masses of impactor and plug; V_0 and V_l are initial and minimum initial impact velocity capable of causing a piercing (ballistic limit). Various empirical formulas can be used to determine the value of V_l, such as Jacob de Marre's equation (see below).

4.2.8 Viscous Crater Formation (Puncture)

In the early stage of penetration, during puncture, the material of the barrier is displaced in the direction of lowest resistance, i.e. to the surface, forming a crater. The developed plastic deformations localize near the crater, forming a region with an unsteady external boundary. The state of the material outside this region can be characterized as elastic and, within the context of idealized models, as rigid. The pressures in the barrier at average impact velocities are low, so the material is displaced rather than compressed at the lowest energy level.

The diameter of the craters can be larger than that of the impactor as a result of the radial inertial motion of the particles. Friction between the impactor and the crater walls has almost no effect on the penetration process. In the case of static penetration, however, the role of friction is significant.

Specific resistance F to the penetration of the cone impactor into the surface layers of a metal semi-infinite plate at any average velocity and into the deep layers at velocities $V > V_{cr}$ is determined by the empirical relation

$$F = H + k\rho V^2, \tag{4.1}$$

where $k = \sin^2\alpha$ – coefficient of influence of the shape for the cone head; α – the angle of the cone half-section; H – dynamic hardness of the plate material, measured when the cone impactor penetrates the surface layers of the barrier; ρ – density of the barrier material. The critical velocity value is determined by the relation

$$V_{cr} \approx \left(\frac{H_B}{k\rho}\right)^{1/2},$$

H_B – the Brinell hardness of the plate material.

Ratio (4.1), due to its simplicity and clear physical sense, is of great importance in practical applications and needs to be modernized taking into account a number of factors (e.g. weakening effect of free surfaces of the barrier, viscous effects).

4.2.9 Ballistic Limit

One of the tasks encountered in the study of impact phenomena is to determine the limiting velocity below which the thrown object is no longer able to penetrate a barrier or any protective device. This is of primary importance in the design of protective structures, evaluation of the armor effectiveness of military vehicles, and in all cases where the impact may cause damage. This velocity is commonly referred to as the critical impact velocity, ballistic limit velocity, or ballistic limit.

The existing methods of determining the ballistic velocity limit can be divided into deterministic and probabilistic ones. In the first case, the limiting velocity is determined based on physical principles (conservation laws and the equation of state of materials), but the corresponding partial differential equations are too complex and we have to make simplifications, which usually require the introduction of one or two empirical constants.

In the probabilistic approach, models are built based on a considerable amount of data, including impact velocity and either the residual velocity of the projectile or the fact that it does or does not pierce the barrier. The resulting critical velocity is usually denoted by V_{50} and is the impact velocity at which the probability of piercing a barrier is 50%.

In the simplest case, V_{50} is determined by averaging six impact velocities, which include the three lowest velocities at which the projectile penetrated the target through and the three highest velocities at which it only partially penetrated the target. This requires that the difference between the lowest partial penetration velocity and the highest full penetration velocity not exceed 46 m/s. In practice, time and economic reasons limit the amount of data available and are usually insufficient to reliably determine critical impact velocities in either a deterministic or a probabilistic approach.

The deterministic approach to determining the ballistic limit is attempting to predict the behavior of a projectile during penetration by developing models based on conservation laws and assumptions about the mechanical properties of the system. In developing models, many difficulties are avoided by assuming that the projectile is a non-deformable or perfectly solid body. Using these assumptions, one can conclude that dependences of the following kind are obtained:

$$V_r = \begin{cases} 0, & 0 \le V_s \le V_l, \\ \alpha\left(V_s^p - V_l^p\right)^{1/p}, & V_s > V_l, \end{cases}$$

where V_r is the residual velocity of the projectile; V_s is the impact velocity, and V_l is the limiting ballistic velocity. For non-deformable projectiles, $p = 2$, α and V_l are empirical constants obtained by regression.

Figure 4.11a shows a typical dependence of the probability of piercing a target on impact velocity for some combination of projectile and target characteristics.

Such graphs are obtained by shooting a large number of projectiles and statistically processing the data. Any velocity value in the area of mixed results can be regarded as a limiting velocity. For example, limiting velocities V_{10} and V_{90} are those impact velocities at which the probability of piercing a target is

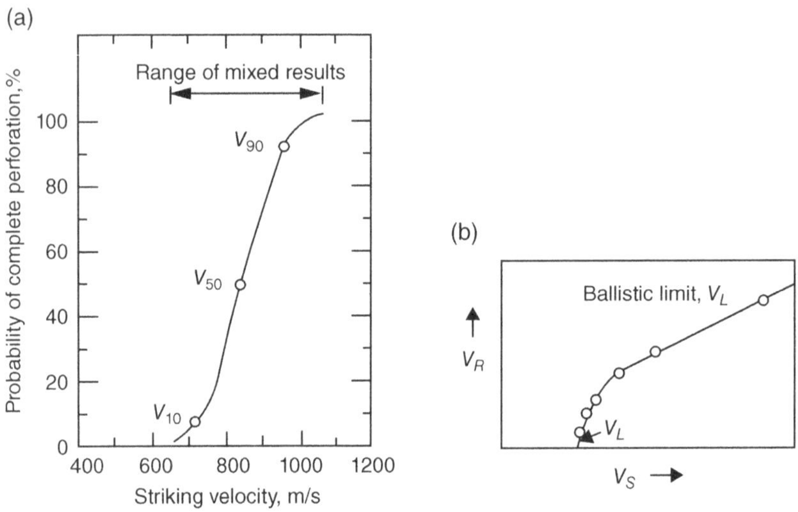

Figure 4.11 Ballistic limit determination by (a) statistical measurements and (b) residual velocity measurement. *Source*: From Zukas et al. [3]. Reprinted with permission of Wiley.

10 and 90%, respectively. Usually, V_{50} is taken as the limiting velocity, because near this point the probability distribution curve has the maximum slope, and it allows to find V_{50} with maximum accuracy.

The ballistic velocity limit curve in Figure 4.11b divides the regions of piercing and non-piercing with an accuracy that cannot be achieved in reality. In fact, near this curve, the projectile penetration of a target should be considered from a probabilistic rather than a deterministic point of view.

Before the ballistic velocity limit can be measured, appropriate definitions must be introduced. First, we must define complete piercing. Clearly, the concept is largely subjective and is determined by the task at hand. As an example of possible variants, here are three definitions.

1) **Complete piercing** is the piercing of a target so that light passes through it freely.
2) **Complete piercing** is the piercing of a target so that at least half of the projectile passes through it.
3) C**omplete piercin**g of a target is one in which a thin aluminum plate placed behind it is also pierced.

Under the same impact conditions, the limiting velocity corresponding to the second definition is usually greater than that corresponding to the first definition. The limiting velocity according to the third definition is close to the velocity according to the second definition, except for the case when there is a spall from the rear surface of the barrier.

Note that only the second definition relates piercing to the behavior of the projectile. Introducing the definition of complete piercing, it is necessary to find the limiting ballistic velocity in terms of achieving success on impact. This velocity is usually taken as V_{50}. An impact with a velocity greater than V_{50} may not result in a complete piercing, but the probability of the latter increases with increasing velocity.

The probabilistic value V_{50} as the limiting ballistic velocity is used frequently and widely spread. It is quite convenient to use in both experimental and theoretical studies. Its main disadvantage is the need to carry out a fairly large number of experiments to obtain a statistically justified value of V_{50}, the results of which are not fully used.

A deterministic V_l value has several obvious advantages, the most important of which is more complete use of the available information. Instead of simply stating the fact of piercing – non-piercing in the case of piercing, weighted velocity values are used, thus providing a more accurate and economical result. It is common to obtain a real $V_r(V_s)$ curve and simultaneously with it a ballistic velocity limit V_l.

This not only facilitates projectile engineering and comparison of experimental data with the results of analytical and numerical studies but also allows us to consider the piercing of a target consisting of several consecutive plates.

Jacob de Marre's equation is widely used to estimate the limiting ballistic velocity for piercing armor in the average velocity range:

$$V_l = K\frac{d^{0.75}b^{0.7}}{q^{0.5}\cos\psi_0}\ (\text{m/s}), \tag{4.2}$$

where K is the armor-piercing coefficient ($K = 1900$–2100 for sharp-headed projectiles, $K = 2100$–2300 for projectiles with blunted head); d is the caliber of the projectile, (dm); b is the thickness of armor pierced, (dm); q is the mass of the projectile, (kg); ψ_0 is the arriving angle of the projectile to the target, measured from the normal to the armor surface.

In modern engineering applications, numerical calculations are used to estimate the V_l or thickness of the barrier to be pierced at a given velocity. At the same time, the role of experimental studies of piercing and penetration remains important and largely determinative. They remain the basis for obtaining a physical understanding of the interaction between the munition and the target, as well as the determination of empirical relationships that describe this process in whole or in part (obtaining formulas to calculate the total depth and time of penetration of the munition into a semi-infinite barrier under known interaction conditions; for a finite thickness barrier – formulas to determine the velocity required to pierce the barrier and the duration of the process).

At the same time, based on physical assumptions about the process of impact–penetration interaction, it is possible to establish the law of resistance of solid media to the movement of a munition in them, the knowledge of which would allow obtaining not only empirical but also more complete in meaning semi-empirical formulas of evaluation of impact–penetration interaction. The accuracy of such formulas depends on the correct selection of coefficients introduced into the resistance law, determined experimentally, and taking into account the factors and assumptions used in the interpretation of the physical picture of the process under consideration.

Of a large number of empirical formulas available for calculating the depth of penetration of a munition into soil and concrete, the most common are the following formulas: Zabudsky's, Berezan, and ANII.

Zabudsky's formula for the current penetration depth is

$$h = \frac{q\sin\theta_0}{2ABFS}\ln\left(\frac{1+BV_s^2}{1+BV^2}\right)\ (\text{m}), \tag{4.3}$$

where q is the mass of the munition, (kg); θ_0 is the angle of the munition meeting the surface of the ground, measured from its surface; A and B are coefficients depending on ground properties (see Table 4.3);

Table 4.3 Coefficients taking into account the properties of different media.

Media	$A \cdot 10^{-6}$	$B \cdot 10^6$	$K_b \cdot 10^6$
Loose soil, freshly poured	4.5	60	13–17
Regular soil	—	—	11–13
Medium-density clay	10.2	35	7–10
Frozen soil	—	—	3.5
Loam	—	—	6–8
The sand of natural humidity	4.3	20	4.5–7
Water saturated sand	4.3	20	6–9
Sandy clay	—	—	5–7
Limestone	—	—	1.8–2
Granite (rocky soil)	—	—	1.4–1.7
Wood (conifers)	11.4	10	—
Masonry	31	15	2–2.5
Medium-strength concrete	—	—	1.2–1.5
Reinforced concrete	—	—	0.7–0.9

Source: From Vodopyanov [6].

$F = (1.15–1.20)h_{mh}/d$ – shape factor of the ammunition head; h_{mh} – the height of the munition head, (m); d – the caliber of the munition, (m); S – midsection of the munition, (m^2); V_s, V – velocity of approach of the munition to the barrier and the current velocity of the munition inside the barrier, respectively, (m/s).

Zabudsky's formula satisfactorily describes the real process of penetrating blunt-headed BPs with meeting velocities $V_s < 400$ m/s.

The Berezan formula is an empirical formula derived from firing tests conducted in 1912 on Berezan Island in the Black Sea near Ochakov:

$$h = K\frac{q}{d^2}V_s \cos\psi_0 \text{ (m)}, \tag{4.4}$$

where K – coefficient characterizing the properties of the barrier; q – a mass of the projectile, (kg); d – projectile caliber, (m); V_s – velocity of the meeting of the projectile with the barrier, (m/s); ψ_0 – the angle of the meeting (from the normal) of the projectile with the barrier.

The Berezan formula is applicable for a relatively small velocity range and gives low accuracy. Calculation of penetration of conventional weapons into soils, rocks, and other materials for non-deformable artillery projectile bodies at meeting velocities up to 400 m/s is recommended using the Berezan formula, transformed by ANII (Russian Artillery Research Institute). This formula is often referred to as the ANII formula:

$$h = \lambda K_b\frac{q}{d^2}V_s\frac{\cos(n\psi_0)}{\sqrt{\cos\psi_0}} \text{ (m)}, \tag{4.5}$$

where $\lambda = \lambda_1 \cdot \lambda_2$ is a coefficient taking into account the shape of the projectile head, $\lambda_1 = 0.5 + 0.4\sqrt[3]{(H_h/d)^2}$, $\lambda_2 = 2.8\sqrt[3]{d} - 1.3\sqrt{d}$; K_b is a coefficient characterizing the properties of the barrier (see Table 4.3); $\psi_0 \leq 90^\circ/(2n-1)$ – angle (from the normal) of the projectile meeting the barrier; n – coefficient taking into account possible curvature of the projectile trajectory in barrier: $n = 1.82$ – for long-range shape projectiles, $n = 2.62$ – for short-range shape projectiles; H_h – the height of head part (m); d – projectile caliber (m).

For supersonic impact velocities, when the munition velocity V_s exceeds the sound speed of the barrier material a_0, the following formula is acceptable:

$$h = \lambda \frac{q}{d^2} K\left(\frac{V_s}{a_0}\right) 10^4 \text{ (m)},$$
$$\text{here } \lambda = (0.45...1.0)\frac{H_h}{d}; \quad K\left(\frac{V_s}{a_0}\right) = 24\frac{V_s}{a_0} \exp 6.65\left(\frac{V_s}{a_0}\right). \tag{4.6}$$

The next section will show which law of resistance to penetration corresponds to each of the above formulas.

Example 4.1.

Calculation of the penetration depth of an OF-472 projectile into the frozen ground when firing the D-74 cannon.

Initial data: projectile caliber $d = 122$ mm; projectile mass $q = 27.3$ kg; height of its head $H_h = 378$ mm; projectile velocity meeting with the barrier $V_s = 338$ m/s; angle of meeting $\psi_0 = 0°$.

$$\text{calculation}: \lambda_1 = 0.5 + 0.4\sqrt[3]{\left(\frac{H_h}{d}\right)^2} = 0.5 + 0.4\sqrt[3]{\left(\frac{0.378}{0.122}\right)^2} = 1.35;$$
$$\lambda_2 = 2.8\sqrt[3]{d} - 1.3\sqrt{d} = 2.8\sqrt[3]{0.122} - 1.3\sqrt{0.122} = 0.935$$
$$\lambda = \lambda_1 \cdot \lambda_2 = 1.35 \cdot 0.935 = 1.262;$$
$$h = \lambda K_b \frac{q}{d^2} V_s \frac{\cos(n\psi_0)}{\sqrt{\cos\psi_0}} = 1.262 \cdot 3.5 \cdot 10^{-6} \cdot \frac{27.3}{0.122^2} \cdot 338 \cdot \frac{\cos(1.82 \cdot 0^\circ)}{\sqrt{\cos 0^\circ}}$$
$$= 2.74 \text{ (m)}.$$

4.2.10 Peculiarities of a High-velocity Impact

In modern weapon systems and under modern conditions, when in addition to conventional warfare, it is necessary to consider the effects of penetrating munitions on rocket components and space objects, impact velocities can reach values of the order of several kilometers per second. This range of velocities refers to the so-called high-velocity impact.

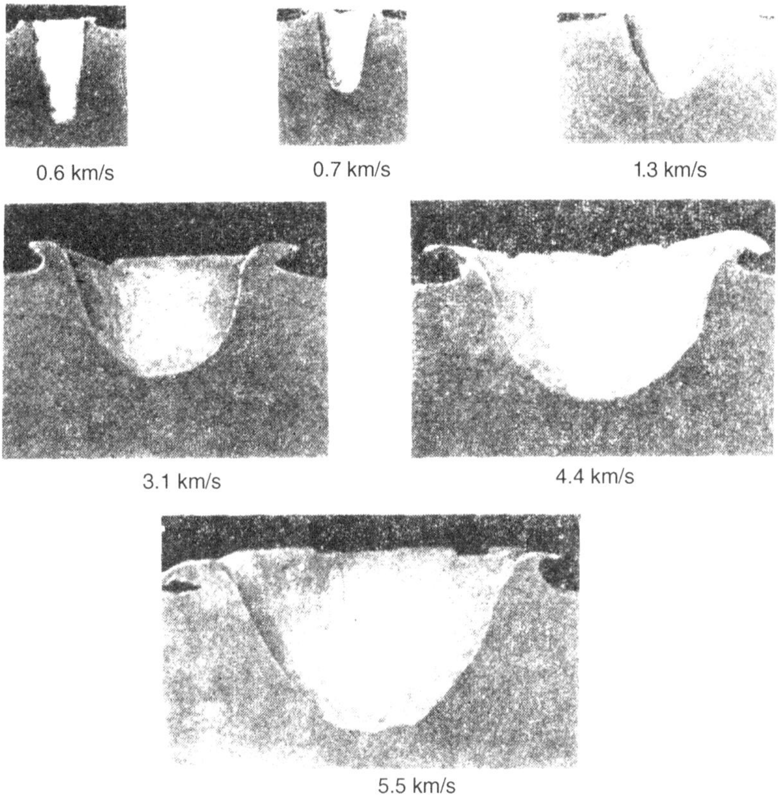

Figure 4.12 Photographs of crater cross-sections illustrating the change in crater shape as the impact velocity increases. *Source:* From Eichelberger, et al. (1967) / with permission of The American Society of Mechanical Engineers.

As noted in the above classification, the behavior of the impactor's interaction with the barrier changes with increasing velocity. This changes the shape of the resulting cavity (crater). It is illustrated in Figure 4.12, which shows how the shape of the crater in lead barriers changes when compact steel impactors of the same mass at different velocities are impacted.

The change in the behavior of interaction between the impactors and the barriers in the transition from velocities typical of conventional small arms and artillery weapons to high-velocity impact is also evidenced in Figure 4.13. In the area of low impact velocities, the depth of penetration increases very rapidly with increasing velocity, especially in the case of the impact of strong projectiles with soft barriers.

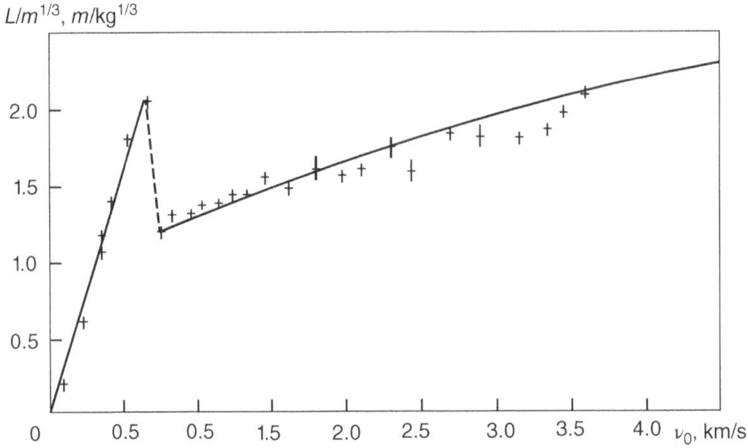

Figure 4.13 Dependence of crater depth on impact velocity in the case of steel projectiles colliding with lead barriers. *Source*: From Eichelberger and Gehring [7].

The depth of penetration reaches a maximum at the velocity determined by the physical and mechanical properties of the materials used, at which the deformation of the projectile begins and falls sharply as the velocity increases further until the velocity is reached, which can be called the high-velocity threshold. A further increase in impact velocity leads to a monotonic increase in the penetration depth.

Thus, a high-velocity impact forms a characteristic crater, the volume of which, due to the plastic flow of the material, is many times greater than the volume of the impacting body. At very-high-impact velocities, conditions sufficient for partial evaporation of the substance during unloading are realized. Data of experiments, in which the impact of glass and steel balls on barriers made of various structural materials in the form of dependences of the ratio of crater depth h to the diameter of the impactor d_0, is shown in Figure 4.14.

It follows, in particular, that the final parameters of the crater are strongly influenced by the strength characteristics of the barrier material, the values of which are comparable with the inertial forces at the late stages of the process. If the crater depth in D16AT, AMG6, and B65 targets, which have very similar strength properties, is described practically by the same dependence, then the crater depth h in the less-strong AD1M alloy is significantly higher at the same impact velocity. This also holds true when the impact velocity increases.

In the case of compact impactors, it is usually assumed that craters in isotropic plastic barriers have a shape close to a hemispherical shape during high-velocity impacts. However, quite frequent deviations from this shape are observed, appearing at relatively low-impact velocities and strong differences in density and/or

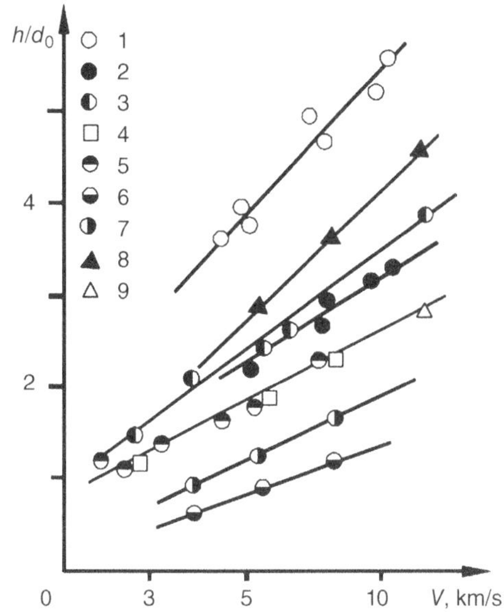

Figure 4.14 Dependencies of the crater depth on the impact velocity: steel balls on barriers: 1 – D16; 2 – St.10; glass ones: 3 – AD1M; 4 – AMG6; 5 – D16AT; 6 – St. H18N10T; 7 – copper M1; 8 – lead; 9 – B65. *Source*: From Merzhievskii and Titov [8]. Reprinted with permission of Springer Nature.

strength of impactor and target materials. This is illustrated in Figure 4.15, where the shape of the crater is characterized by the ratio of its depth h to its radius r, measured at the level of the initial surface of the barrier.

In targets made of aluminum alloys with flow stress significantly lower than the corresponding characteristics of steel and copper, the impact of glass particles makes craters almost hemispherical already with $V \sim 1$ km/s for AD1-M and 2 km/s in other cases. For the same particles in copper barriers, this occurs when $V \approx 5$ km/s, and in steel barriers – in the whole investigated velocity range $h < r$. At the same time, for the impact of a steel particle on a duralumin barrier at

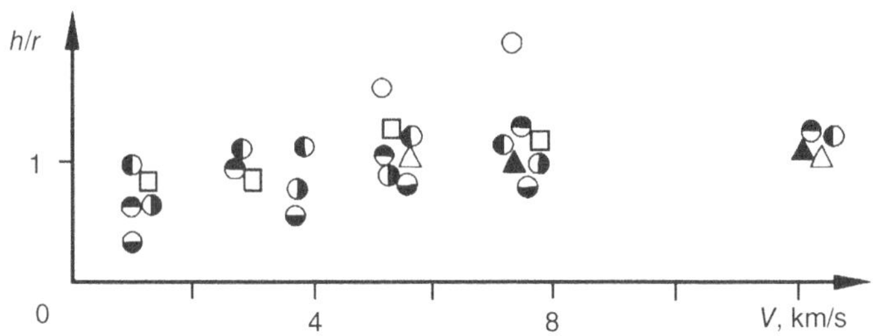

Figure 4.15 Characteristics of the shape of craters. The notations are the same as in Figure 4.14. *Source*: From Merzhievskii and Titov [8]. Reprinted with permission of Springer Nature.

$V = 7.3$ km/s, $h/r > 2$. Comparison of all the data shows that in the range of high-impact velocities, the shape of the crater is determined by the ratio of densities and strength parameters of the particle and target materials.

4.2.11 Damaging Effect of the Impactors on the Living Force

The question of the damaging effect of impactors on the living force deserves special consideration. We will consider rather compact impactors of relatively small-size bullets, small fragments, etc. In the first approximation, the damaging effect will depend on the ratio of the impactor parameters (mass, shape, velocity) and anatomical and physiological properties of the body part to be damaged. Modern representations of the wound ballistics draw the following picture of the development of damages in the tissues under the action of impactors [9]. When the impactor penetrates, a pulsating cavity is formed, the duration of which can be dozens of times longer than the time of the bullet's passage through the entire wound track.

The dimensions of the cavity, duration, and number of pulsations, value of pressure on the surrounding tissues depend, in the first approximation, other conditions being equal, on the value of energy absorbed by the tissues. So, for a certain tissue at an impactor velocity ~ 400 m/s, there are two pulsations of cavity duration $2 \cdot 10^{-2}$ second, at ~730 m/s there are five pulsations of 0.2 seconds, and at ~990 m/s there are eight 0.25 second pulsations.

The described dynamics of pulsating cavity changes and pressure oscillations in the process of penetrating wound formation shows that it is a wave-like process in which the first pulsation corresponds to the shock wave propagation, and the subsequent alternation of compression and expansion phases are associated with the appearance of compression and rarefaction waves, the effect of which determines the destruction in the tissues surrounding the wound track. The process of pulsating cavity formation in the tissues has a hydrodynamic behavior. It should be noted that in the rarefaction phase, as well as in the liquid, cavitation can occur in the zone of tensile stresses, which will lead to additional tissue destructions. Corresponding effects are observed in experiments and real conditions, especially in organs containing a significant amount of fluid.

The destruction of dense bone formations by the impactor produces a flow of cone-shaped bone fragments with their base pointing forward along the direction of the wound track. These fragments can play the role of secondary damage elements. The phenomenon corresponds to the laws of brittle fracture. Let us also note the possibility of additional organ damage as a result of the effect of hydraulic impact – a compression wave propagating along the blood vessels. The listed features of cavity formation in the tissues of a living organism appear to varying degrees depending on the parameters of the impactor (bullet, fragment), such as its mass, shape, velocity, rotation torque, strength properties.

It should be emphasized that the damaging effect of impactors on a living force is not limited only to the mechanical destruction of the surrounding organs. In the mechanism of formation of a gunshot wound under the action of modern wounding projectiles, distinguished by an extremely high damaging effect, a pronounced contusion syndrome comes to the fore, which occurs due to the repeated passage of compression and rarefaction waves through tissues with high liquid content. The resultant disturbances in microcirculation and the nervous apparatus, extending far beyond the wound track, cause those complex disturbances in the injured tissues, which manifest themselves as a neurodystrophic process.

4.3 Formulation of Penetration Problems and Ways to Solve Them

The description of penetration and piercing processes should be based first of all on clear notions and mechanisms of impactor–barrier interaction. At the same time, to fully describe the impact dynamics, it is necessary to take into account the shape of the impacting bodies, propagation of elastic, plastic, and shock waves, the hydrodynamic flow of materials, finite deformations, thermal effects, friction effects, emergence and propagation of fracture zones, and several other phenomena.

Thus, a complete theoretical calculation is very cumbersome, and the accuracy of the results obtained is not guaranteed due to the difficulty of correctly describing the behavior of the materials under the impact conditions. This is the reason why a significant, if not the main, role in the study of penetration and piercing is given to experimental and experimental–theoretical methods. To understand the observed features of the phenomenon, it is necessary to identify the main interaction mechanisms.

In the simplest case, the penetration of a body into a continuous medium, assuming that it penetrates along the normal and moves straight ahead, can be described by the equation of motion expressing Newton's law:

$$M\frac{d^2y}{dt^2} = -F,$$

where y is the distance traveled by the body, t is time, M is the mass of the body, F is the resistance force of the medium.

$$\text{Initial conditions:} \quad t = 0, \quad y = 0, \quad \frac{dy}{dt} = V_S.$$

Now, based on the specific conditions of the impact, it is necessary to choose an expression for F. Most often it is assumed that the resistance force F can be represented as a sum of three forces

$$F = F_1 + F_2 + F_3, \tag{4.7}$$

where F_1, the dynamic resistance force, caused by the inertia of the medium particles, is taken as proportional to the square of the penetration velocity V; F_2 – viscous resistance force resulting from overcoming the friction between the medium particles, proportional to the penetration velocity; F_3 – static resistance force, whose value depends on the strength of the barrier and does not depend on the velocity of the impactor. Thus, the force can be represented as

$$F = AV^2 + BV + C, \tag{4.8}$$

where A, B, C are positive constants that depend on the properties of the barrier material and the shape of the impactor.

The problem of penetration in this form has a long history [1]. Apparently, the first to consider such a problem was Euler, who put

$$F = C. \tag{4.9}$$

There are also known variants:

$$F = AV^2 + C; \tag{4.10}$$

$$F = AV^2 + BV; \tag{4.11}$$

$$F = BV + C. \tag{4.12}$$

The analysis of empirical formulas (4.3)–(4.5) shows that if the resistance force is represented as

$$F = \frac{\pi d^2}{4}\left(AV^2 + C\right),$$

which corresponds to (4.10), we can obtain a formula to determine the depth of penetration of the munition into the barrier

$$h = \frac{2m}{\pi d^2 A}\ln\left(1 + \frac{A}{C}V^2\right),$$

close in structure to Zabudsky's formula (here m is the mass of the ammunition). If the barrier is considered as a medium, the resistance of which depends only on the frictional force proportional to velocity, then

$$F = \frac{\pi d^2}{4}BV$$

(variant (4.12)), and the formula for determining the penetration depth will be

$$h = \frac{4}{\pi B}\frac{m}{d^2}V.$$

Thus, Berezan and Zabudsky's formulas used to estimate the depth of penetration into an object of the same kind correspond to different physical representations of the interaction process, which leads to a different record of the law of resistance of the barrier medium during the penetration of the munition.

Assuming the resistance force of the barrier to be equal to the static resistance

$$F = \frac{\pi d^2}{4} C,$$

we get the formula for determining the depth of penetration

$$h = \frac{2}{\pi C} \frac{m}{d^2} V^2,$$

which can be transformed into Jacob de Marre's formula.

The resistance coefficients of barriers determined experimentally can be used not only in the given empirical formulas for homogeneous barriers but also for multilayer barriers with a heterogeneous structure.

The modern level of science development, with the availability of X-ray and photographic equipment for registration of high-speed processes, allowed to establish the physical structure of several empirical coefficients. Thus, the coefficient of inertial resistance of the barrier turned out to be equal to the product of the density of the material by the shape coefficient of the munition's head part. The static resistance coefficient for a metal barrier is proportional to the dynamic hardness or dynamic yield strength; for a concrete barrier, it is proportional to the cubic strength.

Thus, the above empirical formulas received a more rigorous justification and transferred to the category of semi-empirical formulas. A typical example of this is the LFTI (Ioffe Leningrad Institute of Physics and Technology) formula for calculating the penetration depth

$$h = \frac{l_0}{2\chi_0} \frac{\rho_{02}}{\rho_{01}} \ln\left(1 + \chi_0 \frac{\rho_{01} V^2}{H}\right), \tag{4.13}$$

where l_0 is the length of the ammunition; ρ_{01}, ρ_{02} are the material densities of the barrier and the ammunition, respectively; H is the dynamic hardness of the barrier material (for metals, determined at the penetration velocity of 10–100 m/s); χ_0 is the shape factor of the munition head.

For a munition with a conical head $\chi_0 = \sin^2\delta$,

for a munition with an ogival head $\chi_0 = \dfrac{8(R/d) - 1}{24(R/d^2)}$,

where δ – the angle of taper; R – radius of ogival; d – the caliber of the ammunition.

Completing the analysis of empirical formulas, it should be noted that, along with the considered formulas, some are obtained by an a priori representation of expressions for calculating the path of the munition in the barrier, as the product of functions that take into account the design parameters of the munition, physical and mechanical properties of the barrier, and the conditions of their interaction. These formulas have a sufficiently high accuracy within the conditions for which they are obtained, but they differ little from empirical ones and are insufficient for solving problems arising in the development of real ammunition designs.

4.4 Shock with Long Rods

The collision of elongated impactors (rods) with barriers requires a separate consideration due to the specifics of their interaction with the barriers. It can be said that this process is, in a sense, intermediate between the impact of compact impactors and penetration of shaped charge jets. Therefore, the main approach to developing engineering models of long rod penetration is to modify the hydrodynamic model built to describe the penetration of a shaped charge jet.

This is facilitated by the fact that, in the process under consideration, we can distinguish three main stages, which are unequal in their contribution but different in terms of the mechanics of the process. The first brief stage, the onset of impact, is characterized by the development of wave interaction in the area of initial contact, in which the field of velocities and stresses in the contact zone of the colliding bodies is formed. The second stage, which is the longest and determines the final result, consists of the practically stationary penetration of the impactor into the barrier.

It is at this stage that the application of the hydrodynamic approach becomes justified. At the final stage, the impactor braking occurs, i.e. the process again becomes nonstationary, in which the impact conditions change significantly, first of all, the value of the penetration resistance force changes. From the point of view of the final result, this stage, as well as the first one, is not decisive. Such a division into stages is also preserved in the case of piercing an object, only the third stage, in this case, is associated with the peculiarities of the impactor's exit to the rear free surface.

The penetration of a long rod into a metallic target has been treated in a very preliminary manner by R.F. Bishop et al. [10] and more completely by V.P. Alekseevskii [11] and A. Tate [12]. These analyses apply to impact velocities generating stresses much higher than the dynamic yield stress of either target or projectile. This treatment is applicable to shaped charge jets and long-rod penetrators and is commonly known as the "eroding rod" model [13]. Birkhoff et al. considered, simply, a semi-infinite target impacted by a long rod of length L at a velocity V.

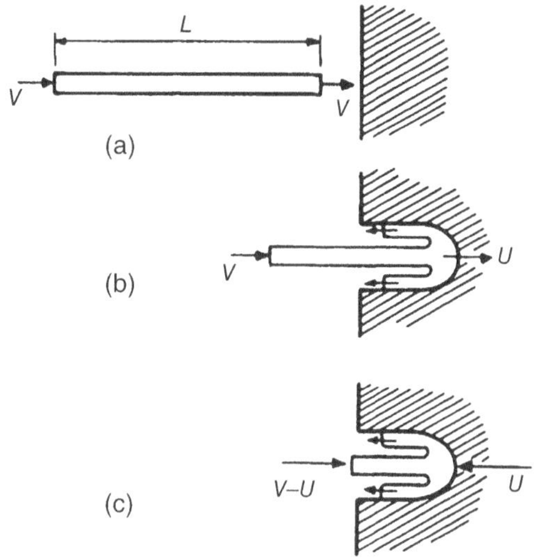

Figure 4.16 High-velocity penetration of projectile into the target (a–c); projectile is consumed as it penetrates. *Source*: From Meyers [13]. Reprinted with permission of Wiley.

The densities of the projectile and target are ρ_p and ρ_t, respectively. Figure 4.16 shows the advance of the tip of the projectile during penetration; its velocity is U. By changing the coordinate system so that the stagnation point (interface between target and projectile) is at the origin, one has the target moving from right to left at a velocity U and the projectile moving from left to right at a velocity $V - U$.

Birkhoff et al. assumed that the pressure P created at the interface was much higher than the strength of the materials, which could therefore be treated as fluids by applying Bernoulli's equation:

$$P = \frac{1}{2}\rho_p(V-U)^2 = \frac{1}{2}\rho_t U^2 \tag{4.14}$$

From this expression, we obtain

$$U = V\left[\frac{1 \pm \left(\rho_t/\rho_p\right)^{1/2}}{1-\left(\rho_t/\rho_p\right)}\right] \tag{4.15}$$

The penetration depth p is equal to the time that the projectile takes for penetration divided by the velocity at which it is penetrating, $V - U$:

$$p = \frac{LU}{V-U}$$

From Eqs. (4.14) and (4.15), one can eliminate U and obtain P:

$$p = \frac{LV\left[\dfrac{1 \pm (\rho_t/\rho_p)^{1/2}}{1-(\rho_t/\rho_p)}\right]}{V - V\left[\dfrac{1 \pm (\rho_t/\rho_p)^{1/2}}{1-(\rho_t/\rho_p)}\right]} \tag{4.16}$$

$$p = L\left(\frac{\rho_p}{\rho_t}\right)^{1/2}$$

Equation (4.16) is significant: it predicts that the depth of penetration p is equal to the projectile length L if the two materials have the same density. It also predicts that the penetration increases with the ratio ρ_p/ρ_t (ratio of projectile and target densities). The fact that the penetration depth is independent of impact velocity V is also significant. As expected, Eq. (4.16) is an oversimplified representation of the penetration process.

The predictions of Eq. (4.16) are not entirely realistic and, therefore, the incorporation of material strength has been implemented. This has been done by Alekseevskii in 1966 and by Tate in 1967. Resistances to the plastic flow of the projectile and target are incorporated into the treatment. For the projectile, the material is unconstrained, and this resistance can be taken as the dynamic yield stress $(\sigma_{yd})_p$. For the target, the material is laterally constrained. Thus, the stress at which it flows is, to a first approximation, the stress to produce a hole through a material. There are elastoplastic solutions to this problem. Tate assumed this value to be R_t, where R_t was assumed by Bishop et al. to be 4.5 $(\sigma_{yd})_t$, the target flow stress is calculated as

$$R_t = 4.5(\sigma_{yd})_t$$

Applying Bernoulli's equation, which equates the pressures on the two sides of the stagnation point, and incorporating the strength of the material, yields

$$P = \frac{1}{2}\rho_t U^2 + R_t = \frac{1}{2}\rho_p (V - U)^2 + (\sigma_{yd})_p \tag{4.17}$$

This is a modification of Bernoulli's equation. The difference in pressures is correlated with $(\sigma_{yd})_p$ and R_t. The solution of Eq. (4.17), coupled with the equations for the conservation of mass and momentum, provides a realistic prediction of the penetration. Figure 4.17 shows Wright's plot of the Alekseevskii–Tate equation for three different cases:

$$(\sigma_{yd})_p \succ (\sigma_{yd})_t;\ (\sigma_{yd})_p = (\sigma_{yd})_t;\ (\sigma_{yd})_p \prec (\sigma_{yd})_t.$$

The predictions are in qualitative agreement with experimental results. The sigmoidal shape of the penetration curve predicts a saturation penetration at a

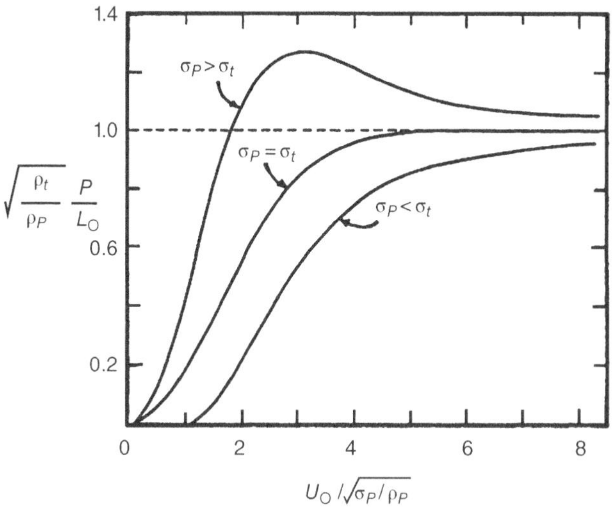

Figure 4.17 Predictions of the Alekseevskii-Tate eroding rod model; both velocity U_0 and penetration P have been scaled to provide nondimensional axes. *Source*: From Wright [14]. Reprinted with permission of Springer.

maximum velocity. It also predicts that the curve is shifted to the right for increasing target strength and to the left for increasing penetrator strength, in agreement with observed results. If the mass of the projectile is constant, increasing the L/D ratio should increase penetration.

Since in some cases barriers made of high-strength ceramic materials (ceramic armor) can be used, the problem of interaction of impactors with brittle media requires separate consideration. As experimental studies show, when rods penetrate high-modulus brittle materials, the depth of penetration turns out to be much smaller (1.5–2 times) than predicted by formulas obtained based on the modified hydrodynamic theory.

It turns out that the maximum deviation of the parameters of the process of penetration into ceramics from the calculated ones is observed at the penetration velocities exceeding the crack propagation velocity in the studied materials. At lower penetration velocities, the effect is reduced. This is because at high velocity, the front of cracks emerging from the walls of the collapsing cavity moves slower than the contact surface of the impactor and the barrier, i.e. the impactor interacts with an intact material having a high yield strength. This is the physical nature of the high resistance of ceramic materials to penetration.

At lower impact velocities, the impactor will basically penetrate a highly fragmented medium practically devoid of strength with a density close to the monolith, and the penetration conditions are close to those described by hydrodynamic theories. This is illustrated by Figure 4.18, where the experimental data on

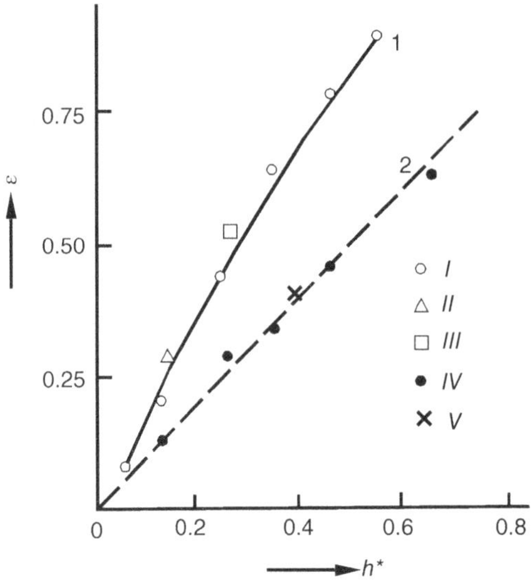

Figure 4.18 Dependence of the relative deformation of the impactor on the specified thickness of the plate: 1 – solid plate; 2 – fragmented medium; I – glass, II – sitall, III – silicon carbide; previously fragmented: IV – glass, V – lead glass. *Source:* From Ziljberbrand et al. [15].

penetration into brittle ceramic materials are compared with the calculation according to the hydrodynamic model.

The figure needs an explanation. The point is that the fracture zone in a brittle material, when it interacts with an impactor, can extend far beyond the penetration depth of the impacting body. This does not allow direct determination of the penetration depth of the impactor as it is done in the case of plastically deformed metal barriers, so when considering the penetration of a plastic body into a brittle medium, an analog of the penetration depth of the impactor should be chosen.

If the barrier is a pierceable plate, then its thickness h in the first approximation can be considered as an element of a cavern of a given depth h. In this case, the measure of the impact of the plate on the impactor can be the value of its shortening when penetrating to a depth h. In accordance with this, Figure 4.18 shows the dependences of the relative shortening of the impactor on the specified thickness of the brittle plate. The data for a solid plate is compared with the data for an impact in a pre-fractured medium. As can be seen from the figure, brittle materials (curve 1) indeed reveal a significant deviation from the hydrodynamic dependence 2, i.e. the above-described mechanism of impactor introduction into brittle (ceramic) materials is confirmed.

As follows from the above analysis, increasing the density and strength of the impactor material leads to an increased penetration depth and, consequently, to an increased penetration efficiency. To exploit this effect for practical purposes, armor-piercing projectiles use cores made of depleted uranium alloys, which significantly increase penetration depth and pierceable thickness. At the same time, the production and use of such cores are associated with economic, political, and environmental problems. For this reason, they try to replace uranium alloys with tungsten composites while maintaining the penetration capability of the impactors. Research shows that it is possible to create cores made of tungsten composites that are as effective as those made of uranium alloys.

Additional possibilities based on the understanding of the physical mechanisms of penetration are used to improve the effectiveness of armor piercing. In particular, the above consideration is based on the second stage of penetration, which can be regarded as a stationary process. A fruitful idea of increasing the piercing depth was the use of a nonstationary stage, in which the maximum pressure in the contact zone is realized, resulting in a significant increase in the penetration depth. Such conditions are provided by a special design of impactors. At present, the possibility of using several types of impactors is being considered.

4.4.1 Segmented Impactors

They are a set of cylinders with a length equal to the diameter, separated from each other by the distance *S*, and *S* can vary from zero to several diameters of the impactor. Depending on the impact velocity, the optimum element spacing is selected. The guiding principle is to choose such a distance between segments so that each subsequent segment arrives at the bottom of the cavern forming at the moment of the full release of the previous one. Without dwelling on the detailed analysis of the occurring physical processes, we shall indicate that the penetration depth increases by 20–40% as compared with an equivalent solid impactor. Structurally, the segments are separated by intermediate elements (Figure 4.19).

4.4.2 Telescopic Impactors

One of the variants of such impactors is a combination of a solid homogeneous impactor connected to a tubular part, the inner diameter of which is equal to the outer diameter of the solid part. When the tubular part hits, a cavern is formed

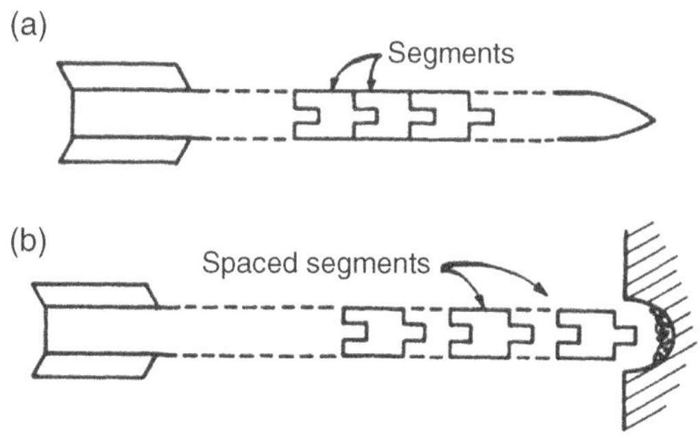

Figure 4.19 (a) Segmented rod with segments adjacent to each other during launching. (b) Segmented rods during impact. *Source*: From Tate [16]. Reprinted with permission of Elsevier.

into which the solid part flies through without resistance and at almost the original velocity, which then interacts with the barrier like an ordinary rod. The effect of increasing the depth is also achieved by using the unsteady stage and can reach tens of percents.

Telescopic segmented impactors are a combination of the previous two types. In this case, the impactor segments are telescopic impactors.

It should be noted that the search for ways to increase armor-piercing continues.

4.5 Peculiarities of Collision with Thin Targets (Screens)

The impact interaction of bodies with relatively thin barriers is of particular interest when considering the impact interaction of bodies with the barriers. The fact is that when a thin barrier is pierced, wave processes develop in the body of the impactor, leading to its deceleration and destruction. This effect is used to protect the object from being pierced. Thin plates, for example, are elements of dynamic protection of tanks, are also used to protect space objects. To understand the processes occurring, let's consider the simplest case of the impact of a straight circular cylinder on a thin screen. Figure 4.20 shows an approximate wave pattern shortly after the impact.

From the interface I, shock waves S_1 and S_2 propagate in opposite directions, with rarefaction waves R_1 and R_2 moving in the direction of the symmetry axis due to the limited diameter of the impactor. The formation of these rarefaction waves leads to the ejection of the impactor and the screen substances in the opposite direction. After the shock wave S_2 is reflected from the free surface, a rarefaction wave R_3 emerges, in which the profile of the lower screen surface takes the form shown in the figure.

Further development of wave processes, accompanied by the interaction of rarefaction waves, can lead to the fact that if the tensile stress at any point of the impactor or screen exceeds the tensile strength of the material, a crack will develop in this place. The resulting crack itself becomes the source of new rarefaction waves in which fracture can occur again, and so on. Thus, the screen and the impactor are fractured.

The process is accompanied by fragments flying in the radial direction and a significant decrease in fragment velocity to values lower than the initial velocity of the impactor. Similarly, the interaction with the screen occurs with other shapes of impactors. Naturally, each particular case has its own specifics of interaction. An X-ray diagram of the process similar to the considered one is shown in Figure 4.21, where a cloud of flying fragments formed as a result of the interaction of a compact cylindrical impactor with a screen is clearly visible.

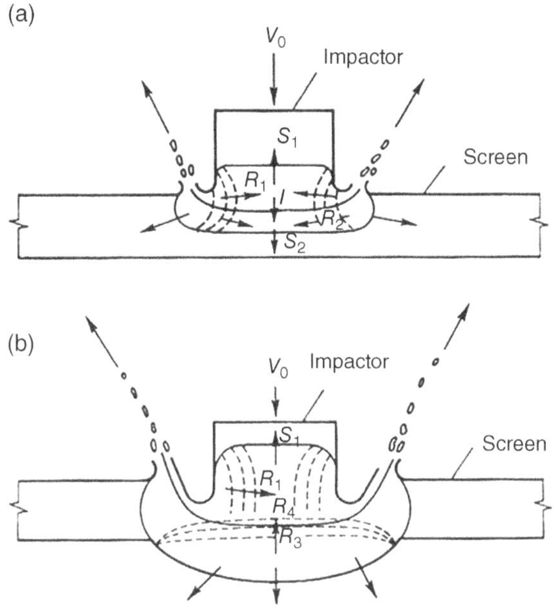

Figure 4.20 (a, b) Scheme of the interaction of the impactor with the thin screen. *Source*: From Kinslow ed. [17]. Reprinted with permission of Elsevier.

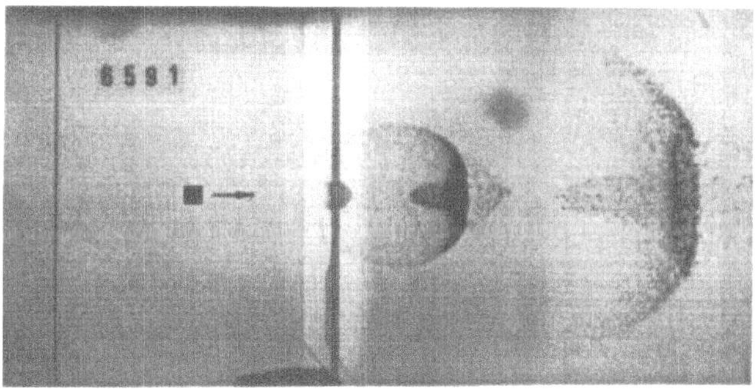

Figure 4.21 The X-ray image of the interaction of a compact cylindrical impactor with a screen. *Source:* From Balagansky et al [1]. Novosibirsk State Technical University [in Russian].

The picture of the interaction and its final result for the rod is shown in Figure 4.22.

In the case of a rod impact, it is established that the greater protective effect is shown when the impact is not along the normal, but at an angle to the screen

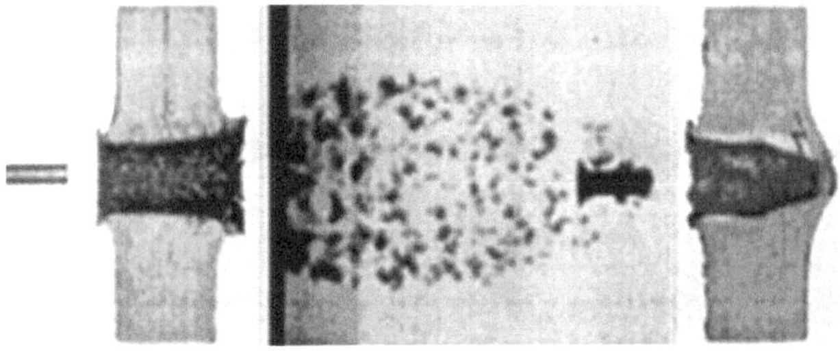

Figure 4.22 The X-ray image of a barrier piercing with a rod. *Source:* From Balagansky et al [1]. Novosibirsk State Technical University [in Russian].

surface, because in this case there is a rotation of the impactor, the consequence of which is the twisting of the residue of the rod penetrating behind the screen. At large deviations of the impact direction from the normal to the surface, a ricochet of the impactor is possible, in which there are practically no fragments penetrating behind the screen (except for possible spalling from the rear surface of the screen).

In the range of very-high-impact velocities, conditions can be realized in which partial or complete evaporation of the impacting body material will occur.

4.6 Self-assessment Questions

1) What targets are penetrating ammunition designed to damage?
2) List the main kinds of penetrating ammunition.
3) What are the basic requirements for penetrating ammunition?
4) What is penetration?
5) What is meant by piercing?
6) What physical processes occur in the barrier and impactor during a collision?
7) How are impact conditions classified by impact velocity?
8) How are impact conditions classified according to the material mechanical properties of the impactor and the barrier?
9) How are impact conditions classified by geometric characteristics of the impactor and the barrier?
10) What is shown in the phase diagram characterizing the collision of the impactor with the barrier?
11) What is the mechanism of plug formation during piercing?
12) What are the piercing particularities of the puncture mechanism?
13) What is the ballistic limit?
14) How is the deterministic ballistic limit specified?
15) How is the probabilistic ballistic limit specified?

16) How is complete penetration specified?
17) What does Jacob de Marre's formula define?
18) What do the Zabudsky's, Berezan, and ANII formulas specify?
19) What is meant by a high-velocity impact?
20) How does the shape of the crater change under a high-velocity impact?
21) What particularities are observed in the formation of a cavity in a human body under the penetration of an impactor?
22) What are the components that make up the resistance force of the medium to the penetration of the impactor?
23) What stages can be distinguished in the process of penetrating a rod into a target?
24) What is the modification of the hydrodynamic theory needed to calculate the penetration of the rod?
25) How does changing the velocity of the impactor affect the mechanism of penetration into ceramic barriers?
26) What kinds of cores are used to increase armor penetration?
27) What constructive methods are used to increase armor penetration?
28) What are the peculiarities of the impactor's interaction with thin barriers?
30) Under what conditions can the partial or complete evaporation of the impactor and the target material be realized?

References

1 Balagansky, I.A. and Merzhievsky, L.A. (2004). *Effects of Means of Destruction and Ammunition*. Novosibirsk: Novosibirsk State Technical University [in Russian].

2 Zhuk, A.B. (1993). *Reference Book on Small Arms*. Moscow: Military Publisher [in Russian].

3 Zukas, J.A., Nicholas, T., Swift, H.F. et.al. (1982). *Impact Dynamics*. New York: Wiley.

4 Backman, M.E. (1976). Terminal ballistics. Naval Weapons Center Report NWC-TP-5780, NWC China Lake, CA.

5 Backman, M.E. and Goldsmith, W. (1978). The mechanics of penetration of projectiles into targets. *International Journal of Engineering Science* **16**: 1–99.

6 Vodopyanov, M.Y. (2002). *Theory and Calculation of Artillery Projectiles: Laboratory Practice*. St. Petersburg: Baltic State Technical University [in Russian].

7 Eichelberger, R.J. and Gehring, J.W. (1967). High velocity impact. In: *Kurzzeitphysik* (ed. K. Vollrath and G. Thomer). Wien: Springer.

8 Merzhievskii, L.A. and Titov, V.M. (1987). High-speed collision. *Combustion, Explosion, and Shock Waves* **23**: 589–604.

9 Popov, V.L. and Dyskin, E.A. (1994). *Wound Ballistics*. St. Petersburg: Military Medical Academy [in Russian].

10 Bishop, R.F., Hill, R., and Mott, N.F. (1945). The theory of indentation and hardness tests. *Proceedings of the Physical Society* **57**: 147–159.

11 Alekseevskii, V.P. (1966). Penetration of a rod into a target at high velocity. *Combustion, Explosion, and Shock Waves* **2**: 63–66.

12 (a) Tate, A. (1967). A theory for the deceleration of long rods after impact. *Journal of the Mechanics and Physics of Solids* **15**: 387–399;
(b) Tate, A. (1969). Further results in the theory of long rod penetration. *Journal of the Mechanics and Physics of Solids* **17**: 141–150.

13 Meyers, M.A. (1994). *Dynamic Behavior of Materials*. New York: Wiley.

14 Wright, T.W. (1983). A survey of penetration mechanics for long rods. In: *Computational Aspects of Penetration Mechanics* (ed. J. Chandra and J.E. Flaherty), 85–106. Springer.

15 Ziljberbrand, E.L., Zlatin, N.A., Kozhushko, A.A. et al. (1989). On the mechanism of interaction of a plastic impactor with a brittle solid medium. *Journal of Technical Physics* **10**: 54–58. [in Russian].

16 Tate, A. (1990). Engineering modeling of some aspects of segmented rod penetration. *International Journal of Impact Engineering* **9**: 327–341.

17 Kinslow, R. (1970). *High-velocity Impact Phenomena*. New York: Academic Press.

5

Numerical Simulation of High-speed Processes

5.1 Introduction. Basic Concepts

At present, it is impossible to project and assess the damaging effect of weapons and munitions without the use of numerical simulation. Although the decisive role is assigned to experimental verification of design solutions, numerical simulation makes it possible to provide calculations of various options with significantly less time and material resources. In some cases, numerical simulation is the only way to validate engineering solutions [1].

Numerical modeling requires:

1) physical models of material behavior under conditions of intensive dynamic loads realized under the action of weapons and munitions;
2) correct mathematical models, in which physical models are represented as mathematical formulas and (or) equations;
3) numerical methods, using which these mathematical problems are solved;
4) software, which implements the solution of problems and presentation of the results in a form convenient for analysis.

Each of the above represents a wide field for research and development. When creating physical models of materials, they are based on their real properties, which are studied in special experiments. The basic data used to construct models is obtained in the form of strain diagrams establishing the relationship between the applied forces (stresses) and the deformations caused by them. This data is increasingly complemented by microstructural studies that allow us to study the mechanisms of irreversible changes in the material, leading to a loss of strength and destruction.

On the other hand, to build physical models, it is necessary to abstract from specific details of material behavior and to identify and generalize basic elements inherent in at least some classes of materials. Often such generalizations are formulated in the form of hypotheses about the structure and properties of

Damaging Effects of Weapons and Ammunition, First Edition. Igor A. Balagansky.

substances. The fundamental hypothesis that allows us to apply a unified approach to describing the behavior of various materials, including those in different aggregate states, is the hypothesis of **continuity** of structure of real bodies. Based on this hypothesis, an independent scientific discipline, **continuum mechanics**, was formulated [2].

We know that all bodies are a set of atoms and molecules. The main mass of matter is concentrated in the nuclei of atoms, which are orders of magnitude smaller than the size of atoms, molecules, and interatomic distances (for example the nuclear radius of a hydrogen atom is 10^{-13}cm, while the radius of a hydrogen molecule is ~10^{-8}cm). Thus, the volumes occupied by bodies are much larger than the volumes in which the substance itself is concentrated, i.e. all bodies are essentially "composed of emptiness."

At the same time, a very small volume of space occupied by particles always contains a large number of particles. This allows to consider the body as a medium filling the space continuously. This hypothesis allows one to use the mathematical apparatus of continuous functions, differential, and integral calculus in further transition to mathematical models. The models considered below are based on the hypothesis about the **continuity** of the structure of real bodies formulated in this way, and for the analysis of explosion and impact processes, such models are mainly used.

Formally, the calculation of any processes within the continuous medium model can be summarized as the calculation of displacements of elements of the medium under the action of loads to be applied. Any displacement is relative, it is realized in relation to something, i.e. to a certain reference system. As a reference system, it is convenient to choose a coordinate system that establishes a one-to-one correlation between numerical characteristics and points in space.

In general, for three-dimensional space, each point is assigned a triplet of numbers (x_1, x_2, x_3), which are called point coordinates. If you fix two of the three coordinates, for example, $x_1 = \text{const}, \quad x_2 = \text{const}$, then all possible values of the third coordinate represent a line in space, which is called a **coordinate line**. Three coordinate lines can be drawn through each point in space. If all coordinate lines are straight lines, then the corresponding coordinate system is called a **rectilinear** one, otherwise, it is a **curvilinear** one.

The basic equations of continuum mechanics are most simply written in a rectilinear coordinate system. However, curvilinear coordinates play an important role in continuum mechanics, in particular, in the numerical solution of specific problems.

Two basic approaches are used when choosing the frame of reference: the **Euler** approach and the **Lagrange** approach. In the former case, the movement of elements of the medium is considered relative to a stationary observer; in the latter one, the change of coordinates of each fixed point of the medium is tracked, i.e. the system moves together with the particles. In other words, in the Lagrangian approach, the history of movement of individual points of a continuous medium is tracked, and in the Eulerian case, we observe what happens at different moments of time in a given geometrical point of space, through which different particles of a continuous medium pass. The general laws of continuum mechanics in each of these approaches have a different form.

Each of the approaches has its advantages and disadvantages, including the construction of numerical solution methods. From a mathematical point of view, the Euler approach differs from the Lagrange approach only in that in the former, the variables are coordinates of points in space, and in the latter, the coordinates that individualize a point in the environment. In both cases, an additional coordinate is time.

The state of a continuous medium at each point at a given time is characterized by a set of parameters including velocity, acceleration, density or components of the strain tensor, pressure or components of the stress tensor, temperature or entropy, etc. Then, from Euler's point of view, it is necessary to know these parameters at a given point of space, and from Lagrange's point of view – at a given individual point of a continuous medium.

To describe changes in the state of a continuous medium, we introduce a Eulerian coordinate system x_1, x_2, x_3, t and a Lagrangian coordinate system ξ_1, ξ_2, ξ_3, t. To know the motion of a continuous medium is to know the motion of all its points, and to know how the state of the medium changes is to know how its parameters change depending on the coordinates. At the same time, it is necessary to formulate rules of individualization of separate points of the medium, perfectly identical from the geometrical point of view.

One of the ways to individualize points of a continuous medium is to set points with values of their initial coordinates. Then the coordinates of points at the initial moment of time $t = t_0$ will be ξ_1, ξ_2, ξ_3, and coordinates of points at an arbitrary moment will be x_1, x_2, x_3. The law of motion, in this case, is described by the functions

$$\begin{aligned} x_1 &= x_1(\xi_1, \xi_2, \xi_3, t), \\ x_2 &= x_2(\xi_1, \xi_2, \xi_3, t), \\ x_3 &= x_3(\xi_1, \xi_2, \xi_3, t), \end{aligned} \tag{5.1}$$

or

$$x_i = x_i(\xi_1, \xi_2, \xi_3, t), \quad i = 1, 2, 3.$$

In the formulated law of motion, the independent variables are Lagrange variables. Considering (5.1) as a system of equations and solving it with relation to ξ_1, ξ_2, ξ_3, we obtain

$$\begin{aligned} \xi_1 &= \xi_1(x_1, x_2, x_3, t), \\ \xi_2 &= \xi_2(x_1, x_2, x_3, t), \\ \xi_3 &= \xi_3(x_1, x_2, x_3, t), \end{aligned} \tag{5.2}$$

i.e. we switch to Euler variables as independent variables. With x_1, x_2, x_3 fixed, the relations (5.2) specify those points ξ_1, ξ_2, ξ_3, of the continuous medium that at different moments arrive at the given point of space. Then if the velocity

$$\vec{u} = \vec{u}(\xi_1, \xi_2, \xi_3, t),$$

acceleration

$$\vec{a} = \vec{a}(\xi_1, \xi_2, \xi_3, t),$$

temperature

$$T = T(\xi_1, \xi_2, \xi_3, t),$$

and other parameters of the medium are given in terms of Lagrange, then (5.2) make it possible to find these parameters as functions of Euler variables. Thus, if the description of the process in Lagrange variables is known, to describe it in Euler variables it is necessary to obtain (5.2) from the law of motion (5.1), i.e. to solve (5.1) concerning the variables ξ_1, ξ_2, ξ_3.

On the contrary, if the process parameters are known from the Euler's point of view, then we can proceed to a Lagrangian description of the process. Let us consider this using the Cartesian coordinate system x_1, x_2, x_3, as an example, then we know $\vec{u} = \vec{u}(x_1, x_2, x_3, t)$. Let the vector $\vec{u}$ have components u_1, u_2, u_3, which are also functions of coordinates x_1, x_2, x_3. Since the velocity components are derivatives of the corresponding coordinates x_1, x_2, x_3 over time t at constant values of ξ_1, ξ_2, ξ_3, the relations hold:

$$\frac{dx_1}{dt} = u_1(x_1, x_2, x_3, t),$$
$$\frac{dx_2}{dt} = u_2(x_1, x_2, x_3, t),$$
$$\frac{dx_3}{dt} = u_3(x_1, x_2, x_3, t),$$

which can be viewed as a system of three ordinary differential equations with respect to x_1, x_2, x_3. Solving this system, we find x_1, x_2, x_3 as functions of t and three arbitrary constants whose values are determined by the values of x_1, x_2, x_3 at the initial moment, i.e., are the parameters that individualize a point in the continuous medium – that is Lagrange variables.

Thus, as a result of solving this system of differential equations, we find the law of motion (5.1), which allows us to pass from Euler variables to Lagrange variables in all formulas for $\vec{u}, \vec{a}, T$ and other parameters characterizing the state of the medium and the process occurring in it. Consequently, the transition from Euler variables to Lagrange variables with a known velocity field consists of the integration of the system of three ordinary differential equations.

It remains to note that defining motion and state of a continuous medium from the Lagrange and Euler points of view are mechanically equivalent to each other.

5.2 The System of Equations of Continuum Mechanics

The main laws describing the behavior of a continuous medium under various kinds of influences are the classical laws of conservation of mass, momentum, and energy. Let us show how they are formulated in continuum mechanics on the example of the law of conservation of mass. The notion of **material bodies** as bodies possessing the property of **inertia** is introduced. The property of inertia is characterized by **mass**, which can be introduced both for the whole body – m, and for any of its parts – m_i.

By definition, the mass of the whole body m is equal to the sum of the masses m_i of all parts composing the body. The notion of **individual or material volume** as a volume consisting of the same particles of the medium is also introduced. All of the above laws are formulated for an individual (material) volume. These laws should be regarded as empirically established laws of nature, true within the context of Newton's mechanics.

The law under discussion is that for any **individual (material), volume mass remains unchanged**, i.e.

$$m = \text{const}.$$

This relation can be rewritten in differential form:

$$\frac{dm}{dt} = 0. \tag{5.3}$$

Let's introduce the average density of the medium according to the formula

$$\rho_a = \frac{\Delta m}{\Delta W},$$

where ΔW is the volume occupied by the mass Δm. We define the density of the medium at a point as the limit ρ_a when the corresponding volume is reduced to this point,

$$\rho = \lim_{\Delta W \to 0} \rho_a = \lim_{\Delta W \to 0} \frac{\Delta m}{\Delta W}.$$

In continuum mechanics, we almost always consider the density ρ instead of mass m. For a small volume ΔW, we can assume that $\Delta m \approx \rho \Delta W$, then for a finite volume $m = \int_W \rho d\tau$, where the integral is taken over the moving individual volume. Thus, knowing the density, one can find the mass. The density for an individual particle may not be conserved since the volume of the particle may change during motion. According to the above, (5.3) can be rewritten in the form

$$\frac{d}{dt} \int_W \rho d\tau = 0.$$

In the more general case, when there is a mass flow across the boundary of the volume, the law of conservation of mass takes the form

$$\frac{\partial}{\partial t}\int_W \rho d\tau = -\int_S \rho \vec{u}\,\vec{n} dS,$$

where S is the surface bounding the considered volume. The procedure for formulating the other laws is similar.

A slightly different approach to the formulation of the basic integral conservation laws is also possible when the change in time of mass, momentum, and energy is considered not in an individual volume, but in a fixed (time-independent) volume w. In this case, the integral conservation laws of mass, momentum, and energy for a continuous medium are expressed by formulas:

$$\begin{aligned}
&\frac{\partial}{\partial t}\int_w \rho d\omega = -\int_S \rho \vec{u}\,\vec{n} dS,\\
&\frac{\partial}{\partial t}\int_w \rho \vec{u} d\omega = -\int_S \left[\rho \vec{u}\left(\vec{u}\ \vec{n}\right) + \sigma \vec{n}\right] dS,\\
&\frac{\partial}{\partial t}\int_w \rho\left(E + \frac{u^2}{2}\right) d\omega = -\int \left[\rho\left(E + \frac{u^2}{2}\right)\vec{u}\vec{n} + \left(\sigma \vec{u}\right)\vec{n}\right] dS.
\end{aligned} \tag{5.4}$$

Here E is the specific internal energy; $\sigma = \|\sigma_{ik}\|$ – stress tensor; $\vec{n}$ – external normal to S, in accordance with which a minus sign is taken before the integrals in the right-hand sides of the equation.

The above equations have a simple physical sense. The mass conservation equation states that the change of mass of a substance in a fixed volume w is equal to its quantity flowing over the same time through the bounding w surface S. The same fact, but for the projections of momentum on the coordinate axes and total energy of matter reflect the other conservation equations.

Converting the surface integrals in the smooth flow regions based on the Gauss–Ostrogradsky theorem into the volume integrals, we conclude that the integral conservation equations are equivalent to the following system of differential equations:

$$\begin{aligned}
&\frac{\partial \rho}{\partial t} + \operatorname{div} \rho \vec{u} = 0,\\
&\frac{\partial \rho \vec{u}}{\partial t} + \operatorname{div}\left(\rho \vec{u} \times \vec{u} + \sigma\right) = 0,\\
&\frac{\partial \rho\left(E + \frac{u^2}{2}\right)}{\partial t} + \operatorname{div}\left[\rho \vec{u}\left(E + \frac{u^2}{2}\right) + \sigma \vec{u}\right] = 0.
\end{aligned} \tag{5.5}$$

The system of differential equations (5.5) is the basic system describing the behavior of a continuous medium. Depending on the type of medium in consideration (solid, liquid, gas), the system can take different forms related to the peculiarities of the reaction of bodies to external loads. Thus, in the case of gaseous media, the stress tensor is spherical, so the corresponding characteristic becomes the pressure in the medium. In Cartesian coordinates, the system (5.5), taking into account the mass forces acting on the considered area, takes a form:

$$\begin{aligned}
&\frac{\partial\sigma_{11}}{\partial x_1}+\frac{\partial\sigma_{12}}{\partial x_2}+\frac{\partial\sigma_{13}}{\partial x_3}+\rho\left(F_1-\frac{\partial u_1}{\partial t}\right)=0,\\
&\frac{\partial\sigma_{12}}{\partial x_1}+\frac{\partial\sigma_{22}}{\partial x_2}+\frac{\partial\sigma_{23}}{\partial x_3}+\rho\left(F_2-\frac{\partial u_2}{\partial t}\right)=0,\\
&\frac{\partial\sigma_{13}}{\partial x_1}+\frac{\partial\sigma_{23}}{\partial x_2}+\frac{\partial\sigma_{33}}{\partial x_3}+\rho\left[F_3-\frac{\partial u_3}{\partial t}\right]=0,
\end{aligned}\tag{5.6}$$

where $\vec{F}=(F_1,F_2,F_3)$is the vector of mass forces.

In tensor notation, these equations are written in the form

$$\frac{\partial\sigma_{ij}}{\partial x_i}+\rho\left(F_j-\frac{\partial u_j}{\partial t}\right)=0.$$

The choice of the coordinate system is arbitrary and is usually determined by the convenience of analyzing the process in consideration. In some cases, for example the cylindrical coordinate system (r, φ, z) is convenient. In this coordinate system, assuming that $\vec{u}=0$ (equilibrium equations of the medium), the system (5.5) takes the form:

$$\begin{aligned}
&\frac{\partial\sigma_{rr}}{\partial r}+\frac{1}{r}\frac{\partial\sigma_{r\varphi}}{\partial\varphi}+\frac{\partial\sigma_{rz}}{\partial z}+\frac{\sigma_{rr}-\sigma_{\varphi\varphi}}{r}+\rho F_r=0,\\
&\frac{\partial\sigma_{r\varphi}}{\partial r}+\frac{1}{r}\frac{\partial\sigma_{\varphi\varphi}}{\partial\varphi}+\frac{\partial\sigma_{\varphi z}}{\partial z}+\frac{2\sigma_{r\varphi}}{r}+\rho F_\varphi=0,\\
&\frac{\partial\sigma_{rz}}{\partial r}+\frac{1}{r}\frac{\partial\sigma_{\varphi z}}{\partial\varphi}+\frac{\partial\sigma_{zz}}{\partial z}+\frac{\sigma_{rz}}{r}+\rho F_z=0.
\end{aligned}$$

From system (5.5), one can obtain equations of motion of solid bodies in any rectilinear or curvilinear coordinate system.

The given systems of equations (equations of motion) are generally supplemented with equations of conservation of mass (the continuity equation) and energy (energy or entropy equation).

For gaseous media, the system of basic equations is rewritten in the form:

$$\begin{aligned}
&\frac{\partial \rho}{\partial t}+\frac{\partial \rho u_1}{\partial x_1}+\frac{\partial \rho u_2}{\partial x_2}+\frac{\partial \rho u_3}{\partial x_3}=0,\\
&\frac{\partial \rho u_1}{\partial t}+\frac{\partial (p+\rho u_1^2)}{\partial x_1}+\frac{\partial \rho u_1 u_2}{\partial x_2}+\frac{\partial \rho u_1 u_3}{\partial x_3}=0,\\
&\frac{\partial \rho u_2}{\partial t}+\frac{\partial \rho u_1 u_2}{\partial x_1}+\frac{\partial (p+\rho u_2^2)}{\partial x_2}+\frac{\partial \rho u_2 u_3}{\partial x_3}=0,\\
&\frac{\partial \rho u_3}{\partial t}+\frac{\partial \rho u_1 u_3}{\partial x_1}+\frac{\partial \rho u_2 u_3}{\partial x_2}+\frac{\partial (p+\rho u_3^2)}{\partial x_3}=0,\\
&\frac{\partial F}{\partial t}+\frac{\partial (F+p)u_1}{\partial x_1}+\frac{\partial (F+p)u_2}{\partial x_2}+\frac{\partial (F+p)u_3}{\partial x_3}=0.
\end{aligned} \tag{5.7}$$

Here, F is the total specific energy,

$$F=\rho\left(E+\frac{u^2}{2}\right), \quad u^2={u_1}^2+{u_2}^2+{u_3}^2,$$

p is the hydrostatic pressure in the medium.

The described systems of equations have a unique property: due to the generality of laws expressed by them, they are valid for any solid [system (5.6)] or gaseous [system (5.7)] media. On the other hand, each medium, condensed or gaseous, has individual properties that are not described in the formulated general systems.

To specify the media, the general equations are supplemented with the **constitutive equations** usually establishing the relationship between the components of the stress tensor and the components of the strain tensors and strain rates (more complex constitutive equations are also possible) and the temperature or entropy. In the case of gaseous media or ideal liquids, the stress and strain tensors become spherical, i.e. they are characterized by scalar quantities – pressure and specific volume (a characteristic inverse to the density of matter).

Then, following this, the constitutive equations that establish the relationship between pressure, density (specific volume), and temperature (entropy) are called **equations of state** (EOS). These relations (constitutive equations, equations of state) make it possible to describe deformation processes of real materials and substances under various external influences. It is easy to see that the number of written equations in each of the considered systems (5.6) and (5.7) is less than the number of required functions. The constitutive equations (equations of state) complete the corresponding systems, which allows solving the formulated mathematical problems.

The problem of constructing the constitutive equations is the problem of constructing a model of the corresponding medium. The basic models describing the behavior of materials and media under intensive dynamic effects realized during the action of munitions and weapons are considered in the next section.

5.3 Behavior of Materials Under Intense Dynamic Loads

In the process of dynamic deformation of materials under the action of intense short-term loads created, for example, during detonation of an explosive charge (HE), as a result of the impact of solid bodies with high velocities, during impulse action of powerful laser radiation, the material passes through a wide range of states – from elastic to hydrodynamic, and during unloading from such states, evaporation of matter is also possible.

Such complex behavior relates to the presence of strong interaction between atoms or molecules of condensed substances located at close distances from each other, called, in contrast to the interaction of heat nature, elastic. The existence of elastic interaction forces determines the main physical features of the behavior of condensed media under shock-wave loading conditions.

By the present time, quite an extensive experimental material has been collected, and considerable attention of researchers is paid to physical and mathematical construction and justification of models adequately describing peculiarities of essentially nonstationary shock-wave processes. It is possible to formulate some absolute requirements, which should be imposed on the models pretending to describe shock-wave deformation processes:

- applicability over a wide range of material strain rate variations, from fractions to 10^6 to $10^7\,s^{-1}$;
- validity for finite, very large deformations;
- accounting for the possibility of material failure based on criteria corresponding to a range of strain rates and suitable for use in the case of the spatial distribution of state characteristics.

Even this set of requirements, far from being complete, shows the complexity of developing such models. These mathematical problems are practically not amenable to analytical solutions and require the use of computers, which, in turn, forces researchers to develop and implement efficient numerical methods and algorithms.

In this section, we briefly analyze the models most commonly used to solve problems of high-velocity deformation and fracture [3]. A detailed description of the matter models used in numerical simulations can be found in Meyers' excellent book [4].

All currently used models of dynamic deformation of solids can be roughly divided into two large classes: models based on the macroscopic approach to describing the behavior of matter under impulse loading, and models based on the microscopic approach.

In the first case, the matter is considered as a continuous medium, and the laws of conservation of mass, momentum, and energy are supplemented by phenomenological connections between the characteristics of loading and the reaction of matter to these loads – the constitutive equations of the model. Most often, such constitutive equations are one or another approximation of the generalized σ-ε diagram of the material (here, σ and ε refer to the first invariants of the stress and strain tensors).

In the microscopic approach, the constitutive equations are based on the description of the evolution of the defect structure of the crystal lattice. The direction in which the dynamics of dislocation ensemble is believed to be the main mechanism of plastic deformation has received the greatest development.

5.3.1 Elastic Medium

To analyze wave processes at a considerable distance from the area where the load is applied, the use of the elastic model of solids is justified. Classical elasticity theory is the most developed model of deformable solid mechanics. One of the variants for writing the constitutive equations for the isotropic elastic medium is of the form:

$$\begin{aligned} \sigma_{11} &= 3\lambda\theta + 2\mu\varepsilon_{11}, \quad \sigma_{12} = 2\mu\varepsilon_{12}, \\ \sigma_{22} &= 3\lambda\theta + 2\mu\varepsilon_{22}, \quad \sigma_{23} = 2\mu\varepsilon_{23}, \\ \sigma_{33} &= 3\lambda\theta + 2\mu\varepsilon_{33}, \quad \sigma_{13} = 2\mu\varepsilon_{13}, \end{aligned} \tag{5.8}$$

here θ is the average strain, defined by the formula

$$\theta = \frac{1}{3}(\varepsilon_{11} + \varepsilon_{22} + \varepsilon_{33}),$$

λ, μ – Lame's constants.

Instead of Lame's constants λ, μ, Young's modulus E, and Poisson's ratio ν are often used, defined by the formulas

$$E = \frac{\mu(3\lambda + 2\mu)}{\lambda + \mu}, \quad \nu = \frac{\lambda}{2(\lambda + \mu)}.$$

The relations (5.8) expressing the generalized Hooke's law can be solved for the components of the strain tensor.

As a result, we obtain:

$$\begin{aligned} \varepsilon_{11} &= \frac{1}{E}[(1 + \nu)\sigma_{11} - \nu(\sigma_{11} + \sigma_{22} + \sigma_{33})], \quad \varepsilon_{12} = \varepsilon_{21} = \frac{1 + \nu}{E}\sigma_{21}, \\ \varepsilon_{22} &= \frac{1}{E}[(1 + \nu)\sigma_{22} - \nu(\sigma_{11} + \sigma_{22} + \sigma_{33})], \quad \varepsilon_{13} = \varepsilon_{31} = \frac{1 + \nu}{E}\sigma_{23}, \\ \varepsilon_{33} &= \frac{1}{E}[(1 + \nu)\sigma_{33} - \nu(\sigma_{11} + \sigma_{22} + \sigma_{33})], \quad \varepsilon_{23} = \varepsilon_{32} = \frac{1 + \nu}{E}\sigma_{23}. \end{aligned}$$

Let us briefly list the main results on the description of elastic wave propagation.

The most common way to analyze the dynamics of stress waves in an elastic medium is to convert the system of equations of motion using the constitutive equations to wave equations for the scalar and vector potentials of the displacement field. In this case, longitudinal and transverse waves propagating with constant velocities are distinguished:

$$c_1 = \sqrt{\frac{\lambda + 2\mu}{\rho}}, \quad c_2 = \sqrt{\frac{\mu}{\rho}}.$$

The general solutions of the wave equation in the most practically interesting cases of plane and spherically symmetric waves have the form

$$w(r,t) = f_1(r - ct) + f_2(r + ct)$$

and

$$w(r,t) = \frac{1}{r}[\, f_1(r - ct) + f_2(r + ct)],$$

where w is the displacement; r is the spatial variable; f_1, f_2 are arbitrary functions; c is one of the wave velocities given above. It can be shown that any plane wave in an infinite homogeneous isotropic elastic medium can propagate only with velocity c_1 or c_2.

Further, wave processes in bodies of limited size are considered, for which adequate initial boundary value problems are formulated and solved according to the general theory of hyperbolic equations. An important result is obtained in the study of the reflection and refraction of a plane elastic wave on the boundary between two different elastic media.

It turned out that, generally speaking, the nature of the wave changes. When a purely longitudinal or purely transverse wave interacts with a contact boundary, two reflected and two refracted waves occur. This does not happen only in the case of normal incidence of plane waves on a plane boundary when the type of the wave is preserved.

In general, the analysis of elastic behavior can be useful for evaluating the final stages of the processes under consideration, complementing approaches that more closely match the real features of dynamic deformation.

5.3.2 Hydrodynamic Model

In the opposite – as compared to the elastic-limit – case, the consideration is carried out based on the hydrodynamic (gas-dynamic) model of the behavior of matter. The main preconditions for its use are related, on the one hand, to the successes of many fundamental studies carried out in the ideal incompressible fluid approximation (hydrodynamic theory of shaped charges, elements of the high-velocity impact theory) and, on the other hand, to the natural desire to use the detailed developed classical tool for the description of shock-wave phenomena in gases and liquids.

The physical justification for the applicability of the hydrodynamic approximation is that the pressure value under the considered loads exceeds the strength characteristics of the materials by an order of magnitude, or, more strongly, there is a significant excess of the normal components of the stress tensor over the tangential ones. It is not taken into account that the loading of the material is carried out by a shock wave, the tangential stresses at the front of which may reach values significantly exceeding the theoretical strength – a load sufficient to break the bonds of atoms.

Only after their relaxation, the level of tangential stresses decreases to a certain value characteristic of the realized spectrum of strain rates and temperature regime. The process of tangential stress relaxation can significantly affect the formation of the stress–strain state as a whole.

The constitutive equations in the hydrodynamic model are formulated in the form of the equation of state of the form:

$$\begin{aligned} &p = p(\rho, E), \quad p = p(\rho, T), \quad p = p(\rho, S),\\ &p = p(V, E), \quad p = p(V, T), \quad p = p(V, S), \end{aligned}$$

and (or)

$$\begin{aligned} &E = E(\rho, p), \quad E = E(\rho, T), \quad E = E(\rho, S),\\ &E = E(V, p), \quad E = E(V, T), \quad E = E(V, S). \end{aligned}$$

The most widely used equations of state for condensed substances are in the Mie–Grüneisen form, according to which it is assumed that

$$\begin{aligned} &E = E_x(V) + E_T(V, T) + E_e(V, T),\\ &p = p_x(V) + p_T(V, T) + p_e(V, T), \end{aligned}$$

where $E_x(V)$, $p_x(V)$ are elastic components of specific internal energy and pressure; $E_T(V, T)$, $p_T(V, T)$, $E_e(V, T)$, $p_e(V, T)$ are their components corresponding to heat motion of atoms (nuclei) and heat excitation of electrons, respectively. There is a large amount of information about the parameters of equations of states of this type, summarized in special collections and even presented on electronic web pages. At the same time, simplified versions of the equations of state, such as the equation in the form of Tate, are often used:

$$P = (n-1)\rho E + n\beta(S)\left(\frac{\rho}{\rho_0} - 1\right),$$

where ρ_0 is the initial density; $\beta(S)$, n are parameters determined from experiments, n for a large number of metals is close to 4.

Obviously, within the framework of hydrodynamic approximation, it is impossible to calculate processes up to the final stages (for example to calculate the crater size during a high-velocity impact of a compact particle on a massive barrier).

Instead of modeling a phenomenon to the end, hydrodynamic approximation is used to calculate the initial stages with a further transition to the description of plastic and elastic deformation. In this case, it is necessary to establish criteria for a transition from one model to another.

The practice of calculations has shown that even the use of the most general equations of state does not lead to complete elimination of drawbacks of hydrodynamic description even in the simplest cases, for example when modeling the damping of a plane shock wave from the action of an overtaking rarefaction wave. As a result, there is an urgent need to use models that describe irreversible deformations, on the one hand, and take into account the strength properties of real materials, on the other hand.

5.3.3 Elastoplastic, Viscoplastic, and Elastoviscoplastic Models

The numerous theories developed to date are usually divided into two classes depending on the principle underlying the construction of the constitutive equations. In **deformation** theories, the relationship between stresses and strains is established; in **plastic flow** theories, between stresses and strain rates. The practice of calculations shows that the best description of the experimental results for dynamic processes in solids is given by the plastic flow theories.

The first attempts to take into account the strength of materials in calculating high-velocity impact generalized the hydrodynamic approach and proceeded from the rigid-plastic body model, but the Prandtl–Reuss model was most widely used. The constitutive equation of the model is based on the hypothesis that the principal axes of the plastic strain rate tensor ξ_{ij}^{p} and the deviator of stress tensor S_{ij} are coaxial:

$$\xi_{ij}^{p} = \lambda S_{ij}. \tag{5.9}$$

When (5.9) is used, the problem of calculating λ arises, which in different theories is solved based on additional assumptions. The well-known specialist in numerical methods M.L. Wilkins suggested instead of directly calculating the coefficient λ in the Prandtl–Reuss model to carry out the procedure for bringing the stress state, determined from the solution of the equations of the linear theory of elasticity to the yield surface specified by the von Mises condition to the principal axes of the stress tensor having the form

$$(\sigma_1 - \sigma_2)^2 + (\sigma_2 - \sigma_3)^2 + (\sigma_3 - \sigma_1)^2 = 2\,(Y_0)^2,$$

where $\sigma_1, \sigma_2, \sigma_3$ – principal stresses; Y_0 – yield strength. Such a procedure provides an approximate fulfillment of the Prandtl–Reuss equations. This version of the Prandtl–Reuss model has become the basis for several computational programs. It should be noted, however, that the procedure is quite correct only for the case of an ideal elastoplastic body, but at variable yield strength (which corresponds to taking into account strain and rate hardening and temperature softening), the expression for correction has a more complex form than the von Mises condition.

In most works devoted to the numerical solution of dynamic problems of elastic–plastic deformation, the deviator of stress tensor appears in the constitutive equations, and the corresponding calculation programs use the procedure of rotating the principal axes of the stress tensor while rotating the calculation cell as a rigid whole, which corresponds to the approximation of the Jaumann derivative.

The works based on the M.L. Wilkins approach postulate a linear relation between the deviators of stress and strain tensor (Hooke's law) with subsequent bringing of stress to the yield surface, and in a more general formulation, the main part of the constitutive equations is linear. All this leads to limitations on the value and nature of deformations and narrows the range of problems, to which the discussed models are applicable. The remarks made are not completely removed for the generalization of the Prandtl–Reuss equations to the case of finite deformations.

The approach formulated in the nonlinear elasticity theory, which uses the full derivative of the strain tensor, does not have such drawbacks.

Comparison of the results on elastic–plastic models with experimental data shows that these models do not describe the elementary phenomenon of shock-wave processes – the attenuation of the amplitude of an elastic–plastic shock wave during its interaction with a catch-up rarefaction wave. This points to the need for a more accurate account of the mechanisms of irreversible dissipative losses characteristic of solids during dynamic deformation.

Further analysis has shown that most models of elastic–plastic deformation take into account only "hydrodynamic attenuation," which is insufficient for the correct description of shock-wave interaction in solids. This circumstance causes the appearance of complicated viscoplastic and elastic–viscoplastic models, in which the lack of dissipative mechanisms is compensated by additional viscous terms in the constitutive relations. The results obtained with their help refer mainly to the field of one-dimensional nonstationary problems, in which the shock impulse evolution is modeled and interpreted.

It should be pointed out that, when using elastic–plastic models, the main qualitative features of dynamic and shock-wave processes associated with the strength properties of real solids are fairly well reproduced, while for their quantitative agreement, one often has to resort to additional model corrections relying on known experimental results. It is difficult to assess the capabilities of models with additionally introduced viscous dissipation mechanisms because they are rarely used for solving specific dynamic problems.

5.3.4 Dislocation Models

To describe irreversible deformations of metals, a physical, microscopic approach that traces the evolution of the defect structure of the crystal lattice within the framework of certain assumptions is increasingly used. Consideration, where the dynamics of dislocation ensembles are believed to be the main mechanism of plastic deformation, has gained the greatest importance. The basis for the dynamic theory was given by Gilman's experimental data on the mobility and number of dislocations in lithium fluoride single crystals. A significant contribution to the development of the theory was made by L.A. Merzhievsky [5–7].

The constitutive equations connecting microscopic and macroscopic characteristics of the stress–strain state of the material, in the simplest case, have the form

$$\varepsilon_p = bNL, \tag{5.10}$$

where ε_p – plastic deformation; L – average distance traveled by dislocations; N – dislocation density, defined as the total length of all dislocations in the unit volume of material; b – modulus of Burgers vector – maximum plastic deformation. Differentiating (5.10) by time gives the expression

$$\frac{d\varepsilon_p}{dt} = bL\frac{dN}{dt} + bNV_a,$$

which connects the plastic strain rate with the dislocation propagation rate and the average dislocation velocity V_a. The real contribution to the second summand is made only by mobile dislocations, i.e.

$$\frac{d\varepsilon_p}{dt} = bL\frac{dN}{dt} + bN_nV_a, \tag{5.11}$$

where N_n, V_a are the density and average velocity of mobile dislocations. With few exceptions, (5.11) is used in the form of Orowan relations, when the first term on the right-hand side is ignored, assuming that the dislocation reproduction rate is relatively small. Several investigations show that under intensive dynamic and especially shock-wave loading, dislocation multiplication mechanisms additional to quasi-static ones are activated, and their rate can be high.

As a result, when solving specific problems based on the Orowan relation, one has either to introduce an initial density of mobile dislocations overestimated by 2–3 orders of magnitude compared to the observed one or to use unjustifiably large values of the dislocation multiplication factor. In the latter case, the value dN/dt in (5.11) cannot be ignored.

The description of dislocation ensemble kinetics is based on the notion that the dislocation propagation mechanism is dominated by double cross-slip, which gives for N

$$N = N_0 + M\varepsilon_p,$$

and for N_n

$$N_n = (N_0 + M\xi)\exp\left(\frac{-h_0\varepsilon_p}{\sigma}\right),$$

where N_0 is the initial number of dislocations; M is the multiplication coefficient; h_0 is the strain-hardening coefficient; σ is the tangential stress. Approximation of experimental data on dislocation mobility is used for V_a.

Other attempts to develop dislocation kinetics are also known. For example, one of the dislocation models of plastic deformation uses the notion of thermally activated behavior of dislocation motion, which limits its applicability to high-level loads that cause dislocation motion in the above-barrier sliding mode. Sometimes they try to take into account the mechanisms of dislocation nucleation in shock waves and additional ideas about the behavior of dislocation motion.

Here, it should be noted that the proposed models focus on a detailed description of the structure of plane shock waves and rarefaction waves; it is difficult to imagine a way of extending this approach to more complex cases, for example to solve spatial problems.

In the only known work based on the full Eq. (5.11), dislocation kinetics is also developed using the idea of heterogeneous nucleation of dislocations in shock-wave fronts. The presence in it of a sufficient number of parameters, chosen from the condition of the best coincidence with the experimental data, makes it possible to accurately transmit all the features of the shock-wave profiles measured in the experiment into the calculation.

The listed models were developed to solve one-dimensional nonstationary problems of plastic deformation. In general, the ways of generalization of dislocation models to the case of arbitrary dimensionality are clear. Let us illustrate this by an example of a generalization of the Gilman model to the three-dimensional case. Let us restrict ourselves to the consideration of elastic–plastic deformations and, following the established approach, represent the strain tensor ε_{ij} and the strain rates ξ_{ij} as a sum of the elastic and plastic parts:

$$\varepsilon_{ij} = \varepsilon_{ij}^e + \varepsilon_{ij}^p, \xi_{ij} = \xi_{ij}^e + \xi_{ij}^p.$$

To relate the change in the components of the stress tensor $\dot{\sigma}_{ij}$ to the elastic components ξ_{ij}^e, we will accept Hooke's law:

$$\dot{\sigma}_{ij} = \lambda\xi_{kk}^e\delta_{ij} + 2\mu\xi_{ij}^e,$$

$$\xi_{ij}^{e} = \frac{1}{2\mu}\dot{S}_{ij} + \frac{1-2\nu}{E_*}\dot{\sigma}_{kk}\delta_{ij},$$

$$S_{ij} = \sigma_{ij} - \frac{1}{3}\sigma_{kk}\delta_{ij},$$

where S_{ij} is the deviator of the stress tensor. Here, the dot above the symbol means time differentiation; λ, μ are Lamé coefficients; ν, E_* are Poisson's ratio and Young's modulus; δ_{ij} is the Kronecker symbol.

Hooke's law, however, can also be written in terms of stress–strain. Further consideration will be conducted in the principal axes of the tensors characterizing the stress–strain state of the medium, which are common for all tensors due to the assumption of small deformations.

Let us consider the main tangential stresses

$$\tau_1 = \frac{\sigma_2 - \sigma_3}{2}, \quad \tau_2 = \frac{\sigma_3 - \sigma_1}{2}, \quad \tau_3 = \frac{\sigma_1 - \sigma_2}{2}$$

and main tangential plastic strains

$$\gamma_1 = \frac{\varepsilon_2^p - \varepsilon_3^p}{2}, \quad \gamma_2 = \frac{\varepsilon_3^p - \varepsilon_1^p}{2}, \quad \gamma_3 = \frac{\varepsilon_1^p - \varepsilon_2^p}{2}.$$

From the last relations and the conditions of incompressibility of the material under plastic deformation:

$$\varepsilon_1^p = \frac{2}{3}(\gamma_3 - \gamma_2), \ \varepsilon_2^p = \frac{2}{3}(\gamma_1 - \gamma_3), \ \varepsilon_3^p = \frac{2}{3}(\gamma_2 - \gamma_1),$$

$$\xi_1^p = \frac{2}{3}(\dot{\gamma}_3 - \dot{\gamma}_2), \ \xi_2^p = \frac{2}{3}(\dot{\gamma}_1 - \dot{\gamma}_3), \ \xi_3^p = \frac{2}{3}(\dot{\gamma}_2 - \dot{\gamma}_1).$$

Hooke's law can be rewritten as

$$\sigma_i = \lambda\theta + 2\mu(\varepsilon_i - \varepsilon_i^p), \quad \theta = \varepsilon_1 + \varepsilon_2 + \varepsilon_3 = \varepsilon_1^e + \varepsilon_2^e + \varepsilon_3^e,$$

from where, as in one-dimensional formulations, it follows that the increase of stress in the considered medium is due to the growth of total strain, and their relaxation is due to plastic strain.

To develop the relations defining plastic deformation, we assume that it is the result of dislocation motion in the planes of the main tangential stresses, which coincide, due to the assumptions made, with the planes of the main tangential plastic strains and their velocities. The deformations are determined from expressions like (5.10):

$$\gamma_i = bN_iL_i, \quad i = 1, 2, 3, \tag{5.12}$$

where N_i is now the dislocation density in a given sliding plane; L_i is their average displacement in a given plane. Analog of (5.11):

$$\gamma_i = b\,N_i L_i + b\,N_{ni} V_i. \tag{5.13}$$

Applying (5.12) or (5.13), we can use the known relations for N, N_n, V_a, considering, respectively, that

$$N_i = N(\gamma_i), \quad N_{ni} = N_n(\gamma_i), \quad V_i = V(\sigma_i).$$

The value of the average dislocation displacement – the distance a dislocation travels during loading until it stops at some barrier or comes to a free surface – is very sensitive to the loading conditions and structural peculiarities of the material. There is no systematic data on the influence of these or those parameters on the value of L. It can be assumed that as the dislocation structure develops, the average displacement changes in proportion to the size of the dislocation cell formed in the material during deformation, then

$$L_i = a_0/\sqrt{N_i}, \tag{5.14}$$

where the constant a_0 is subject to experimental determination. According to various estimates, $10^{-3}\text{cm} \geq L_i \geq 10^{-5}\text{cm}$.

These relations are the constitutive equations of an elastic–plastic medium with dislocation deformation kinetics. The formulated three-dimensional generalization is illustrative in nature since its practical applicability is strongly limited by the serious assumptions made during its construction.

Recent investigations show that, in addition to dislocations (linear defects), defects of higher dimensions – planar, volumetric ones – may also play a significant role in plastic deformation. Their evolution introduces additional mechanisms, which led to the introduction of the concept of structural levels of solid-state deformation. At present, this new promising direction in modeling dynamic deformation and fracture of solids is still under development. The formulated formal apparatus is not yet suitable for solving specific physical problems.

5.4 Numerical Methods for Solving Dynamic Problems

The complexity of mathematical formulations of specific dynamic problems in the described models requires the use of numerical methods and algorithms implemented on powerful computers for their solution. Essentially nonstationary, wave processes in deformable solids are described by systems of quasi-linear or nonlinear hyperbolic equations, the general theory of the solution of which is far from being completed. This statement also applies to the analysis of difference methods used for the numerical solution of practical problems. One of the first publications on numerical methods was the book [8].

The main problems – stability and convergence of the applied difference scheme – can be rigorously justified for linear homogeneous boundary value problems with constant coefficients, for some cases of variable coefficients, and partial types of quasi-linear systems. The analysis of stability and convergence on the simplest model problems, or comparison with the known (analytical, numerical) solutions, or comparison with the results of the corresponding experiments still remain the main methods of investigation of numerical solutions.

Among the numerical methods for solving these systems of differential equations, there are explicit and implicit ones. The latter are almost always stable, but the peculiarities of wave processes do not allow calculations with any large time step, as this will lead to loss of accuracy. Also, it is known that with a large step, implicit schemes greatly overestimate the dependence area of the numerical solution in comparison to the dependence area of the solution of the original system, while in high-speed processes it is important to pass this area as accurately as possible.

The correct representation of the dependence area requires a condition, which, as a rule, coincides with the Courant condition, which guarantees the stability of explicit schemes. Thus, the main advantages of implicit difference schemes are not realized in this class of problems, and preference is given to explicit difference schemes for solving them.

The peculiarities of the numerical method or its implementation are often related to the choice of the reference system to describe the motion of the medium. Here, the Euler and Lagrange approaches are distinguished, in which the independent quantities (variables) are associated with a stationary coordinate system or with moving parts of the medium, respectively, as discussed above. Mesh-characteristic methods can be implemented in both Eulerian and Lagrangian coordinate systems. Most methods are oriented to one or another coordinate system, by virtue of which a division of numerical schemes into Eulerian and Lagrangian arises.

Both classes have certain disadvantages and advantages. The main advantage of Eulerian schemes is the possibility to calculate processes accompanied by finite, large deformations of the medium. In this case, however, one has to put up with an overestimation at the initial stage of the counting domain size, roughness, and inaccuracy of the resolution of flow details in the vicinity of free and contact boundaries, the necessity to introduce a variable number of counting nodes.

The main advantages of the Lagrangian approach are the logical simplicity of the algorithms due to the invariance of the number of nodes of the difference mesh and the associated preservation of the initial approximation accuracy (at small deformations of the calculation domain), the ease of calculating the boundaries of regions and subdomains and a good description of the flow in their vicinity. Disadvantages begin to appear when it is necessary to compute flows with large deformations, when large mesh distortions occur, up to possible self-intersection of cell boundaries. This requires rearrangement of the mesh with the inevitable use of interpolations, leading to a loss of accuracy.

When choosing between Eulerian and Lagrangian methods, one should be guided by the peculiarities of the problem (range of problems) under consideration. The main criterion may be the characteristic value of the strain rate. At low strain rates, it is natural to use the Lagrangian methods. The technique of their application is well developed, so they are quite often used in calculations of high-velocity processes as well, overcoming considerable computational difficulties.

One of the main difficulties is the re-interpolation of intermediate results on the new mesh. The necessity of its implementation appears when the image of the Lagrangian grid in Eulerian coordinates starts to be significantly distorted. The new mesh must satisfy certain conditions, for example sometimes it is required that the ratio of lengths of diagonals of its cells is close enough to unity. Naturally, re-interpolations distort the flow pattern by smearing the features of the solution. In some cases, the correct reproduction of the integral characteristics of the flow using re-interpolation to a new mesh is provided by techniques that preserve the total mass, momentum, and energy contained in the medium bounded by the computational domain.

The desire to combine the advantages of both approaches led to the emergence and widespread use of hybrid methods. One of the first such methods was the particle-in-cell method [8]. Here, the medium is modeled by mobile Lagrangian particles having mass, which, moving along the cells of a stationary Eulerian mesh, transmit momentum and energy. The method is quite widespread for solving problems of high-velocity deformation both in the hydrodynamic formulation and considering the strength properties of the material based on the viscous compressible fluid model and more complex viscoelastic–plastic approximations.

Its main drawback is the low resolution of the flow pattern in cells containing a small number of particles (two or three). When calculating, for example, the motion of the medium in strong rarefaction waves, material outflows occur, which leads to sharp fluctuations of the calculated physical quantities when even a single particle moves from one cell to another.

Besides, there are also some disadvantages of Eulerian methods: diffusion of contact boundaries, which worsens the description of the flow pattern in their vicinity, uneconomical nature associated with the initial overestimation of the

Eulerian calculation domain, and the need to store information about a large number of Lagrange particles, if the detailed resolution of the flow pattern is required. Some improvement of the method is achieved by introducing, in addition to the Eulerian one, a Lagrangian calculation mesh.

Let us note one of the branches of the general group of particle methods – the method of individual particles developed by V.A. Agureikin and B.P. Kryukov [9]. Like in the particle-in-cells method (F.H. Harlow method), the flow region is covered with a fixed Euler mesh on which the particles move. However, now each particle is assigned internal energy, density, pressure, etc. together with coordinates, mass, and its sort. The calculations are based on the fact that each equation of motion can be written in an equivalent form when the left part is a Lagrangian derivative of some flow characteristic, and the right part contains only spatial derivatives of flow parameters. Let us consider the continuity equation as an example:

$$\frac{d\rho}{dt} = -\rho \operatorname{div} \vec{u}.$$

Its integration is carried out as follows. At a fixed moment t, the velocity value is re-interpolated according to a certain rule from particle centers to mesh nodes. After that, the values div $\vec{u}$ are determined here using finite differences. The latter are re-interpolated to the particle centers again, after which the density of the i-th particle at a new time layer $t + \Delta t$ is calculated by explicit integration of the continuity equation:

$$\rho_i(t + \Delta t) = \rho_i(t) - \rho_i(t) \operatorname{div} \vec{u}\, \Delta t.$$

Similarly, the values of velocity and energy are recalculated. The new coordinates of the particle are calculated using the established values of velocities. The obvious advantage of the method is that the problem of mixed cells is absent here. In Harlow's method, such difficulty arises when one cell contains several particles of a different sort. In this case, the procedure of pressure calculation becomes significantly more complicated. The main advantage of the individual particle method, however, is that particle parameter fluctuations play a much smaller role here.

From the idea of discrete redistribution of physical parameters using particles, it is logical to move to continuous distribution by "smearing" the mass over the whole cell volume and calculating the mass flow over the Euler cell boundaries according to the same regularities as in the case of a discrete set of particles. As a result, the large particle method (the flow method) proposed by O.M. Belotserkovsky and Y.M. Davydov was developed [10]. The method is based on the sequential use of Eulerian and Lagrangian approaches, with approximations in fixed nodes of calculation mesh and subsequent consideration of the interaction of calculation cells as independent large particles of liquid.

We know examples of solving several problems of the discussed class in the gas-dynamic formulation and in the formulation accounting for material strength. Here, the difficulties of realization of the boundary conditions, inherent to the Eulerian methods, remain. In the vicinity of the boundary of the computational domain, the introduction of dummy cells or a special approximation of the initial differential equations is required, which may reduce the possibilities of detailing the flow in the boundary zone.

Further development of the idea of combining the advantages of the Eulerian and Lagrangian approaches was the movable mesh method. In it, the boundaries of the computational domain move together with the particles of the medium (Lagrangian), and a mesh, most often curvilinear and adapting to the specifics of the flow, which is Eulerian relative to the inner cells, is built inside the regions bounded by these movable boundaries. Mass, momentum, and energy can flow through the ribs (cell boundaries).

The main difficulty in implementing the method is the construction of nonuniform meshes within regions with curvilinear boundaries. Algorithms of construction using quasiconformal mappings or based on variational principles require a very long computation time, sometimes even more than it is necessary to solve the equations describing the medium flow. More often, simpler, more economical, and more limited in applicability problem-oriented ways of constructing meshes are used.

Calculating the motion of free and contact boundaries requires special care in the movable mesh method. Often the displacements of the boundary points are determined under the assumption that the flow is close to one-dimensional, and the boundary is sufficiently smooth. At finite deformations of the medium and the essentially nonuniform behavior of the flow, these conditions may be violated. In these cases, either the peculiarities of a particular problem are used or additional, for example physical, considerations are involved, allowing one to still determine the displacement of the boundary nodes.

After choosing the approach to description (Eulerian, Lagrangian, mixed) it is necessary to determine the method of approximation of differential equations. The accuracy of the laws of conservation, the order of approximation of the initial equations, the degree of smearing of strong discontinuities at the end-to-end calculation, etc. largely depend on it.

One of the most common is the finite-difference "cross" scheme. For two-dimensional calculations of high-velocity processes in models considering the strength of materials, it was first applied by M.L. Wilkins (HEMP method) [8]. The scheme has the second order of accuracy, but in the boundary points due to the necessity of inserting dummy cells, the approximation order decreases to the first order. The Prandtl–Reuss relations are also realized here with the first order of accuracy. For calculations of flows with strong gradients of physical parameters (shock waves), tensor artificial viscosity is used. The HEMP method is widespread worldwide. It is often also taken as the basis for calculating processes in materials with complicated rheology.

The Lax–Wendroff method is often used in Eulerian coordinates. It is effective for solving systems of equations that allow writing in a divergent form. The two-step version of the method, in which the conservation laws are automatically satisfied, is the most efficient. The scheme has the second order of accuracy. The scheme's viscosity is insufficient for the end-to-end calculation of discontinuous solutions, so an artificial viscosity of a special type that practically disappears in the smooth part of the solution and is essential in the jumps is introduced. The Lax–Wendroff method was used to solve the problems of impact of impactors with barriers of finite thickness. The necessity of introducing dummy cells to approximate the boundary conditions reduces the accuracy of the scheme in the boundary zone to the first order. The Prandtl–Reuss equations cannot be written down in a fully divergent form, so the stress field has to be determined from the solution of difference equations with "frozen" coefficients.

The finite element method has been widely used in the problems of nonlinear continuum mechanics. It is quite often used in problems of high-velocity deformation as well. As a rule, this method is implemented in combination with a Lagrangian description. In contrast to static problems, in this case, partitioning into elements is carried out only in the spatial domain, while time resolution of a process is produced by solving a system of ordinary differential equations. The main disadvantage of the finite element method is the same as that of any other technique associated with the Lagrangian approach: the distortion of the shape of cells (elements), which can significantly limit the time step and lead to an increase in the calculation error in the zone of large deformations.

One of the fruitful ideas used in constructing difference schemes for solving nonlinear hyperbolic systems is to calculate each time step in two stages (fractional step method, "predictor-corrector" type schemes). When calculating shock-wave motions of material, it is convenient to use characteristic relations at the "predictor" stage, and at the "corrector" stage to recalculate using integral conservation laws written in finite-difference form, thereby guaranteeing their fulfillment. When applied to systems of linear partial differential equations in "predictor–corrector" schemes, the first stage creates a stability reserve of the numerical solution, and the second stage provides an approximation of the original equations.

In a more general case, the use at the first stage of the characteristic relations allows to carry out the correct transfer of dependence areas of the solution (according to the Courant condition, it is enough for the stability of the obtained solution), and the "corrector" stage uses the conservation laws, providing approximation. This approach is implemented for the Prandtl–Reuss model as well. At the "corrector" stage, the equations of linear elasticity theory with the subsequent correction to the yield circle are used here, since the Prandtl–Reuss equations are not written in divergent form even in the case of small deformations.

A more complex modification of the "predictor–corrector" idea as applied to nonlinear equations can be considered the method of disintegration of discontinuities (Godunov's method), which has arisen independently and is of independent importance. Here at the "predictor" stage, it is proposed to use the exact solutions of the problem of disintegration of discontinuities (Riemann problem) occurring at the boundaries of difference cells assuming that the medium parameters are constant inside the cells. The method of disintegration of discontinuities has found broad adoption.

The application of the S.K. Godunov method to the problems of dynamics of nonlinear–elastic and elastic–plastic media is also known. In complicated cases or in more complex constitutive equations of elastic–plastic deformation, it is impossible to find an exact solution to the problems of disintegration of discontinuities. Moreover, the equations of models that take into account the dependence of the yield strength on the strain rate (including the viscoelastic body model of the Maxwell type) generally do not have automodel solutions describing the disintegration of discontinuities. The way out is found in using the solution of the decomposition problem in the sound approximation, a method that has been repeatedly tested in gas dynamic problems.

The practice of calculations shows the preservation of good accuracy in this case. In essence, at the first stage, an approximate solution of the initial system of equations is attracted by the method of characteristics. Usually, the equations of either elasticity theory or gas dynamics, which have self-similar solutions, are used. As a generalization, a method is proposed in which both systems of equations are used at the stage of calculating the disintegration of discontinuities, and the specific choice for a given point of the medium and time moment is made from the analysis of the features of the physical state at the point of time and space in question. This method, extended further to the two-dimensional case, in a combination with moving meshes has been used to solve two-dimensional nonstationary problems. As its merits (the drawbacks are clear from the brief critical analysis above), it is possible to note sufficient correctness in the transfer of the dependence region of the solution, the property of monotonicity preservation, and the absence of the need for the introduction of artificial viscosity.

Smoothed-particle hydrodynamics (SPH) is a computational method used for simulating the mechanics of continuum media, such as solid mechanics and fluid flows. It was developed by Gingold and Monaghan and Lucy in 1977, initially for astrophysical problems [11, 12].

By construction, SPH is a mesh-free method, which makes it ideally suited to simulate problems dominated by complex boundary dynamics, like free surface flows, or large boundary displacement.

5.5 Short Introduction to ANSYS AUTODYN

To date, a significant number of software systems have been developed that implement various methods of numerical integration of the equations describing explosion and shock processes in both two- and three-dimensional formulations. At present, the most well-known systems are, perhaps, LS DYNA and ANSYS AUTODYN. Below is a brief description of ANSYS AUTODYN [13, 14].

ANSYS AUTODYN was initially developed by the American company Century Dynamics and is intended for numerical modeling of nonlinear dynamics of solids, liquids, gases, and their combinations in two- and three-dimensional formulations. This is a powerful tool for interdisciplinary computations in problems where loads change over a short period, as in the case of a high-velocity impact or explosion. ANSYS AUTODYN contains many solvers and computational methods that use Lagrangian and Eulerian approaches to the flow of a continuous medium, an arbitrary Lagrangian–Euler formulation, as well as the method of smoothed particle hydrodynamics (SPH), which forms the foundation of the mesh-free solver.

ANSYS AUTODYN is fully integrated into an easy-to-use graphical interface, the capabilities of which are complemented by the functions available in the ANSYS Workbench. These functions enable quick and easy preparation of a numerical model based on computer-aided design (CAD) geometry, generation of a mesh that is most suitable for analysis, and multiple computations using parameterization. Geometric dimensions, materials, and their physical and mechanical properties, as well as initial conditions (for example the impact velocity), can be used as parameters.

Let us briefly discuss the problems that arise when applying AUTODYN to the numerical simulation of a particular process. The main problem for any software system for numerical simulation of high-speed processes is a reasonable choice of **constitutive relations** for materials, the choice of the **numerical method**.

Let's consider in detail here the recommendations on the choice of **constitutive relations** for the materials involved in the calculation of the bodies [14].

In general, materials have a complex response to dynamic loading.

The following phenomena may need to be modeled:

- Nonlinear pressure response
- Strain hardening
- Strain rate hardening
- Thermal softening
- Compaction (porous materials)
- Orthotropic behavior (e.g. composites)
- Crushing damage (e.g. ceramics, glass, geological materials, concrete)
- Chemical energy deposition (e.g. explosives)

- Tensile failure
- Phase changes (solid–liquid–gas)

No single material model incorporates all these effects.

How do we choose a set of material modeling options for a particular material? In terms of the material itself, it is relatively easy to identify the basic category that a material lies in:

- Liquid or Solid?
- Isotropic or Anisotropic/Orthotropic?
- Inert or Reactive?
- Porous or Not?
- Ductile or Brittle?
- Pressure Dependent Strength (cohesive) or not?

The actual set of models used, however, is highly dependent on the application and the available material data. First, we need to understand how parameters influence response and which parameters are critical for good results. Which material model should we use?

An AUTODYN material model consists of three components:

Equation of state (EOS), strength model, failure model (Tables 5.1–5.3).

You can select materials by specifying a new material with any properties or from the standard material library. The library contains about 270 materials: metals and alloys, explosives and mixtures, ceramics, composites, concrete, plastics, granular materials, water, and air. A large number of equations of state, strength, and failure models are provided for the materials, allowing research under various conditions.

5.5.1 Choice of the Numerical Method

5.5.1.1 Lagrange Solvers

Use meshes that are embedded in the material and move and distort with the material. Three element types: solid, shell, beam.

Provide the most efficient and accurate method for computing structural response.

5.5.1.2 Euler Solvers

Use meshes that are fixed in space. Materials are advected through the mesh as they flow or deform. Avoids problems of mesh distortion and tangling that are prevalent in Lagrange simulations with large flows. Provides the best solution for gas/fluid flow. Provides an alternative solution for the extreme deformation of solids.

Table 5.1 Choice of the equation of state.

Material	Equation of state		
Inert gas			IDEAL GAS
Inert isotropic solid (metals, polymers, concrete, geological material)	Nonporous	Small compressions (<1%)	LINEAR
		Large compressions (>1%)	SHOCK, TILLOTSON, PUFF, SESAME, TWO-PHASE
	Porous	Very small compressions (compaction not important)	LINEAR
		Small compressions (compaction but no shock effects)	POROUS COMPACTION
		Large compression (shock effects important)	P-ALPHA
Inert anisotropic solid (fiber composite, rolled material)		Small compression (<1%)	ORTHOTROPIC WITH LINEAR EOS
		Large compression (>1%)	ORTHOTROPIC WITH POLY/SHOCK EOS
Reactive isotropic solid/gas (high explosive, incendiary powder, reactive gas)	Detonation	Detonates at constant velocity	JWL
		Initiation and propagation of detonation depends on local material conditions	LEE–TARVER
	Deflagration	Porous	POWDER BURN WITH COMPACTION SOLID EOS
		Nonporous	POWDER SLOW BURN WITH LINEAR SOLID EOS

Source: From ANSYS Explicit Dynamics Analysis Guide [14]. Reprinted with permission of Ansys Inc.

AUTODYN has two Euler solvers:

5.5.1.2.1 Multi-material Euler Solver

General Euler solver capable of modeling and tracking the motion of multiple materials (with strength) within the mesh.

Table 5.2 Choice of the strength model.

Material	Strength model		
Inert gas			HYDRO
Inert isotropic solid (metals, polymers, concrete, geological)	Elastic	Rate dependent elasticity	VISCOELASTIC
		Rate independent elasticity (no plasticity)	ELASTIC
	Elastic–plastic	No hardening	VON MISES/BILINEAR ISOTROPIC HARDENING
		Strain hardening, Strain rate sensitive, melting	JOHNSON–COOK, STEINBERG–GUINAN, ZERILLI–ARMSTRONG, PIECEWISE JC/MULTILINEAR ISOTROPIC HARDENING
	Elastic–Brittle/ granular	Pressure hardening only	DRUKER–PRAGER, MO GRANULAR
		Non-porous	JOHNSON–HOLMQUIST
		Porous	RHT
Reactive isotropic solid/gas (high explosive, incendiary powder, reactive gas)	Detonation		Strength usually not important
	Deflagration	Porous	MO GRANULAR STRENGTH
		Brittle	ELASTIC
		Ductile	VON-MISES/BILINEAR ISOTROPIC HARDENING, JOHNSON -COOK

Source: From [14]. Reprinted with permission of Ansys Inc.

5.5.1.2.2 Euler Blast Solver

Designed for efficient and accurate blast solutions:

2nd order; hydrodynamic (no stress tensor); single material, ideal gas.

5.5.1.3 ALE (Arbitrary Lagrange Euler) Solver

Uses solid meshes that are embedded in the material and move and distort with the material. The mesh within any material region can be continuously adjusted in arbitrary and predefined ways (automatic rezoning). The most efficient way of

Table 5.3 Choice of the failure model.

Material	Failure model		
Inert gas			NONE
Inert isotropic solid (metals, polymers, concrete, geological)	Hydrodynamic		HYDRO (PMIN)
	Elastic–plastic	Brittle material/spall	PRINCIPAL STRESS (+ CRACK SOFTENING)
		Ductile material	EFFECTIVE PLASTIC STRAIN, PRINCIPAL STRAIN, PRINCIPAL STRESS
	Elastic–Brittle/ granular	Tensile failure	PRINCIPAL STRESS/STRAIN (+ CRACK SOFTENING)
		Shear damage	JH2, RHT, CUMULATIVE DAMAGE MODEL
Inert anisotropic solid (fiber composite, rolled material (low compression)	Brittle		ORTHOTROPIC DAMAGE
	Ductile		ORTHOTROPIC DAMAGE

Source: From [14]. Reprinted with permission of Ansys Inc.

calculating blast response of structures that do not deform significantly. Mesh distortions cannot easily be avoided if gross deformation or failure of structures occurs.

5.5.1.4 Mesh Free Solver

Uses the Smooth Particle Hydrodynamics Method (SPH). Lagrange particle method (particles are embedded in the material).

No mesh distortion/tangling problems. Provides the most accurate solution for modeling fracture and fragmentation of brittle materials. Applied very successfully to problems involving high-velocity impacts and impacts into ceramics and concrete. Not recommended to model ductile materials. "Numerical" failure occurs before the onset of physical failure.

5.6 Numerical Modeling Example

Recall that some examples were given earlier. The first chapter presented the results of numerical calculation of the explosive loading process of a spherical shell carried out by the method of individual particles. The second chapter presented the results of numerical simulation of the formation of a hypervelocity cumulative jet from a ceramic tube.

To illustrate the capabilities of numerical simulation, we consider here the problem of forming a Mach stem in an explosive charge pre-compressed by a leading wave without a shock front [15].

5.6.1 Experimental Data

The scheme of the experimental assembly is shown in Figure 5.1. Active and passive HE charges were shaped as cylindrical cartridges with a diameter and height of 40 mm. On the rear end of the passive HE charge, an identification steel specimen was mounted to detect the presence or absence of detonation. Inert inserts of silicon carbide were shaped as 20 mm × 20 mm square prisms of varying heights Δ and were located between active and passive HE charges without any clearance on the way of initiating shock wave and with partial overlap of HE cross-sections. Inert inserts were mounted using the vacuum lubricant "Ramzai" to prevent any clearance between HE charges and prisms.

The phenomenon of energy focusing in explosion systems that include high modulus elastic elements was observed in a passive HE charge of cast TG-40/60 without a cumulative shape under shock-wave loading by the explosion of an active HE charge in water after preliminary compression by a leading wave in a silicon carbide insert. The phenomenon manifested itself as a hole in the identification steel specimen with a depth of about 10 mm and a diameter of about 5 mm (see Figure 5.2). The physical reason for this focusing process is the propagation of a convergent detonation wave into highly compressed HE and the formation of a Mach stem on the axis of the HE charge.

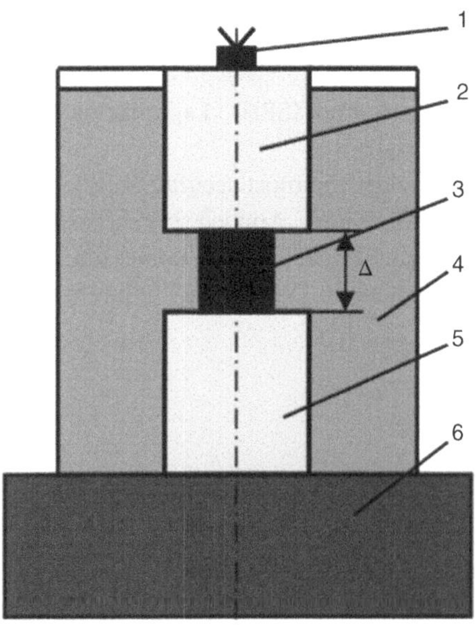

Figure 5.1 Experimental assembly scheme: 1 – electric detonator; 2 – active HE charge; 3 – inert insert (shaped like a 20 mm × 20 mm square prism); 4 – degassed water; 5 – passive HE charge; 6 – identification steel specimen. *Source:* From Balagansky et al. [15]. Reprinted with permission of American Institute of Physics.

Figure 5.2 A typical surface of a steel identification specimen with a hole caused by the effect of detonation products of a passive HE charge. *Source:* From Balagansky, et al [15]. (2011). pg 110: 123516 / with permission of American Institute of Physics.

5.6.2 Numerical Simulation

For a better understanding of the Mach stem formation phenomenon, we have used numerical simulation with help of ANSYS AUTODYN 11.0. Initial posting corresponded to the experimental assembly for $\Delta = 20$ mm (see Figure 5.1). It is shown in Figure 5.3.

The simulation was performed in 2D-axis symmetry posting on an Eulerian mesh of 2600×800 cells.

The detonation wave was initiated as a plane wave and propagated from the left to the right boundary of the Eulerian mesh. Boundary conditions on the left, right, and upper boundaries were defined in terms of AUTODYN as "Flow out." We used values of constants for inert matter from the AUTODYN EOS library.

The behavior of a Composition B passive HE charge with a density of 1.63 g/cm^3 was calculated from JWL EOS and Lee–Tarver kinetic constants. Values of the EOS and kinetic constants are given in Table 5.4.

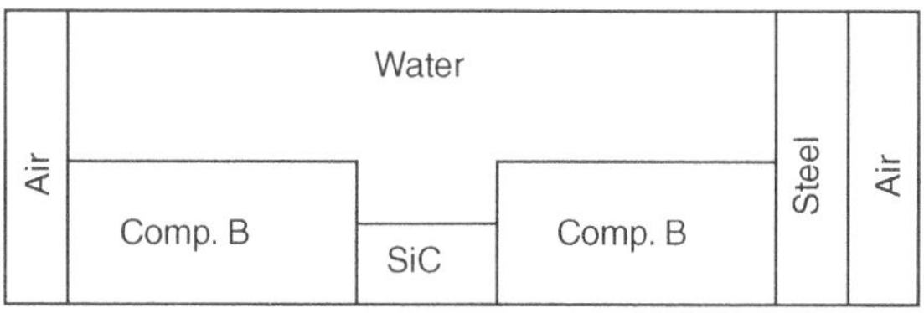

Figure 5.3 Initial posting with dimensions of 13.0 cm × 4.0 cm. The bottom boundary is the symmetry axis. The direction of wave propagation is from left to right. *Source:* From Balagansky et al. [15]. Reprinted with permission of American Institute of Physics.

Table 5.4 Ignition and growth parameters for cast Composition B.

Unreacted JWL	Product JWL
$A = 1479$ Mbar	$A = 5.308$ Mbar
$B = -0.05261$ Mbar	$B = 0.0783$ Mbar
$R_{1u} = 12$	$R_1 = 4.5$
$R_{2u} = 12$	$R_2 = 1.2$
$\omega_u = 0.912$	$\omega = 0.34$
$C_v = 2.487 \times 10^{-5}$ Mbar/K	$C_v = 1.0 \times 10^{-5}$ Mbar/K
$T_o = 298K$	C–J energy/unit volume $E_{og} = 0.081$ Mbar
Shear modulus = 0.035 Mbar	C–J detonation velocity $U_D = 7.576$ mm/μs
Yield strength = 0.002 Mbar	C–J pressure $P_{CJ} = 0.265$ Mbar
$\rho_0 = 1.63$ g/cm^3	Reaction zone width $W_{reac} = 2.5$
Von Neumann spike rel vol. $c_0 = 0.7$	Max change in reaction ratio $\Delta F_{max} = 0.1$
C–J energy/unit volume $E_{0,\ u} = -0.00504$ Mbar	
Reaction rates	
$a = 0.0367$	$x = 7.0$
$b = 0.667$	$y = 2.0$
$c = 0.667$	$z = 3.0$
$d = 0.333$	$F_{igmax} = 0.022$
$e = 0.222$	$F_{G1max} = 0.7$
$g = 1.0$	$F_{G2min} = 0.0$
$I = 4.0 \times 10^6 \mu s^{-1}$	$G_1 = 140/\text{Mbar}^2/\mu s$
Maximum relative volume in tension = 1.1	$G_2 = 1000/\text{Mbar}^3/\mu s$

Source: From Balagansky et al. [15]. Reprinted with permission of American Institute of Physics.

The Lee–Tarver ignition and growth model [16, 17] has been used to describe the behavior of the Composition B acceptor HE charge. The model uses two Jones–Wilkins–Lee (JWL) equations of state, one for the unreacted explosive and another one for its reaction products, in the temperature-depended form

$$p = Ae^{-R_1 V} + Be^{-R_2 V} + \frac{\omega C_V T}{V},$$

where p is pressure, V is the relative volume, T is temperature, ω is the Grüneisen parameter, C_v is the average heat capacity, and A, B, R_1, and R_2 are constants.

The reaction rate law contains three terms to model the ignition of hot spots by shock compression, the slow growth of reaction from these isolated hot spots, and the rapid completion of the reaction as the hot spots coalesce.

$$\frac{d\alpha}{dt} = I(1-\alpha)^b\left(\frac{\rho}{\rho_0} - 1 - a\right)^x + G_1(1-\alpha)^c\alpha^d p^y + G_2(1-\alpha)^e\alpha^g p^z,$$

where α is the fraction reacted (degree of HE decomposition), t is time, ρ is the current density, ρ_0 is the initial density, p is pressure, and I, G_1, G_2, a, b, c, d, e, g, x, y, and z are constants. The parameter a is the critical compression that is used to prohibit ignition until a certain degree of compression (or a certain input pressure) has been reached. This parameter provides the ability to calculate HE desensitization after precursor wave loading.

The results of the simulation at different times with the initial time of 5 μs and a time step of 1 μs are shown as material flow fields and pressure contours in Figure 5.4.

A complex interaction process of shock, detonation, and compression waves develops in the assembly. We can see the Mach stem formation in passive HE charge.

The detonation wave profile with C-J pressure of 22.4 GPa in the active HE charge can be seen in Figure 5.5.

Figure 5.6 demonstrates the graph of the axial stress in the silicon carbide insert versus axial distance at the time of 6.75 μs.

Figure 5.7 Demonstrates the pressure value in the Mach stem at $t = 15.75$ μs. The value is equal to 1.25 Mbar.

The physical reason for the Mach stem formation is the propagation of a convergent detonation wave into highly compressed HE.

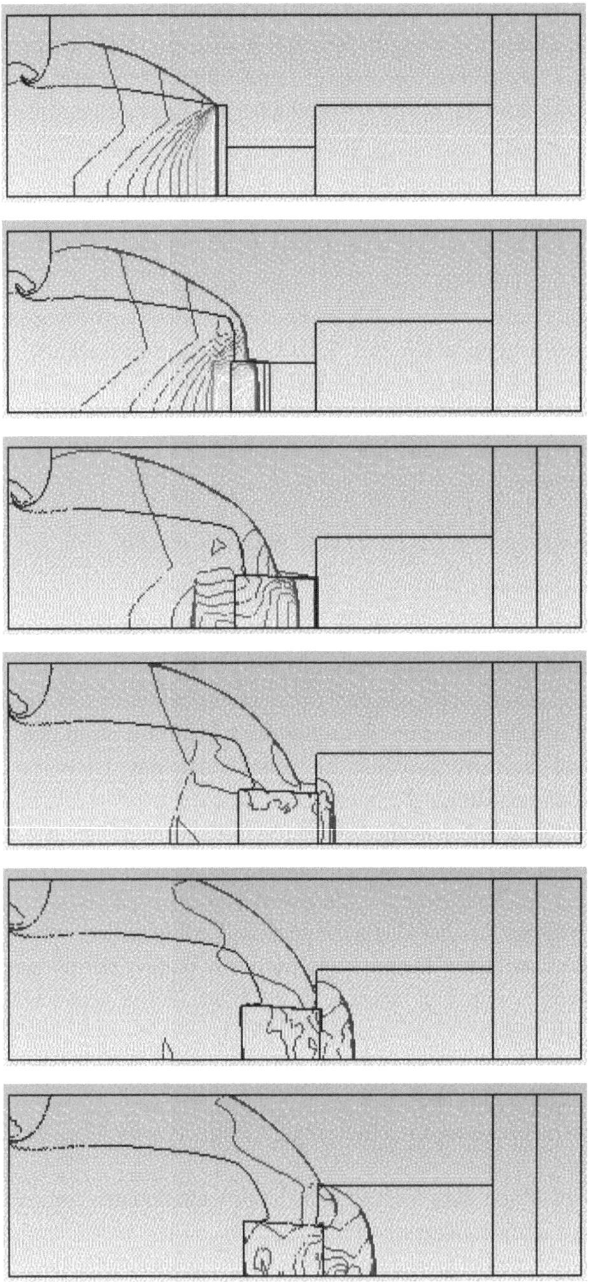

Figure 5.4 Material flow fields and pressure contours at different times with a time step of 1 μs. Calculated area was $x = 13$ cm (2600 cells), $y = 4$ cm (800 cells). *Source:* From Balagansky et al. [15]. Reprinted with permission of American Institute of Physics.

Figure 5.4 (Continued)

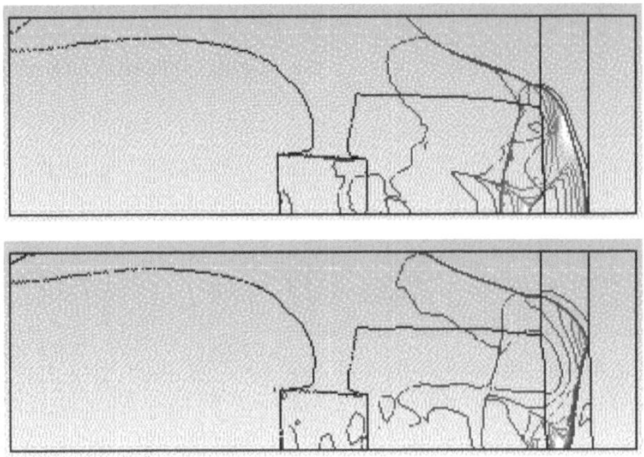

Figure 5.4 (Continued)

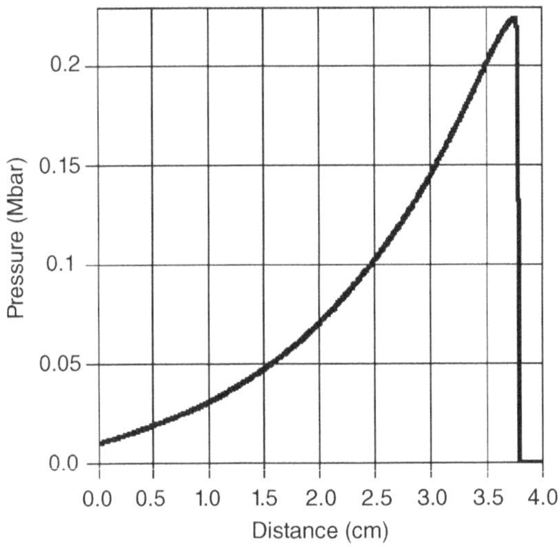

Figure 5.5 Pressure profile on the symmetry axis of active HE charge at the time of 5 μs from x_1 = 1 cm to x_2 = 5 cm. *Source:* From Balagansky et al. [15]. Reprinted with permission of American Institute of Physics.

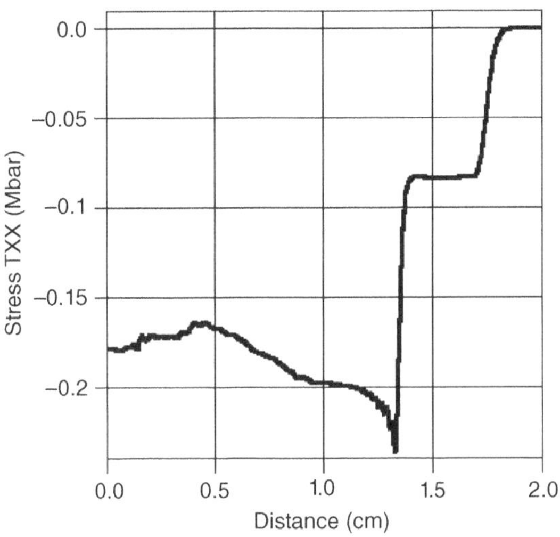

Figure 5.6 Graph of the axial stress in the silicon carbide insert versus axial distance at t = 6.75 μs from x_1 = 5 cm to x_2 = 7 cm. *Source:* From Balagansky et al. [15]. Reprinted with permission of American Institute of Physics.

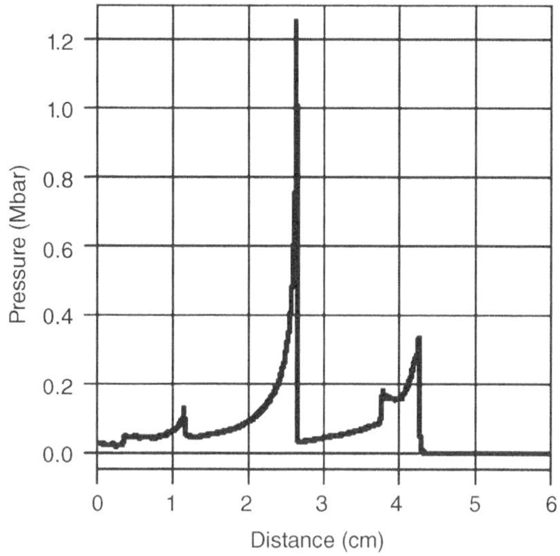

Figure 5.7 Pressure graph in the passive HE charge versus axial distance at $t = 15.75\ \mu s$ from $x_1 = 7$ cmto $x_2 = 13$ cm. *Source:* From Balagansky et al. [15]. Reprinted with permission of American Institute of Physics.

5.7 Self-assessment Questions

1) What is needed for the numerical simulation of explosion and impact processes?
2) Which fundamental hypothesis is most often used in building physical models of deformation of materials?
3) What types of coordinates are used in the modeling of explosion and impact processes?
4) What is the basis of Euler's approach to the choice of the system of reference?
5) What is the basis of Lagrange's approach to the choice of the system of reference?
6) How to make a conversion from the Eulerian to the Lagrangian coordinate system?
7) How to make a conversion from the Lagrangian to the Eulerian coordinate system?
8) How is the density of a substance defined?
9) How are the equations of movement of gaseous media different from the equations of movement of bodies with a strength?
10) What are the constitutive equations?
11) What states are realized in a solid in the process of deformation under the action of intense dynamic loads?

12) What are the requirements for models describing the behavior of continuous media under intensive dynamic loads?
13) What are the main approaches used to build models describing the behavior of continuous media under intensive dynamic loads?
14) What is the area of application of the elastic medium model in the class of problems under consideration?
15) What constants characterize an isotropic elastic body?
16) At what velocities do small perturbations in an isotropic elastic body propagate?
17) What is the area of application of hydrodynamic models in the considered class of problems?
18) On what physical concepts is the Mie–Grüneisen equation of state based?
19) On what hypothesis are deformation theories of plasticity based?
20) On what hypothesis are the theories of plastic flow based?
21) What are the advantages and disadvantages of Eulerian computational schemes?
22) What are the advantages and disadvantages of Lagrangian numerical schemes?

References

1 Balagansky, I.A. and Merzhievsky, L.A. (2004). *Effects of Means of Destruction and Ammunition*. Novosibirsk: Novosibirsk State Technical University [in Russian].

2 Sedov, L.I. (1970). *Mechanics of Continuous Media*, vol. **1**. Moscow: Nauka [in Russian].

3 Godunov, S.K. (1978). *Essential of Continuous Mechanics*. Moscow: Nauka [in Russian].

4 Meyers, M.A. (1994). *Dynamic Behavior of Materials*. New York: Wiley.

5 Merzhievsky, L.A. (1980). Method for calculation of viscoelastic fluid flow. In: *Dynamics of Continuous Media*, vol. **45**, 141–151. Novosibirsk: Institute of Hydrodynamics [in Russian].

6 Merzhievskii, L.A. and Shamonin, S.A. (1980). Construction of the time dependence of the relaxation of tangential stresses on the state parameters of a medium. *J. Apl. Mech. Tech. Phys* **21**: 716–724.

7 Merzhievsky, L.A. (1983). Three-dimensional generalization of the elastic-plastic model of D.D. Gilman. In: *Dynamics of Continuous Media*, vol. **59**, 158–163. Novosibirsk: Institute of Hydrodynamics [in Russian].

8 Alder, B., Fernbach, S., and Rotenberg, M. (ed.) (1964). Methods in computational physics. In: *Fundamental Methods in Hydrodynamics*, vol. **3**. New York and London: Academic Press.
9 Agureikin, V.A. and Kryukov, B.P. (1986). The method of individual particles for the calculation of flows of multicomponent media with large deformations. *Numer. Methods Contin. Mech.* **17** (4): 17–31. [in Russian].
10 Belotserkovsky, O.M. and Davydov, Y.M. (1982). *Method of Large Particles in Gas Dynamics*. Moscow: Nauka [in Russian].
11 Gingold, R.A. and Monaghan, J.J. (1977). Smoothed particle hydrodynamics: theory and application to non-spherical stars. *Mon. Not. R. Astron. Soc.* **181** (3): 375–389.
12 Lucy, L.B. (1977). A numerical approach to the testing of the fission hypothesis. *Astron. J.* **82**: 1013–1024.
13 Century Dynamics (2005). Autodyn. Explicit Software for Nonlinear Dynamics. Theory manual.
14 ANSYS Explicit Dynamics Analysis Guide. (2016). ANSYS, Inc. Release Southpointe.
15 Balagansky, I.A., Hokamoto, K., Manikandan, P. et al. (2011). Mach stem formation in explosion systems, which include high modulus elastic elements. *J. Appl. Phys.* **110**: 123516.
16 Lee, E.L. and Tarver, C.M. (1980). Phenomenological model of shock initiation in heterogeneous explosives. *Phys. Fluids* **23** (12): 2362–2372.
17 Tarver, C.M., Hallquist, J.O., and Erickson, L.M. (1985). Modeling short pulse duration shock initiation of solid explosives. In: *Proceedings of 8th Symposium on Detonation, Albuquerque, USA (15–19 July 1985)*. Silver Spring: Naval Surface Weapons Center.

Appendix A

A.1 Combined Median Errors for Artillery Shooting After Full Preparation of the Source Data

Table A.1 Combined median errors for artillery shooting after full preparation of the source data.

Type of artillery system	Firing distance, % of D_m	E_{x0}, %D	E_{y0}, angle fraction	B_d, %D	B_s, angle fraction
Rifled guns with $V_0 > 400$ m/s	20	2.76	7.3	0.82	2.2
	30	1.64	5.0	0.59	2.1
	40	1.16	3.9	0.51	1.8
	50	0.99	3.3	0.46	1.8
	60	0.91	3.0	0.45	1.7
	70	0.86	2.8	0.46	1.6
	80	0.84	2.8	0.46	1.7
	90	0.85	3.0	0.46	1.7
	100	0.86	3.2	0.47	1.7
Rifled guns with $V_0 < 400$ m/s	20	3.69	15.1	0.71	3.0
	30	2.01	9.7	0.53	2.6
	40	1.53	7.6	0.48	2.3
	50	1.29	6.0	0.48	2.2
	60	1.17	5.1	0.47	2.0
	70	1.09	4.3	0.48	2.0
	80	1.04	4.0	0.49	1.9
	90	1.03	3.7	0.51	1.9
	100	1.03	3.7	0.55	1.9
Mortars	40	2.49	15.6	1.14	9.1
	50	1.83	12.9	0.94	6.7
	60	1.50	10.7	0.85	6.0
	70	1.33	8.6	0.81	5.9
	80	1.20	7.9	0.77	5.2
	90	1.08	7.7	0.77	5.0
	100	1.02	5.3	0.77	4.0

Source: From Fendrikov and Yakovlev [1].

When using the data of this table, it should be kept in mind that the firing distance is expressed as a percentage of the maximum distance corresponding to the most advantageous charge number of the artillery system. Suppose, for example, that a division fires 122 mm howitzers at a range of 8 km. For these conditions, the combined median errors are found as follows.

According to the tables of the firing of 122-mm howitzers for a distance of 8 km, the second charge is chosen at which the maximum range $D_m = 9.72$ km. Therefore, the range $D = 8$ km is 82% of D_m, and since $V_0 > 400$ m/s for the second charge, we obtain from Table A.1 $E_{x0} = 0.84 \cdot 0.01D = 0.84 \cdot 80 \approx 67$ m. Similarly, we find $E_{y0} = 2.84 \cdot 0.001D = 2.84 \cdot 8 \approx 23$ m. $B_d = 0.46 \cdot 80 \approx 37$ m; $B_s = 1.7 \cdot 8 \approx 14$ m.

Absolute errors in efficiency calculations due to the use of combined median errors instead of characteristics of the actual error system are quite insignificant – less than 0.01–0.02 when hitting both elementary targets and area and group objects.

A.2 Values of Normalized Laplace Function

Table A.2 Values of the normalized Laplace function at different values of the argument.

x	0	1	2	3	4	5	6	7	8	9
0.0	0.0000	0054	0108	0161	0215	0269	0323	0377	0430	0484
0.1	0538	0591	0645	0699	0752	0806	0859	0913	0966	1020
0.2	1073	1126	1180	1233	1286	1339	1392	1445	1498	1551
0.3	1603	1656	1709	1761	1814	1866	1918	1971	2023	2075
0.4	2127	2179	2230	2282	2334	2385	2436	2488	2539	2590
0.5	2641	2691	2742	2793	2843	2893	2944	2994	3044	3093
0.6	3143	3192	3242	3291	3340	3389	3438	3487	3535	3584
0.7	3632	3680	3728	3775	3823	3870	3918	3965	4012	4059
0.8	4105	4152	4198	4244	4290	4336	4381	4427	4472	4517
0.9	4562	4604	4651	4695	4739	4783	4827	4870	4914	4957
1.0	5000	5043	5085	5128	5170	5212	5254	5295	5337	5378
1.1	5419	5459	5500	5540	5581	5620	5660	5700	5739	5778
1.2	5817	5856	5894	5932	5970	6008	6046	6083	6120	6157
1.3	6194	6231	6267	6303	6339	6375	6410	6445	6480	6515
1.4	6550	6584	6618	6652	6686	6719	6753	6786	6818	6851
1.5	6883	6915	6947	6979	7011	7042	7073	7104	7134	7165
1.6	7195	7225	7255	7284	7313	7342	7371	7400	7428	7457
1.7	7485	7512	7540	7567	7594	7621	7648	7675	7701	7727

Table A.2 (Continued)

x	0	1	2	3	4	5	6	7	8	9
1.8	7753	7778	7804	7829	7854	7879	7904	7928	7952	7976
1.9	8000	8023	8047	8070	8093	8116	8138	8161	8183	8205
2.0	8227	8248	8269	8291	8312	8332	8353	8373	8394	8414
2.1	8433	8453	8473	8492	8511	8530	8549	8567	8585	8604
2.2	8622	8639	8657	8674	8692	8709	8727	8742	8759	8775
2.3	8792	8808	8824	8839	8855	8870	8886	8901	8916	8930
2.4	8945	8959	8974	8988	9002	9016	9029	9043	9056	9069
2.5	9082	9095	9108	9120	9133	9145	9157	9169	9181	9193
2.6	9205	9216	9228	9239	9250	9261	9272	9282	9293	9303
2.7	9314	9324	9334	9344	9354	9363	9373	9382	9392	9401
2.8	9410	9419	9428	9437	9445	9454	9462	9471	9479	9487
2.9	9495	9503	9511	9518	9526	9533	9541	9548	9555	9562
3.0	9569	9567	9583	9590	9596	9603	9609	9616	9622	9628
3.1	9634	9640	9646	9652	9658	9663	9669	9674	9680	9685
3.2	9691	9696	9701	9706	9711	9716	9721	9725	9730	9735
3.3	9739	9744	9748	9753	9757	9761	9765	9769	9773	9777
3.4	9781	9785	9789	9793	9796	9800	9803	9807	9810	9814
3.5	9817	9820	9824	9827	9830	9833	9836	9839	9842	9845
3.6	9848	9851	9853	9856	9859	9861	9864	9866	9869	9871
3.7	9874	9876	9879	9881	9883	9885	9888	9890	9892	9894
3.8	9896	9898	9900	9902	9904	9905	9907	9909	9911	9913
4.0	9930	9943	9953	9962	9970	9976	9980	9984	9987	9990
5.0	9992	9994	9995	9996	9997	9997	9998	9998	9999	9999

Source: From Wentzel [2].

A.3 Basic Systems of Units and the Relationships Between Them

The concept of a system of units of physical quantities was first introduced by the German mathematician K. Gauss in 1832. His idea was to choose several quantities that are independent of each other. These quantities are called basic quantities, and their units are the basic units of the system of units. The basic units are chosen so that, using the formulas expressing the relationship between physical quantities, it is possible to form units of other quantities. Gauss called the units obtained by using formulas and expressed through the basic units "derived units."

Using his idea, Gauss constructed a system of units of magnetic quantities. The basic units of this system were as follows: millimeter – the unit of length, milligram – the unit of mass, and second – the unit of time. Although Gauss's system was not widespread due to the small size of the basic units, Gauss's ideas proved to be very fruitful. All later systems of units were based on his principles.

Today three systems of units are most commonly used in practice: the CGS, the MKGFS, and the SI [3].

A.3.1 The CGS System

The CGS system has been around for more than a century and is extremely widely used in both scientific research and several engineering fields. It is based on the *LMT* system of units. The basic units of the system: centimeter (cm) – the unit of length *L*, gram (g) – the unit of mass *M*, second (s) – the unit of time *T*. In the CGS system, using these three basic units, derived units of mechanical and acoustic quantities are set up.

With the use of the unit of thermodynamic temperature, kelvin (K), and the unit of light intensity,– candela (cd), the CGS system extends to the field of thermal and optical quantities. Using the unit of luminous flux, lumen (lm), and the unit of quantity of substance, mole (mol), the CGS system extends to the fields of light engineering and molecular physics, etc.

To measure electromagnetic quantities based on the CGS system using its three basic units (centimeters, grams, seconds), three systems of units are obtained: the electrostatic system of units – the CGSE system, the electromagnetic system of units CGSM, and the symmetrical system of units CGS.

When constructing the GSCE system, the unit of electric charge is introduced as the first derived of the electric unit using Coulomb's law as the defining equation. The absolute permittivity is treated as a dimensionless electrical quantity. Therefore, the square root of the speed of light in a vacuum appears explicitly in some equations that relate electromagnetic quantities.

The symmetric system of units is a combination of the GGSE and GGSM systems. In this system, the units of the GGSE system are used as units of electrical quantities, and the units of the GSSM system are used as units of magnetic quantities. As a result of the combination of the two systems, the speed of light in a vacuum appears explicitly in some equations linking electrical and magnetic quantities. The main advantage of the GGS system is its logicality and consistency. The main disadvantage of the SGS system is that the vast majority of its units are not directly applied in practice.

A.3.2 The MKGFS System of Units

The MKGFS system of units is based on the *LFT* system of quantities. Its basic units: meter (m) – the unit of length *L*, kilogram-force (kgf) – the unit of force *F*, second (s) – the unit of time *T*. Kilogram-force is the force equal to the weight of a body with the mass of 1 kg at a normal acceleration of free fall $g_0 = 9.806\,65\ \mathrm{m/s^2}$.

This unit of force, as well as some derived units of the MKGFS system, turned out to be convenient for their application in engineering. Therefore, the system is widely used in mechanics, heat engineering, and several other areas. The main drawback of the MKGFS system is a limited area of possible application – only for mechanical measurements.

The disadvantages of this system are many, for example such a seemingly minor one as the absence of a special name for the unit of mass (usually this unit is denoted by TME – a technical unit of mass), and, because of this, the eternal confusion between kilogram-force and kilogram – unit of mass. The "weight" terminology (such as "weight fluid flow," etc.) common in the respective fields is completely unacceptable.

A.3.3 The International System of Units

The International System of Units is based on the *LMTIΘJN* system of quantities. The basic SI units are meter (m) – unit of length L, kilogram (kg) – unit of mass M, second (s) – unit of time T, ampere (A) – unit of intensity of current I, kelvin (K) – unit of thermodynamic temperature Θ, candela (cd) – unit of light intensity J, mole (mol) – unit of substance quantity N. Additional units are plane angle – unit of angle radian (rad), solid angle – unit of solid angle steradian (sr).

Derived units are force, newton (1 N = 1 kg m/s^2); energy, joule (1 J = 1 N m); power, watt (1 W = 1 J/s); electric charge, coulomb (1 C = 1 A s); electric potential, volt (1 V = 1 J/C); electrical resistance, ohm (1 Ω = 1 V/A); electrical capacitance, farad (1 F = 1 C/V); magnetic flux, Weber (1 Wb = 1 V s); magnetic induction, Tesla (1 T = 1 Wb/m^2); inductance, henri (1 H = 1 V s/A); frequency, Hertz (1 Hz = 1 s^{-1}); temperature, degrees Celsius (°C); luminous flux, lumen (1 lm = 1 kd sr); illuminance, lux (1 lx = 1 lm/m^2).

The international system has undoubted advantages over all existing systems of units. It is universal, i.e. it covers all fields of measurement. Its seven basic units allow generating units for any physical quantities in all fields of science and branches of technology. The international system is a coherent system, which makes it possible to simplify the calculation formulas by freeing them from conversion coefficients. Both the basic and the vast majority of the derived SI units are convenient in their size for practical application.

Many derived units of the International System (watt, ampere, volt, ohm, etc.) were widely used long before its introduction. The main difficulty in constructing the SI system was the need to "stitch" electrical and magnetic units together with mechanical ones. This was achieved by introducing two constants, the electrical (ε_0) and magnetic (μ_0). As a result, to a certain extent, the clarity of the equations of the electromagnetic field and, in particular, of the theory of relativity was lost.

A.3.4 Non-SI Units

Despite certain advantages, which the use of units defined by one or another system provides, various units that do not fit into any system are still widespread. Many of them cannot be abandoned due to the convenience of their application in certain fields, while others have been preserved due to historical traditions.

Among off-system units, decimal multiples and fractional units are to be singled out in the first group. The name of these units is formed using appropriate prefixes (deci-, centi-, milli-, deca-, hecto-, kilo-, etc.). The list of these prefixes and the corresponding designations are given in Table A.3.

The second group of off-system units is made up of units constructed from the basic units of the system not according to the decimal principle. These include, first of all, units of time: minute, hour; units of length: foot, inch, etc.

Finally, the third group includes units that are not related to the units of the established systems. Such units include, in particular, the unit of heat quantity – calorie, the units of pressure – normal atmosphere and millimeter of the mercury column, etc. The relations of some off-system units with SI units are given in Table A.4.

The temperature in the SI is measured in kelvin. You can convert temperature from one temperature scale to another using a simple ratio

$$\frac{(T-273.15)\mathrm{K}}{5}=\frac{t\,^\circ\mathrm{C}}{5}=\frac{t\,^\circ\mathrm{R}}{4}=\frac{(t-32)^\circ\mathrm{F}}{9},$$

where K is the kelvin scale; °C is the Celsius scale; °R is the Reaumur scale; °F is the Fahrenheit scale.

Table A.3 Multipliers and prefixes for the formation of decimal multiples and fractions in the international system of units.

Multiplier	Prefixes	Prefix designation	Multiplier	Prefixes	Prefix designation
10^{18}	exa	E	10^{-1}	deci	d
10^{15}	peta	P	10^{-2}	centi	c
10^{12}	tera	T	10^{-3}	milli	m
10^{9}	giga	G	10^{-6}	micro	μ
10^{6}	mega	M	10^{-9}	nano	n
10^{3}	kilo	k	10^{-12}	pico	p
10^{2}	hecto	h	10^{-15}	femto	f
10^{1}	deka	da	10^{-18}	atto	a

Source: From Chertov [3].

Table A.4 Relationship of some nonsystem units with SI units.

Name of unit	Designation	Relationship to SI unit
Length		
Angstrom	Å	1 Å = 10^{-10} m
Micron	μ	1 micron = 10^{-6}m
Inch	in.	1 inch = 0.025 4 m
Foot	ft	1 foot = 0.304 8 m
Cable	—	1 cable = 185.2 m
Nautical mile	n. mile	1 nautical mile = 185 2 m
Land mile	mile	1 land mile = 160 9.344 m
Area		
Are	ar	1 ar = 100 m^2
Hectare	ha	1 hectare = 10^4 m^2
Acre	acre	1 acre = 4 046.86 m^2
Volume		
Imperial dry barrel	—	1 imperial dry barrel = 0.163 65 m^3
US oil barrel	—	1 US oil barrel = 0.158 988 m^3
US dry barrel	bbl (US)	1 US dry barrel = 0.115 628 m^3
Imperial gallon	gal (UK)	1 imperial gallon = 4.546 09 · 10^{-3} m^3
US liquid gallon	gal (US)	1 US liquid gallon = 3.785 43 · 10^{-3} m^3
US dry gallon	—	1 US dry gallon = 4.405 · 10^{-3} m^3
Liter	l, L	1 l = 10^{-3} m^3
Imperial pint	pt (UK)	1 imperial pint = 5.682 61 · 10^{-4} m^3
US liquid pint	lig pt (US)	1 US liquid pint = 4.731 79 · 10^{-4} m^3
US dry pint	dry pt (US)	1 US dry pint = 5.506 14 · 10^{-4} m^3

(*Continued*)

Table A.4 (Continued)

Name of unit	Designation	Relationship to SI unit
Speed		
Kilometer per hour	km/h	1 km/h = 0.277 78 m/s
Mile per hour	mile/h	1 mile/h = 0.447 04 m/s
Sea knot	kn	1 kn = 0.514 444 m/s
Mass		
Grain	gr	1 gr = $6.479\,891 \cdot 10^{-5}$ kg
Carat	ct	1 carat = $2 \cdot 10^{-4}$ kg
Ton	t	1 ton = 1 000 kg
Imperial ton	ton	1 imperial ton = 1 016.05 kg
Imperial short ton	sh ton	1 imperial short ton = 907.185 kg
Apothecaries' ounce	oz apoth	1 apothecaries' ounce = $31.103\ 5 \cdot 10^{-3}$ kg
Russian apothecaries' ounce	—	1 Russian apothecaries' ounce = $29.86 \cdot 10^{-3}$ kg
International avoirdupois ounce	oz	1 international avoirdupois ounce = $28.349\,5 \cdot 10^{-3}$ kg
International troy ounce	oz t	1 international troy ounce = $31.103\,5 \cdot 10^{-3}$ kg
Imperial pound	lb	1 imperial pound = 0.453 592 37 kg
Pound in the Russian system of measures	—	1 pound in the Russian system of measures = 0.409 512 41 kg
US pound	lb (US)	1 US pound = 0.453 592 427 7 kg
Pressure		
Atmosphere	atm	1 atm = $1.013\,25 \cdot 10^{5}$ Pa
Technical atmosphere	at	1 at = $0.980\,665 \cdot 10^{5}$ Pa
Millimeter of mercury	mmHg	1 mmHg = 133.322 Pa

Table A.4 (Continued)

Name of unit	Designation	Relationship to SI unit
Work, energy, heat		
Kilowatt-hour	kW h	$1\ \text{kW h} = 3.6 \cdot 10^{6}\ \text{J}$
Electron volt	eV	$1\ \text{eV} = 1.602\,177\,33 \cdot 10^{-19}\ \text{J}$
International calorie	cal	$1\ \text{cal} = 4.186\,8\ \text{J}$
Power		
Metric horsepower	hp(M)	$1\ \text{hp(M)} = 735.499\ \text{W}$
Imperial horsepower	hp(I)	$1\ \text{hp(I)} = 745.700\ \text{W}$

Source: From Chertov [3].

In the existing sources, one can find all three systems of units, CGS, MKGFS, and SI, as well as various off-system units, so it is necessary to be able to convert the units of quantities from one system to another. To facilitate such conversion, tables of relations between units of various physical quantities are provided below. Conversion coefficients are given with accuracy sufficient for engineering practice.

A.3.4.1 Relationships Between Units of Different Physical Quantities

Table A.5 Relationships between units of speed.

Unit	m/s	m/min	cm/s	km/h	kn
1 m/s	1	60	100	3.6	1.94
1 m/min	$1.67 \cdot 10^{-2}$	1	1.67	$6 \cdot 10^{-2}$	$3.24 \cdot 10^{-2}$
1 cm/s	10^{-2}	0.6	1	$3.6 \cdot 10^{-2}$	$1.94 \cdot 10^{-2}$
1 km/h	0.278	16.7	27.8	1	0.540
1 kn	0.514	30.9	5 1.4	1.85	1

Source: From Sena [4].

Table A.6 Relationships between the units of angular velocity.

Unit	rad/s	Round/s	Round/min	°/s
1 rad/s	1	0.159	9.55	57.3
1 round/s	6.28	1	60	$3.6 \cdot 10^2$
1 round/min	0.105	$1.67 \cdot 10^{-2}$	1	6
1 ° /s	$1.75 \cdot 10^{-2}$	$2.78 \cdot 10^{-3}$	0.167	1

Source: From Sena [4].

Table A.7 Relationships between units of mass.

Unit	kg	g	TME	t
1 kg	1	10^3	0.102	10^{-3}
1 g	10^{-3}	1	$1.02 \cdot 10^{-4}$	10^{-6}
1 TME	9.81	$9.81 \cdot 10^3$	1	$9.81 \cdot 10^{-3}$
1 t	10^3	10^6	102	1

Source: From Sena [4].

Table A.8 Relationships between units of force.

Unit	N	dyn	kgf
1 H	1	10^5	0.102
1 dyn	10^{-5}	1	$1.02 \cdot 10^{-6}$
1 kgf	9.81	$9.81 \cdot 10^5$	1

Source: From Sena [4].

Table A.9 Relationships between units of pressure.

Unit	Pa	dyn/cm^2	kgf/cm^2	bar	atm
1 Pa	1	10	$1.02 \cdot 10^{-5}$	10^{-5}	$9.87 \cdot 10^{-6}$
1 dyn/cm^2	0.1	1	$1.02 \cdot 10^{-6}$	10^{-6}	$9.87 \cdot 10^{-7}$
1 kgf/cm^2	$9.81 \cdot 10^4$	$9.81 \cdot 10^5$	1	0.981	0.968
1 bar	10^5	10^6	1.02	1	0.987
1 atm	$1.01 \cdot 10^5$	$1.01 \cdot 10^6$	1.03	1.01	1

Source: From Sena [4].

Table A.10 Relationships between units of work and energy.

Unit	J	erg	kgf·m	kcal	kW·h
1 J	1	10^7	0.102	$2.39 \cdot 10^{-4}$	$2.78 \cdot 10^{-7}$
1 erg	10^{-7}	1	$1.02 \cdot 10^{-8}$	$2.39 \cdot 10^{-11}$	$2.78 \cdot 10^{-14}$
1 kgf·m	9.81	$9.81 \cdot 10^7$	1	$2.34 \cdot 10^{-3}$	$2.72 \cdot 10^{-6}$
1 kcal	$4.19 \cdot 10^3$	$4.19 \cdot 10^{10}$	$4.27 \cdot 10^2$	1	$1.16 \cdot 10^{-3}$
1 kW·h	$3.6 \cdot 10^6$	$3.6 \cdot 10^{13}$	$3.67 \cdot 10^5$	$8.6 \cdot 10^2$	1

Source: From Sena [4].

A.4 Properties of High Explosives

A.4.1 Explosion

The concept of "explosion" was supposedly first defined by M.V. Lomonosov. In modern language, his definition sounds as follows: *Explosion is a very rapid release of large amounts of energy and large volumes of gases.*

In modern natural science, the concept of "explosion" has acquired a much broader meaning. Without claiming completeness and finality of formulation, we can say that nowadays an explosion is understood as a process of a very rapid physical or chemical transformation of a system, accompanied by the transfer of its potential energy into mechanical work. The work performed in an explosion is the result of the rapid expansion of gases or vapors, regardless of whether they existed before or were formed during the explosion.

Already the definition of M.V. Lomonosov, despite its simplicity, contains several fundamental points. Let us dwell on two of them, which seem to be the most important. First, what is meant by "very fast"? The speed of a process is a relative concept and should be measured against the speed of other processes. For example, it can be argued that the release of energy under explosion is much faster than under combustion of the same substance. Second, "a large amount of energy." It is most logical to compare this amount to the energy somehow contained in the environment.

Explosions can be caused by a variety of physical or chemical phenomena, for example:

- Explosions of a steam boiler or a balloon filled with compressed gas
- Explosions of a ball lightning or a conductor carrying a high voltage electric current
- Nuclear and thermonuclear explosions
- Explosions of explosives.

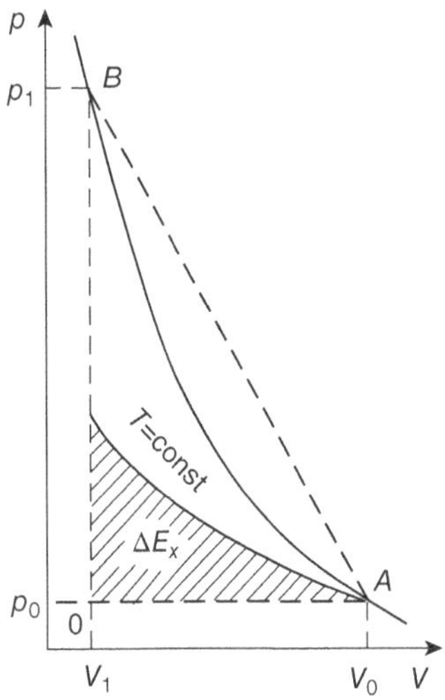

Figure A.1 Hugoniot shock adiabat. *Source:* From Balagansky and Merzhievsky [5].

A.4.2 Shock Wave

The term "shock wave" refers to a narrow zone (several molecule free paths thick for gases, several interatomic distances in crystalline bodies, and particle sizes for granular materials) in which there is a sharp increase in pressure, density, and velocity of matter. Beyond this zone, the substance is in a state close to thermodynamically equilibrium. In the most general case for a plane shock wave, the parameters before the front and after the front are related by relations expressing the laws of conservation of mass, momentum, and energy:

$$\begin{aligned} \rho_0(D-u_0) &= \rho_1(D-u_1), \\ p_1 - p_0 &= \rho_0(D-u_0)(u_1-u_0), \\ E_1 - E_0 &= \frac{p_1+p_0}{2}(v_0-v_1). \end{aligned} \tag{A.1}$$

Here ρ, density; $v=1/\rho$, specific volume; p, pressure; u, D, mass velocity and shock wave velocity; E, specific internal energy; index 0 refers to the state before the front; 1, behind the front of the shock wave.

The last of these relations, linking kinematic and thermodynamic parameters, is called the shock adiabat or Hugoniot adiabat. The relations have a unique property: they are valid for shock waves in any medium. To specify the medium, it is necessary to add the equation of state of the medium, for example in the form $p = p$ (ρ, E). Using the equation of state, one can specify the shock adiabat for a given substance. For an ideal gas with a constant adiabatic index k, the shock adiabat takes the form

$$\frac{p_1}{p_0} = \frac{(k+1)\rho_1-(k-1)\rho_0}{(k+1)\rho_0-(k-1)\rho_1}, \quad \text{or} \quad \frac{\rho_1}{\rho_0} = \frac{(k+1)p_1+(k-1)p_0}{(k+1)p_0+(k-1)p_1}. \tag{A.2}$$

The plot of the shock adiabat (curve AB) is shown in Figure A.1. The total energy increment per unit mass of matter during shock compression corresponds to the area of a rectangular triangle OAB. If one were to imagine that compression took place at a constant temperature, then the change of state of the substance would follow the isotherm $T = \text{const}$, and then the increase in internal energy of the substance ΔE_x would depend only on elastic compression. However, in real substances, shock compression always occurs in accordance with the curve AB passing above the isotherm, which indicates that the substance subjected to shock compression is heated.

We emphasize that the shock adiabat is the geometric location of the points corresponding to the states of the matter reached during the transition from the initial state (p_0, v_0) to the state at the shock-wave front (p_1, v_1). The intermediate points do not lie on the shock adiabat. The loading goes along the straight line *AB* (the dashed line in Figure A.1). This line is referred to in the Russian literature as the Mikhelson line, and in the English-language literature as the Rayleigh line.

A.4.3 Combustion Deflagration

Combustion is a complex, rapid chemical transformation accompanied by the release of a significant amount of heat and a bright glow. In most cases, combustion is the exothermic oxidation of fuel (a substance capable of combustion) by an oxidizer (oxygen, including air oxygen, nitrous oxide, etc.).

Combustion also includes other processes associated with rapid transformation and their thermal or chain acceleration, including the decomposition of explosives. According to the velocity of propagation of transformation, combustion is divided into deflagration, which proceeds at subsonic speeds, and detonation, which propagates at supersonic speeds. It should be noted that, initially, the term "deflagration" referred only to a combustion mode in which a compression wave was formed and transformed into a shock wave; in this case, deflagration transforms into detonation.

A.4.4 Detonation

Detonation is the fastest form of explosive transformation propagating at supersonic velocity. It is a hydrodynamic process of propagation through the matter of a detonation wave, which is a complex of a shock wave and an exothermic reaction zone located behind it. In the simplest Chapman–Jouguet detonation model, the detonation wave front (transformation zone) is assumed to consist of a shock front and a parallel plane (the Chapman–Jouguet plane) located behind it, in which an instantaneous release of explosive energy occurs.

In terms of gas dynamics, the Chapman–Jouguet plane is the plane of transition from the supersonic flow of matter ahead of the detonation wave front to the subsonic flow behind the detonation wave front. On the Chapman–Jouguet plane, the flow is sonic. Approximate formulas for calculating density, pressure, mass velocity, sound velocity, and detonation velocity in the Chapman–Jouguet plane are often used:

$$\rho_H = \rho_0 \frac{k+1}{k}; \quad p_H = \frac{\rho_0 D^2}{k+1}; \quad u_H = \frac{D}{k+1};$$
$$c_H = \frac{kD}{k+1}; \quad D = \sqrt{2Q_v(k^2-1)}, \tag{A.3}$$

where Q_v is the energy released under the detonation transformation. For detonation, products of high-density condensed explosives often take $k = 3$.

The basic properties of some common explosives are given in Table A.11.

Table A.11 The properties of explosives.

Name	Makeup	Density ρ_0, g/cm^3	Heat of explosion Q_v, kJ/kg	Isoentropy index of the detonation products k	The steady-state detonation velocity D, km/s	Pressure at the Chapman–Jouguet point p_H, GPa
TNT	2,4,6-Trinitrotoluene	1.06 1.64	3570 4330	1.66 3.16	5.25 6.95	11.0 19.0
RDX	Cyclotrimethylene trinitramine	1.00 1.80	5000 5530	2.48 2.98	5.98 8.75	10.8 34.7
HMX	Cyclotetramethylene tetranitramine	1.90	5460	3.00	9.10	39.3
PETN	Pentaerythritol tetranitrate	1.00 1.77	5440 5870	2.45 2.64	5.48 8.30	8.7 33.5
TATE	1,3,5-Triamino-2,4,6-trinitrobenzene	1.895	3062	2.72	7.86	31.5
Comp B	60 RDX/40 TNT	1.713	4860	2.76	8.03	29.4
Pentclite	50 PETN/50 TNT	1.65	—	2.84	7.47	24.7
NG	Nitroglycerine	1.59	6220	2.83	7.58	24.6
NM	Nitromethane	1.128	4640	2.17	6.29	14.1
C-4	91 RDX/9 polyisobutylene	1,64	5030	—	8,19	—
DYNA	Dioxyethylnitramine dinitrate	1.49	5458	—	7580	—

Source. From Balagansky and Merzhievsky [5].

A.5 Shock Adiabats of Substances

In the experimental study of plane shock waves in condensed matter, the velocities of shock waves (D) and the free surface are usually measured, from which the mass velocity (u) is determined (in the free surface approximation, the velocity u is equal to half of the free surface velocity). For various materials in the pressure range from 10 to 100 GPa, the dependencies between shock wave velocity and mass velocity fit well into the linear dependence

$$D = a + \lambda u. \tag{A.4}$$

By now, the constants a and λ have been determined for many materials. Some of these data are given in Table A.12. Having these experimental data, it is possible to calculate the shock adiabats, determine the equations of state, and calculate the

Table A.12 Values of the coefficients a and λ for different substances.

Substance	Density ρ_0, g/cm^3	Bulk speed of sound c_0, km/s	a, km/s	λ	Pressure range p, kbar	Velocity range u, km/s
1. Elements						
Aluminum	2.71		5.330	1.35	20–2000	
Beryllium	1.845	7.934	7.975	1.091	144–290	
Vanadium	6.1	5.180	5.108	1.210	207–1270	
Bismuth	9.8	4.050	2.000	1.340	350–3500	
Tungsten	19.17		4.005	1.268	400–2110	
Iron	7.85		3.800	1.580	400–5000	
Gold	19.3	2.330	3.150	1.470	590–5200	
Indium	7.27		2.370	1.61	218–414	
Iodine	4.93		1.350	1.7	50–700	
Cadmium	8.64		2.650	1.48	360–3500	
Potassium	0.86	4.630	2.000	1.172	33–860	
Cobalt	8.82		4.748	1.33	246–1640	
Lithium	0.53		4.450	1.115	<500	
Magnesium	1.72	3.980	4.780	1.16	60–400	
Copper	8.90	5.192	3.915	1.495	<4000	
Molybdenum	10.2		5.157	1.238	260–1670	
Sodium	0.97	4.630	2.550	1.262	<970	

(Continued)

Table A.12 (Continued)

Substance	Density ρ_0, g/cm^3	Bulk speed of sound c_0, km/s	a, km/s	λ	Pressure range p, kbar	Velocity range u, km/s
Nickel	8.86	4.427	4.646	1.445	240–1520	
Niobium	8.6	2.761	4.447	1.212	250–491	
Tin	7.28	3.800	2.640	1.476	175–1400	
Palladium	11.95	3.640	3.793	1.922	267–541	
Platinum	21.37	4.630	3.671	1.405	300–598	
Rhodium	12.42		4.680	1.645	284–562	
Mercury	13.55	2.029	1.450	2.200	230–473	
Lead	11.34		2.028	1.517	200–1410	
Sulfur	2.1		2.200	1.00	68–204	
Silver	10.49		3.300	1.54	460–4100	
Antimony	6.69	1.830	2.000	1.60	252–1200	
Thallium	11.84	3.410	1.859	1.515	218–1550	
Tantalum	16.46	4.847	3.374	1.155	277–557	
Titanium	4.51	2.059	4.779	1.088	175–1080	
Thorium	11.68		2.132	1.278	207–1430	
Uranus	18.9	5.150	2.550	1.504	342–6570	
Chromium	7.10	3.030	5.217	1.465	238–1400	
Zinc	7.14	3.74	3.050	1.559	190–1430	
Zirconium	6.49		3.770	0.93	260–417	
2. Alloys and mixtures						
VNZh-90	17.1		3.832	1.497		0.6–3.6
AMG-6	2.64		5.508	1.233		0.6–6.1
D-16	2.78		5.386	1.284		0.7–6
Steel 45	7.85		3.940	1.58		0.8–3.1
Steel 40X	7.83		4.010	1.5		0.8–3.2
12X18H10T	7.89		4.553	1.482		0.2–2.8
Tungsten carbide			4.920	1.339	<2000	
Brass	8.41		3.800	1.42	224–1800	

Table A.12 (Continued)

Substance	Density ρ_0, g/cm^3	Bulk speed of sound c_0, km/s	a, km/s	λ	Pressure range p, kbar	Velocity range u, km/s
3. Various substances						
Paraffin	0.91		1.810	2.31	22.5–66.3	
			3.320	1.24	66.3–260	
Plexiglass	1.18	2.440	2.590	1.51	3–76	
			3.100	1.32	130–1300	
			2.740	1.35	17–2000	
Quartz	2.65	3.690	3.710	1.24	39–185	
Granite	2.6		2.435	1.525		2–5.1
Limestone			3.750	1.44		0.8–8
Water	1.0		1.500	2.0	<20	
			1.700	1.7	10–114	
Methyl alcohol	0.80		1.800	1.5	48–112	
Nitromethane	1.14		2.00	1.38	20–88	
TNT	1.614	2.57	2.39	2.05		
RDX	1.8		2.87	1.61		
HMX	1.891		3.07	1.79		
PETN	1.67		2.83	1.91		
40 TNT/60 RDX	1.7		2.95	1.58		

Source: From Baum et al. [6].

temperature at the shock front. In particular, the relationship between the parameters at the shock front $p(u)$, $v(u)$, $p(v)$, $p(\rho)$, $(v = 1/\rho)$ is determined from the following dependencies:

$$\begin{aligned} p &= \rho_0 u(a + \lambda u); \\ v &= v_0 \frac{a + (\lambda - 1)u}{a + \lambda u}; \\ p &= \frac{a^2}{v_0}\left(1 - \frac{v}{v_0}\right)\left[1 - \lambda\left(1 - \frac{v}{v_0}\right)\right]^{-2}; \\ p &= a^2\rho\left(\frac{\rho}{\rho_0} - 1\right)\left[\lambda - (\lambda - 1)\left(\frac{\rho}{\rho_0}\right)\right]^{-2}. \end{aligned} \tag{A.5}$$

For explosives, the shock adiabat parameters are true for those pressures until the chemical decomposition of the explosive behind the shock front begins.

References

1 Fendrikov, N.M. and Yakovlev, V.I. (1971). *Methods of Calculation of Combat Efficiency of Armament*. Moscow: Voyenizdat [in Russian].
2 Wentzel, E.S. (1964). *Introduction to Operations Research*. Moscow: Soviet Radio [in Russian].
3 Chertov, A.G. (1990). *Physical Quantities*. Moscow: High School [in Russian].
4 Sena, L.A. (1977). *Units of Physical Quantities and Their Dimensions*. Moscow: Nauka [in Russian].
5 Balagansky, I.A. and Merzhievsky, L.A. (2004). *Effects of Means of Destruction and Ammunition*. Novosibirsk: Novosibirsk State Technical University [in Russian].
6 Baum, F.A., Orlenko, L.P., Stanukovich, K.P. et al. (1975). *Physics of Explosion* (ed. K.P. Staniukovich). Moscow: Nauka [in Russian].

Bibliography

This bibliography is provided to enable the studious reader to expand his or her knowledge.

Introduction

Driels, M. (2020). *Weaponeering: Conventional Weapon System Effectiveness (AIAA Education Series)*. AIAA.
Kenneth, C., Dullum, O., Garlasco, M. et al. (2015). *Explosive Weapons in Populated Areas: Technical Considerations Relevant to Their Use and Effects. Special Report*. Armament Research Services (ARES): Perth.

Fragmentation Ammunitions

Grady, D.E. and Kipp, M.E. (1993). Dynamic fracture and fragmentation. In: *High-Pressure Shock Compression of Solids. High-Pressure Shock Compression of Condensed Matter* (ed. J.R. Asay and M. Shahinpoor). New York: Springer.
Meyers, M.A. (1994). *Dynamic Behavior of Materials*. New York: Wiley.

Ammunitions with Shaped Charges

Balagansky, I.A., Bataev, A.A., and Bataev, I.A. (2019). *Explosion Systems with Inert High-Modulus Components: Increasing the Efficiency of Blast Technologies and Their Applications*. Hoboken: Wiley.

Chou, P.C., Carleone, J., and Karpp, R.R. (1976). Criteria for jet formation from impinging shells and plates. *J. Appl. Phys.* **47**: 2975–2982.
Kinelovskii, S.A. and Trishin, Y.A. (1980). Physical aspects of the hollow-charge effect. *Combust. Explos. Shock Waves* **16**: 504–515.
Walsh, G.M., Shreffler, R.G., and Willig, F.G. (1953). Limiting conditions for jet formation in high-velocity collisions. *J. Appl. Phys.* **24**: 349–359.
Walters, W.P. and Zukas, J.A. (1989). *Fundamentals of Shaped Charges*. New York: Wiley.

High-explosive Ammunitions

Baker, W.E., Cox, P.A., Westine, P.S. et al. (1983). *Explosion Hazards and Evaluation*. Amsterdam: Elsevier.
Ben-Dor, G. (2007). *Shock Wave and High Pressure Phenomena*. Berlin Heidelberg: Springer-Verlag.
Cole, R.H. (1948). *Underwater Explosions*. Princeton: Princeton University Press.
Kinney, G.F. and Graham, K.J. (1985). *Explosive Shocks in Air*. Berlin Heidelberg: Springer-Verlag.
Nettlenton, M.A. (1987). *Gaseous Detonation: Their Nature, Effects, and Control*. Chapman and Hall.

Penetrating Ammunitions

Eichelberger, R.J. and Kineke, J.H. (1967). High velocity impact. In: *Kurzzeitphysik* (ed. K. Vollrath and G. Thomer). Wien: Springer.
Kinslow, R. (ed.) (1970). *High-Velocity Impact Phenomena*. New York: Academic Press.
Meyers, M.A. (1994). *Dynamic Behavior of Materials*. New York: Wiley.
Zukas, J.A., Nicholas, T., Swift, H.F. et al. (1982). *Impact Dynamics*. New York: Wiley.

Numerical Simulation of High-speed Processes

Alder, B., Fernbach, S., and Rotenberg, M. (1964). *Methods in Computational Physics. Fundamental Methods in Hydrodynamics*, vol. **3**. New York and London: Academic Press.
ANSYS, Inc (2016). *ANSYS Explicit Dynamics Analysis Guide*. Release Southpointe.
Lee, R. (2019). *LS-DYNA for Engineers. A Practical Tutorial Book*. BW Publications.
Mader, C.L. (2007). *Numerical Modeling of Explosives and Propellants*. CRC Press.
Oran, E.S. (2000). *Numerical Simulation of Reactive Flow*. Cambridge University Press.

Appendices

Trunin, R.F., Gudarenko, L.F., Zhernokletov, M.V. et al. (2001). *Experimental Data on Shock Compression and Adiabatic Expansion of Condensed Matter* (ed. R.F. Trunin). Sarov: RFNC-VNIIEF.

Index

Wiley The manufacturer's authorized representative according to the EU General Product Safety Regulation is Wiley-VCH GmbH, Boschstr. 12, 69469 Weinheim, Germany, e-mail: Product_Safety@wiley.com.

Printed and bound by CPI Group (UK) Ltd, Croydon, CR0 4YY
19/08/2026
02201089-0007